MCBU

Molecular and Cell Biology Updates

Series Editors

Prof. Dr. Angelo Azzi
Institut für Biochemie
und Molekularbiologie
Bühlstr. 28
CH–3012 Bern
Switzerland

Prof. Dr. Lester Packer
Dept. of Molecular
and Cell Biology
251 Life Science Addition
Membrane Bioenergetics Group
Berkeley, California 94720
USA

Vitamin A and Retinoids: An Update of Biological Aspects and Clinical Applications

Edited by   M. A. Livrea

Springer Basel AG

Editor

Prof. Maria A. Livrea
Cattedra di Chimica Biologica
Istituto di Farmacologia e Farmacognosia
Università di Palermo
Via Forlanini 1
I-90134 Palermo
Italy

Library of Congress Cataloging-in-Publication Data

Vitamin A and retinoids : an update of biological aspects and clinical
    applications / edited by Maria A. Livrea
        p.   cm. – (Molecular and cell biology updates)
    Includes bibliographical references and index.
    ISBN 978-3-0348-9574-3   ISBN 978-3-0348-8454-9 (eBook)
    DOI 10.1007/978-3-0348-8454-9
    1. Retinoids – Physiological effect. 2. Retinoids – Therapeutic
    use. 3. Vitamin A – Physiological effect. 4. Vitamin A – Therapeutic
    use. 5. Cancer – Chemotherapy. 6. Skin – Diseases – Chemotherapy.
    I. Livrea, Maria A., 1946– . II. Series.
    QP801.R47V58 2000
    612.3'99 – dc21

Deutsche Bibliothek Cataloging-in-Publication Data

Vitamin A and retinoids : an update of biological aspects and
clinical applications / ed. by M. A. Livrea. - Basel ; Boston ; Berlin :
Birkhäuser, 2000
    (Molecular and cell biology updates)
    ISBN 978-3-0348-9574-3

© 2000 Springer Basel AG
Originally published by Birkhäuser Verlag in 2000
Softcover reprint of the hardcover 1st edition 2000

Printed on acid-free paper produced from chlorine-free pulp. TFC ∞
Cover illustration: Reproduced with friendly permission of the International Life Sciences Institute,
        Washington DC 20036-4810 (page 30)

ISBN 978-3-0348-9574-3

9 8 7 6 5 4 3 2 1

Table of contents

List of contributors

Brigitte Almond-Roesler, Department of Dermatology, University Medical Center Benjamin Franklin, Freie Universität Berlin, Hindenburgdamm 30, D-12200 Berlin, Germany

Athena Andreadis, E. Kennedy Shriver Center and Department of Neurology, Harvard Medical School, 200 Trapelo Road, Waltham, MA 02452, USA; e-mail: AAndreadis@shriver.org

Amir A. Bajoghli, Department of Dermatology, Boston University School of Medicine, 609 Albany Street, Boston, MA 02118, USA; e-mail: Bajoghli@bu.edu

John S. Bertram, Cancer Research Center of Hawaii, University of Hawaii, 1236 Lauhala Street, Honolulu, HI 96813, USA

William S. Blaner, Department of Medicine and Institute of Human Nutrition, College of Physicians and Surgeons, Columbia University, HHSC 502, 701 W. 168th St., New York, NY 10032, USA; e-mail: WSB2@columbia.edu

Carsten Carlberg, Institut für Physiologische Chemie I, Heinrich-Heine-Universität, Postfach 101007, D-40001 Düsseldorf, Germany; e-mail: carlberg@uni-duesseldorf.de

Roshantha A. S. Chandraratna, Retinoid Research, Departments of Biology and Chemistry, Allergan Pharmaceuticals, 2525 Dupont Drive, PO Box 19534, Irvine, CA 92715, USA; e-mail: chandraratna_rosh@allergan.com

Harvey J. Clewell, ICF/Kaiser International, 602 East Georgia Avenue, Ruston, LA 71270, USA

John L. Clifford, Department of Clinical Cancer Prevention, University of Texas, M. D. Anderson Cancer Center, 1515 Holcombe Boulevard, Houston, TX 77030, USA

Joan M. Creech Kraft, 87 Sudden Valley, Bellingham, WA 98226, USA; e-mail: jmkraft@pacificrim.net

Rosalie K. Crouch, Medical University of South Carolina, 96 Jonathan Lucas Street, P.O. Box 250617, Charleston, SC 29425, USA; e-mail: crouchrk@musc.edu

Marcia I. Dawson, Molecular Medicine Research Institute, 325 East Middlefield Road, Mountain View, CA 94043, USA; e-mail: mdawson@mmrx.org

Laurent Degos, Service d'Hématologie, Hôpital Saint-Louis, F-75475 Paris, France

Ursula C. Dräger, E. Kennedy Shriver Center and Department of Psychiatry, Harvard Medical School, 200 Trapelo Road, Waltham, MA 02452, USA; e-mail: UDrager@shriver.org

Pierre Fenaux, Service des Maladies du Sang, CHU, Place de Verdun 1, F-59037 Lille, France; e-mail: pfenaux.lille@invivo.edu

Gary J. Fisher, Department of Dermatology, University of Michigan Medical Center, 1910 Taubman Center, Ann Arbor, MI 48109, USA; e-mail: gjfisher@umich.edu

Franca Formelli, Istituto Nazionale per lo Studio e la Cura dei Tumori, Department of Experimental Oncology, Via Venezian 1, I-20133 Milan, Italy; e-mail: formelli@istitutotumori.mi.it

Mary V. Gamble, Department of Medicine and Institute of Human Nutrition, College of Physicians and Surgeons, Columbia University, 701 W 168th Street, New York, NY 10032, USA

Christoph C. Geilen, Department of Dermatology, University Medical Center Benjamin Franklin, Freie Universität Berlin, Hindenburgdamm 30, D-12200 Berlin, Germany

Barbara A. Gilchrest, Department of Dermatology, Boston University School of Medicine, 609 Albany Street, Boston, MA 02118, USA; e-mail: bgilchre@bu.edu

Fabien Guidez, Leukaemia Research Fund Centre at the Institute of Cancer Research, Chester Beatty Laboratories, 237 Fulham Road, London SW3 6JB, UK

Peter D. Hobbs, SRI International, 333 Ravenswood Avenue, Menlo Park, CA 94025, USA

Laura Isnardi, Laboratory of Pre-Clinical Oncology, Advanced Biotechnology Center, Largo Rosanna Benzi 10, I-16132 Genoa, Italy

Ling Jong, SRI International, 333 Ravenswood Avenue, Menlo Park, CA 94025, USA

Sewon Kang, Department of Dermatology, University of Michigan Medical Center, 1910 Taubman Center, Ann Arbor, MI 48109, USA; e-mail: swkang@umich.edu

Elliott S. Klein, Retinoid Research, Department of Biology, Allergan Pharmaceuticals, 2525 Dupont Drive, PO Box 19534, Irvine, CA 92715, USA; e-mail: klein_elliott@allergan.com

Norman I. Krinsky, Department of Biochemistry, Tufts University School of Medicine, 136 Harrison Avenue, Boston, MA 02111, USA; e-mail: nkrinsky_mna@opal.tufts.edu

Scott M. Lippman, Department of Clinical Cancer Prevention, University of Texas, M. D. Anderson Cancer Center, 1515 Holcombe Boulevard, Houston, TX 77030, USA

Reuben Lotan, Department of Thoracic/Head and Neck Medical Oncology, University of Texas, M. D. Anderson Cancer Center, 1515 Holcombe Boulevard, Houston, TX 77030, USA; e-mail: rlotan@notes.mdacc.tmc.edu

Jian-Xing Ma, Department of Ophthalmology, Medical University of South Carolina, 96 Jonathan Lucas Street, Charleston, SC 29425, USA

Susan T. Mayne, Department of Epidemiology and Public Health, Yale University School of Medicine and Yale Cancer Center, New Haven, CT 06520, USA; e-mail: susan.mayne@yale.edu

Peter McCaffery, E. Kennedy Shriver Center and Department of Psychiatry, Harvard Medical School, 200 Trapelo Road, Waltham, MA 02452, USA; e-mail: PMcCaffery@shriver.org

Joseph L. Napoli, Department of Nutritional Sciences, University of California, 119 Morgan Hall, Berkeley, CA 94720, USA; e-mail: JLN@uclink4.berkeley.edu

James A. Olson, Department of Biochemistry, Biophysics and Molecular Biology, 3252 Molecular Biology Building, Iowa State University, Ames, IA 50011, USA; e-mail: jaolson@iastate.edu

Constantin C. Orfanos, Department of Dermatology, University Medical Center Benjamin Franklin, Freie Universität Berlin, Hindenburgdamm 30, D-12200 Berlin, Germany

Patsie Polly, Institut für Physiologische Chemie I, Heinrich-Heine-Universität, Postfach 101007, D-40001 Düsseldorf, Germany

Patrizia Raffo, National Institute for Cancer Research, Largo Rosanna Benzi 10, I-16132 Genoa, Italy

A. Catharine Ross, Department of Nutrition and Department of Veterinary Science, 115 Henning Building, Pennsylvania State University, University Park, PA 16802, USA; e-mail: acr6@psu.edu

Richard D. Semba, Ocular Immunology Service, Suite 700, 550 North Broadway, Johns Hopkins School of Medicine, Baltimore, MD 21205, USA; e-mail: rdsemba@jhmi.edu

Salvatore Toma, Department of Oncology, Biology and Genetics, National Institute for Cancer Research, Largo Rosanna Benzi 10, I-16132 Genoa, Italy

John Voorhees, Department of Dermatology, University of Michigan Medical Center, 1910 Taubman Center, Ann Arbor, MI 48109, USA; e-mail: voorhees@umich.edu

Elisabeth Wagner, E. Kennedy Shriver Center and Department of Psychiatry, Harvard Medical School, 200 Trapelo Road, Waltham, MA 02452, USA; e-mail: EWagner@shriver.org

Calvin C. Willhite, State of California, 700 Heinz Street, Suite 200, Berkeley, CA 94710, USA

Arthur Zelent, Leukaemia Research Fund Centre at the Institute of Cancer Research, Chester Beatty Laboratories, 237 Fulham Road, London SW3 6JB, UK

Xiao-kun Zhang, The Burnham Institute, 10901 N. Torrey Pines Road, La Jolla, CA 92037, USA; e-mail: xzhang@burnham-inst.org

To Francesco and Claudio

Preface

Research on vitamin A and retinoids has a very long history, yet major advances are continuously in progress in this field. As we approach the new century, these molecules still are at the center of many different interests, from basic chemical, biochemical and cell biology research, to nutritional aspects and clinical applications. This book presents current knowledge and state-of-the-art information on research in many areas of this huge field of studies. Prominent scientists contributed chapters on the role of vitamin A in nutrition and the metabolism of retinol and retinoic acid, retinoic acid receptors and the interplay with vitamin D, the cell biology of retinoids and their role in the immune system. New synthetic molecules and their use in pharmacology and therapeutics are treated, and the importance of retinoids in the clinical practice of neoplastic and skin diseases is also covered.

I hope that this volume will be of value to all research workers in this field, but also and mainly to young people now approaching these fascinating molecules, who may take suggestions to further retinoid research and produce new and exciting results.

I gratefully and heartily thank all the contributors whose efforts have made this book possible.

Maria A. Livrea
Palermo, August 1999

Vitamin A and retinoids: an update of biological aspects and clinical applications
M.A. Livrea (ed.)
© 2000 Birkhäuser Verlag Basel/Switzerland

Factors affecting blood levels of vitamin A

M.V. Gamble and W.S. Blaner

Department of Medicine and Institute of Human Nutrition, College of Physicians and Surgeons, Columbia University, HHSC 502, 701 W. 168th St., New York, NY 10032, USA

Introduction

The predominant retinoid in the fasting circulation is retinol, all of which is bound to its specific plasma transport protein, retinol-binding protein (RBP) [1, 2]. Although retinol accounts for approximately 95 to 99% of all retinoid in the circulation, other retinoids also are present. Fasting human and rodent blood contains very low levels of both all-*trans*- and 13-*cis*-retinoic acid (approximately 0.2 to 0.7% of those of retinol) [3], as well as low levels of retinyl esters in lipoprotein fractions, particularly very low-density lipoproteins (VLDL) and low density lipoproteins (LDL) [4]. Soluble glucuronides of both retinol and retinoic acid are also detectable in the circulation of humans and rodents [5], as are provitamin A carotenoids like β-carotene

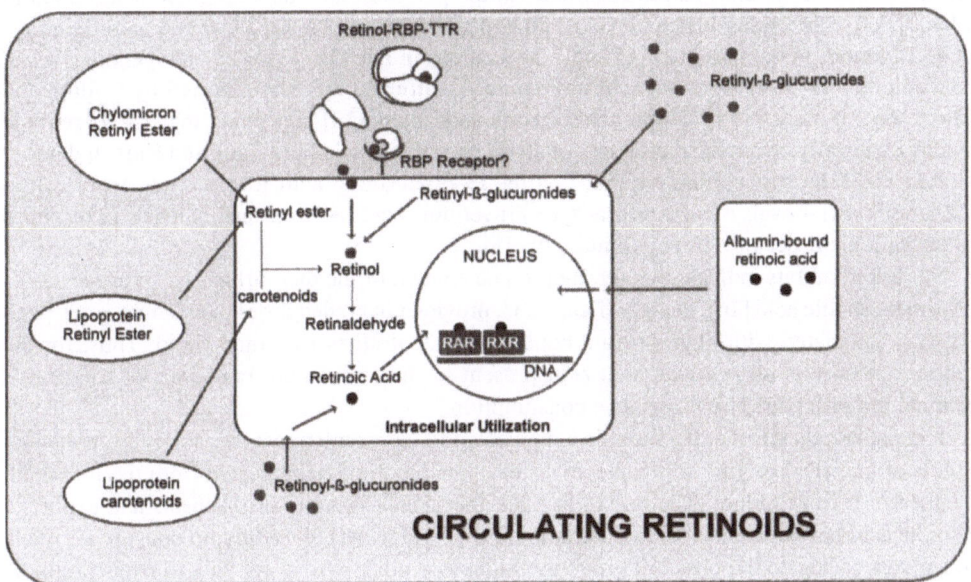

Figure 1. Overview representation of the major retinoid forms found in the circulation. This scheme takes into account the different retinoid forms that may be encountered in the circulation and the carriers responsible for retinoid transport through the circulation.

which can be converted by enzymes present in many tissues to retinoids [3]. These pathways are summarized diagramatically in Figure 1.

Following consumption of a retinoid-rich meal, the circulation can also contain large amounts of retinyl esters in chylomicrons and/or their remnants [6, 7]. Postprandial retinyl ester concentrations can easily exceed fasting retinol levels by several fold [7, 8]. Moreover, consumption of a retinoid-rich meal brings about an increase in circulating concentrations of all-*trans*- and 13-*cis* retinoic acid and of glucuronides of both retinol and retinoic acid [7, 9].

For the purposes of this brief review, we will focus primarily on the factors affecting retinol bound to RBP. In addition, we will summarize literature regarding chylomicron retinyl ester transport as well as circulating retinoic acid. Although it is believed that the predominant route through which tissues acquire retinoid is through delivery of retinol by RBP, it is likely that other circulating retinoids, although redundant, also serve as physiologically significant sources of retinoid for tissues. This is clearly evidenced by the phenotype of RBP knockout mice that totally lack RBP. Adult RBP-deficient mice maintained on a control diet are viable and fertile, suggesting that these alternative delivery pathways can be significant for the delivery of retinoids to tissues (unpublished observations).

Circulating retinoic acid

A small percentage of dietary vitamin A is converted to retinoic acid in the intestine (or may arrive as such in the diet) and is absorbed via the portal circulation as retinoic acid bound to albumin [3]. The fasting plasma level of all-*trans*-retinoic acid is very low and is in the range of 4–14 nmol/l in humans [10, 11] and 7.3–9 nmol/l in rats [12, 13]. The endogenous plasma concentration of 13-*cis*-retinoic acid in vitamin A-sufficient rats was reported by Cullum and Zile to be 3.0 nmol/l [12]. These investigators and others [13] concluded that 13-*cis*-retinoic acid is a naturally occurring metabolite of all-*trans*-retinoic acid [12]. Tang and Russell demonstrated in 26 human volunteers that fasting 13-*cis*-retinoic acid levels range from 3.7 to 7.2 nmol/l [14]. Levels of all-*trans*- and 13-*cis*-retinoic acid rise over fasting levels in response to physiologic doses of retinyl palmitate [9, 15].

Studies in rabbits indicate that the β-carotene content of the diet influences serum levels of all-*trans*-retinoic acid [16]. Dietary β-carotene, provided in graded doses over a nine-week period, was associated with higher serum concentrations of all-*trans*-retinoic acid. Thus, for the rabbit, not only is all-*trans*-retinoic acid present in the circulation, but levels of this active retinoid are influenced by β-carotene consumption.

It is not known whether the retinoic acid present in the circulation arises solely from the diet (i.e. is of intestinal origin) or if some arises through export of retinoic acid from tissues which synthesize it from retinol. The possibility that the kidney is a site of synthesis and export of retinoic acid has been raised in the literature [17, 18]. However, currently no data are available to support or negate the possibility that the kidney or other tissues are able to export retinoic acid to the circulation for delivery to other tissues.

Circulating retinoic acid can be taken up efficiently by cells, and no cell surface receptor specific for retinoic acid is known. All-*trans*-retinoic acid, although fully ionized in free solution at pH 7.4, is uncharged when within a lipid environment [19, 20]. In the uncharged state, all-*trans*-retinoic acid moves rapidly between the outer and inner leaflets of the plasma membrane. Thus, all-*trans*-retinoic acid is able to traverse cellular membranes and rapidly enter cells.

Using steady state tracer kinetic techniques in rats, Kurlandsky et al. determined how much of the all-*trans*-retinoic acid present in a given tissue was derived from the circulation [21]. Some of the findings of this study are summarized in Table 1. For the liver and brain, greater than 75% of the all-*trans*-retinoic acid present in these tissues was derived from the circulation (88.4% for brain and 78.2% for liver). The seminal vesicles, epididymis, kidney, epididymal fat, perinephric fat, spleen, and lungs derived, respectively, 23.1%, 9.6%, 33.4%, 30.3%, 24.5%, 19.0%, and 26.7% of their tissue all-*trans*-retinoic acid pools from the circulation. Only 2.3% and 4.8%, respectively, of the retinoic acid present in the pancreas and eyes was contributed by the circulation. For all rats studied, the testes did not take up any (<1%) all-*trans*-retinoic acid from the circulation. Kurlandsky et al. also reported a fractional catabolic rate for all-*trans*-retinoic acid in plasma of 30.4 pools/hr and an absolute catabolic rate for all-*trans*-retinoic acid of 640 pmol/h [21]. These rates are very rapid compared to those of the only other naturally occurring retinoid that has been studied under normal physiological conditions, all-*trans*-retinol [22].

Table 1. Tissue all-*trans*-retinoic acid concentrations and percentage of all-*trans*-retinoic acid present in the tissue contributed from the circulating all-*trans*-retinoic acid pool in rats[1,2]

Tissue	All-*trans*-retinoic acid pmol/g tissue	Plasma contribution %
Liver	11.3 ± 4.7	78.2 ± 28.2
Brain	6.8 ± 3.3	88.4 ± 21.9
Testis	10.7 ± 2.7	0.7 ± 0.9
Seminal vesicles	12.0 ± 7.0	23.1 ± 13.5
Epididymis	4.2 ± 1.6	9.6 ± 6.1
Kidney	8.3 ± 4.0	33.4 ± 17.1
Pancreas	29.3 ± 16.3	2.3 ± 2.5
Epididymal Fat	15.7 ± 12.3	30.3 ± 29.9
Perirenal Fat	12.7 ± 8.7	24.5 ± 14.5
Spleen	12.7 ± 12.0	19.0 ± 10.1
Eyes	125 ± 37.3	4.8 ± 3.6
Plasma[3]	1.8 ± 0.7	

[1] All values are taken from [23].
[2] Each value is given as the mean ± 1 standard deviation for separate measures employing 8 individual 400–450 g male Sprague-Dawley rats.
[3] The plasma concentration of all-*trans*-retinoic acid is given as pmol/ml plasma.

Chylomicron retinoid

Together with triglycerides, cholesteryl esters, phospholipids and other dietary lipids, dietary retinol is incorporated as retinyl ester into chylomicron particles that are secreted by the enterocyte into the lymphatic system. Once in the general circulation, chylomicrons interact with lipoprotein lipase (LPL) bound to the luminal surface of the vascular endothelium where rapid lipolysis of triglycerides occurs. LPL catalyzed hydrolysis gives rise to free fatty acids and smaller lipoprotein particles termed chylomicron remnants. These remnants acquire apolipoprotein E (apoE) in the plasma or in the space of Disse within the liver [23–26]. ApoE acquisition is essential for chylomicron remnant and thus retinyl ester clearance by the liver [6]. The importance of this role of apoE is apparent in apoE-deficient mice, which clear post-prandial cholesterol and retinyl ester very slowly [27–29]. These mice also show very high circulating levels of total cholesterol and retinyl esters, even in the fasting state [27–29]. The exact mechanisms responsible for uptake of chylomicron remnants by the liver are still unsettled. Several distinct cell surface receptors that are able to bind apoE-containing lipoproteins are probably involved in the receptor-mediated uptake of chylomicron remnants by hepatocytes. Among those are the low-density lipoprotein receptor, the LDL receptor related protein and the lipolysis-stimulated receptor [6]. In addition, heparan sulfate proteoglycans, located on the surface of hepatocytes, may be responsible for the initial interaction of remnants with the cells of the liver [6].

In humans, molar concentrations of chylomicron retinyl esters can exceed those of retinol-RBP by 3- to 4-fold during the 3 to 6 h following consumption of a bolus test dose of retinol [8] or of a retinol-rich meal [7]. For healthy humans, postprandial clearance of chylomicron retinoids normally occurs within 6 to 8 h [4, 8]. However, in some disease states in which clearance of postprandial lipids is delayed, the rate of chylomicron retinoid clearance can be much slower than that of healthy individuals. Chylomicron retinoid clearance in rats and mice is more rapid than in humans, with circulating chylomicron retinyl ester levels reaching a maximum within 2 to 4 h after consumption of retinol [30].

Data accumulated from studies carried out over the past 35 years consistently show that approximately 75% of chylomicron retinyl ester is removed from the circulation by the liver and the remaining 25% by extrahepatic tissues including skeletal muscle, adipose tissue, heart, spleen and the kidney [30, 31]. Neither the physiological significance nor the biochemical basis for the observation that some chylomicron or chylomicron remnant retinyl ester is taken up by extrahepatic tissues has been the subject of systematic investigation until very recently. These recent investigations suggest that LPL may play an important role in facilitating uptake of post-prandial vitamin A by extrahepatic tissues [31]. At present we do not understand the physiological significance of postprandial retinoid uptake by extrahepatic tissues, but it is tempting to speculate that it represents an important delivery pathway through which some tissues can acquire retinoids.

RBP-bound retinol

RBP-bound retinol (RBP-ROH), the predominant form of retinoid in the fasting circulation, has a half-life of approximately 12 h [32]. RBP-ROH concentrations in well nourished Western populations range between 2 to 3 µmol/l [2, 33]. Since retinol is released from hepatic stores together with RBP, and since RBP has one high affinity binding site for one molecule of retinol,

retinol and RBP exist in the circulation in approximately equimolar concentrations. Furthermore, factors which affect either hepatic stores of retinol or hepatic RBP synthesis or secretion usually affect the concentrations of both retinol and RBP in the circulation [34–36]. Tissues other than the liver, including adipose tissue, kidney, eye and testis are known to synthesize RBP [2], but it is generally accepted that the majority of circulating RBP is synthesized by hepatic parenchymal cells [37–39]. The mechanism whereby concentrations of circulating RBP-ROH are regulated is not well understood, but could involve regulation of RBP synthesis and/or secretion, regulation of RBP-ROH clearance and catabolism, or a combination of these. Studies of retinol homeostasis in animals by many different investigators have suggested that a feedback control mechanism (possibly involving plasma retinoic acid, apo-RBP or a modified form of RBP) regulates mobilization of RBP-ROH from hepatic retinol stores [22, 40–47].

There are many disease states which influence RBP-ROH levels. For example, hypothyroidism is known to decrease [48, 49], and renal impairment to increase [1, 41, 50], circulating RBP-ROH levels. RBP-ROH levels have been reported to be altered in both Type 1 and Type 2 diabetes (see [51] for a recent review). Below we focus on some of the more well characterized factors known to influence circulating levels of RBP-ROH.

Retinol availability: deficiency and excess

Under conditions of either insufficient or excessive dietary retinoid intake an individual's circulating level of retinol is defended irrespective of the abundance of liver retinol stores, until liver stores reach some critical level beyond which the amount of circulating RBP-ROH is affected [35, 42, 52]. The regulatory factors responsible for establishing and maintaining this homeostatic set point within an individual are not well understood. In addition to any individual's normal serum retinol homeostatic set point, the "critical level" of hepatic stores also varies from one individual to another, and it has been suggested that it is influenced by non-retinoid factors such as the protein quality and quantity of the diet [34–36, 47, 53].

When retinol is unavailable within the liver, RBP continues to be synthesized, but is retained and accumulates substantially within the endoplasmic reticulum (ER) of hepatocytes [52]. The mechanism by which retinol enables RBP to be secreted from hepatocytes is not well characterized. Several studies employing isolated cells have demonstrated that retinol-deficiency prevents the transit of newly synthesized RBP from the ER to the Golgi [2, 39, 54]. It has been hypothesized that since retinol binding to RBP increases the affinity of RBP for transthyretin (TTR), that RBP-ROH and TTR are co-secreted as a preformed complex from hepatocytes, and that complex formation may be requisite for optimal secretion of RBP-ROH [55, 56]. Kaji and Lodish have demonstrated in human HepG2 cells that retinol enhances the stability of RBP protein to unfolding under reducing conditions (i.e. in the presence of exogenously added DTT). This stabilizing effect by retinol is dependent upon other intracellular factors, including protein disulfide isomerase and an as yet unidentified ATP-requiring factor [57, 58]. It is possible that RBP translocation through the ER membrane resembles that of apolipoprotein B, where it is believed that availability of the lipid ligand on the luminal side of the ER is involved in translocation of the protein through the translocation channel [59]. Although it has been demonstrated that newly ingested retinol follows the last-in-first-out rule in the liver [8], it is not known how retinol is delivered to newly synthesized RBP in the ER lumen of parenchymal cells for secretion as RBP-ROH. In contrast to recent advances in our understanding of post-translational regulatory pathways for synthesis, secretion and intracellular degradation of other secretory proteins, the mechanisms underlying regulation of RBP secretion have not been extensively studied. Greater understanding of this process would undoubtedly impact upon our understanding of the regulation of circulating retinoid levels.

Dietary-induced retinoid toxicity occurs infrequently since preformed retinoids are not abundant in the food supply with the exception of liver, which is generally not consumed in large quantities. Toxicity induced by excessive supplement use does occur, albeit rarely, i.e. reported incidences averaged less than ten cases per year between 1976 and 1987 [60]. The development of hypervitaminosis A is influenced by the composition of the supplement, the dosing regimen, and possibly interactions with other micronutrients [60]. Alterations in serum retinoids induced by vitamin A toxicity include increased circulating retinyl esters and retinoic acid and a decline in circulating RBP-ROH levels [61]. Most toxicology studies are undertaken to determine the safety of synthetic retinoids. Paradoxically, the toxic effects of many synthetic retinoids resemble the effects seen in vitamin A deficiency, e.g. night-blindness. This is because many synthetic retinoids cause a reduction in circulating RBP-ROH levels, possibly by binding to newly synthesized hepatic RBP and preventing its secretion from the liver [62].

Transthyretin (TTR)

RBP and retinol exist in the circulation bound as a ternary complex to TTR. TTR consists of a tetramer made up of four identical subunits, with a total molecular weight of 55 kDa. Binding of RBP-ROH to TTR prevents renal filtration of the smaller RBP (MW 21 kDa) and retinol [61]. It was originally proposed, based on the observation that plasma RBP levels drop in vitamin A-deficiency whereas those of TTR do not, that RBP must complex with TTR in the circulation after both proteins are independently secreted from the hepatocyte [63]. However, recent studies by several independent research groups have suggested that the RBP-TTR complex is formed within the ER of hepatocytes prior to secretion [55, 56, 64–68].

Deletion of the gene for TTR is not lethal, and in fact, TTR deficient mice are phenotypically normal despite circulating RBP-ROH levels which are only approximately 5% of normal. Circulating all-*trans*-retinoic acid levels in these animals are elevated 2.4-fold [66]. While hepatic total retinol levels are the same as wild type mice, hepatic RBP levels are 60% higher for TTR-deficient than wild type mice, suggesting a partial blockage of RBP secretion [65, 66]. Studies of hepatocytes from the two strains of mice indicate that mRNA levels and RBP release into the medium are not altered. Thus, it seems that the accumulation of hepatic RBP protein is due to either increased RBP synthesis or decreased intracellular degradation of RBP.

Similarly, patients who carry a specific point mutation in one allele of their TTR gene that results in a single amino acid substitution at position 84 have very low circulating levels of RBP-ROH [67, 68]. This mutant TTR (in contrast to at least 50 other reported TTR point mutations) has very low affinity for RBP, and thus could result in impaired secretion of RBP from the liver, increased renal filtration, or both. Nevertheless, TTR may not be essential for RBP secretion for all tissues or cell types, since several studies of non-hepatic cell types have demonstrated that RBP synthesis and secretion can occur in the absence of TTR [56, 69, 70]. For example, BFC-1β adipocytes [70] and primary rat Sertoli cells [69] synthesize and secrete RBP but neither cell type synthesizes TTR.

Interestingly, while the predominant carrier protein for retinoic acid is believed to be albumin, retinoic acid can also bind to TTR with relatively high affinity, although the physiological significance of this is not understood [71, 72]. Other biologically active molecules can also bind to TTR. These include penicillin, salicylates, steroids, flavonoids, and environmental contaminants such as hydroxylated polychlorinated biphenyls (TCB), dibenzodioxins and dibenzofurans [71]. Exposure to TCBs causes greatly reduced circulating retinol levels, and is attributed to disruption of the RBP-TTR complex, resulting in some of the toxic effects of TCB exposure [73].

Retinoic acid

It has long been known that all-*trans* retinoic acid (RA) provided in the diet to rats brings about a decline in RBP-ROH levels of 25 to 50% [47, 74, 75]. Consequently, it has been proposed that circulating RA levels could serve as a feedback signal responsible for regulating RBP secretion from hepatocytes. In cell culture, although RA is able to bind to newly synthesized RBP and can stabilize nascent RBP within the ER, binding of RA does not permit secretion of RBP from HepG2 hepatocytes as does retinol [58]. In principle, retinoid metabolism could mimic that of thyroid or vitamin D metabolism, two instances in which there is an abundant circulating precursor metabolite which is synthesized and released from one organ, taken up by target tissues, and converted to a more active metabolite resulting in feedback regulation.

The acute phase response

The acute phase response (APR) is a systemic event that occurs in response to stress such as trauma, infection or immunological disorders, part of which involves alterations in hepatic protein synthesis, with increases in positive- and decreases in negative acute phase proteins (APP). The time-course and magnitude of these alterations may be related to the cause, intensity and duration of the insult. Since RBP and TTR are both negative APPs, decreases in circulating RBP-ROH concentrations occur in response to stress, irrespective of retinoid nutritional status. This notion is supported by the observation that low RBP-ROH levels occur in children with measles in the U.S. where retinoid nutritional status would be expected to be adequate [76]. Moreover, these low RBP-ROH levels have been shown to improve post-infection in the absence of supplemental vitamin A [77, 78]. Assessment of retinoid nutritional status in children in underdeveloped countries is therefore complicated by a high prevalence of concurrent infections within the populations [77, 79–84]. The biological effects of the stress-induced decline in RBP-ROH seen during the APR are unclear. Lower serum retinol levels are associated with severity of disease [76, 77] and vitamin A supplementation has been demonstrated to reduce measles morbidity and mortality [84]. Nevertheless, these findings are not incompatible with the possibility that retinoid redistribution could serve to favorably enhance local retinoid concentrations at the site of utilization. Thus, it is presently unclear whether the decline in RBP-ROH observed during the APR has an adverse affect on the individual or whether the response reflects the actions taken by the body to combat the stress inducing the APR.

The mechanisms proposed to be responsible for stress-induced reduction in RBP-ROH levels include redistribution of RBP-ROH into the extravascular space [82], reduction in liver protein synthesis due to increased zinc utilization [77] or due to reduced secretion of hepatic TTR [85], increased urinary excretion of RBP-ROH [86] and increased utilization of plasma retinol by peripheral tissues [87]. While it seems likely that several or all of these mechanisms may be contributing simultaneously, it is clear that the APR results in diminished liver RBP and TTR mRNA levels, and this may represent the primary causative effect that reduces RBP-ROH levels in the circulation. Inflammation induced in rats by turpentine oil [88] or lipopolysaccharide treatment [89, 90] results in a decrease in liver RBP mRNA levels. The decrease in RBP mRNA was accompanied by changes in the circulation: parallel decreases in RBP and ROH, and a somewhat delayed decrease in TTR [89]. Interestingly, although kidney RBP and TTR protein concentrations are reduced by inflammation, in contrast to liver, kidney RBP mRNA levels were elevated, thus confirming the earlier observation that regulation of RBP expression is tissue-specific [91].

A principal effector of alterations in protein synthesis during the APR is interleukin-6 (IL-6). RBP and TTR protein levels were shown to be decreased in the circulation in cancer patients receiving subcutaneous IL-6 treatments, suggesting this cytokine may play a regulatory role in RBP secretion during the APR [92]. While a direct inhibitory effect of IL-6 on transcription of the TTR gene has been shown in HepG2 cells [93], such an effect has not been demonstrated for RBP gene transcription. Furthermore, since IL-6-deficient mice are still capable of inducing an APR, other cytokines are clearly involved in inducing the APR as well [94, 95]. TGFβ has been shown to be involved in the negative regulation of RBP synthesis by Hep3B cells, by decreasing RBP mRNA via a post-transcriptional mechanism [96]. While treatment of Hep3B cells with TGFβ diminished the accumulation of RBP mRNA, nuclear run-on assays demonstrated that the rate of transcription of the RBP gene was not altered [96]. Taken together, these studies suggest that synthesis of retinoid carrier proteins is decreased in hepatocytes during the APR, which would be expected to result in a re-distribution of retinol throughout the body.

Another affect of the APR on circulating retinoid levels occurs as a result of reduced lipolysis and delayed clearance of chylomicrons and their component retinyl esters [3, 97]. Recent work by Rosales and Ross demonstrated that retinoid supplementation of marginally retinoid-deficient rats undergoing an APR induced by lipopolysaccharide treatment showed lower concentrations of retinyl esters in the liver and perirenal fat, and delayed clearance of circulating retinyl esters [90].

Clearly, the APR has a profound effect on retinol metabolism and influences not only circulating ROH-RBP levels, but also circulating retinyl ester levels and probably the distribution of retinoids within tissues as well.

Protein malnutrition and protein-calorie malnutrition (PCM)

Both RBP and TTR are sensitive indicators of protein malnutrition, as their blood levels respond rapidly to changes in dietary protein intake due to their relatively short half-life in the circulation compared to albumin [98]. Consequently, decreased RBP synthesis and secretion arising from protein malnutrition will also result in decreased circulating retinol. A number of studies on children suffering from protein and protein-calorie malnutrition in the 1970s reported increases in serum retinol in response to treatment with adequate protein independent of supplemental retinoid administration [35, 36, 53]. Other studies have demonstrated that the increase in circulating RBP in response to treatment with retinoid supplements is blunted in protein-calorie malnutrition, and this was attributed to impairment of hepatic production of RBP [99]. Around the same time, diet studies in rats demonstrated that the quality of protein in the diet also affects RBP-ROH levels, and since hepatic RBP reserves in this study appeared adequate, it was suggested that a diet of poor quality protein exerts its effects by limiting hydrolysis of endogenous retinyl esters rather than limiting RBP synthesis [100, 101]. In addition, xeropthalmia relapse after vitamin A supplementation has been reported to occur more frequently in children with concurrent PCM [102].

The use of RBP and TTR to assess protein malnutrition and PCM is well justified by the recent literature. The fractional synthetic rate of RBP, TTR and apolipoprotein A1 were measured using a constant intragastric infusion of $^2H_3$-leucine in 22 Jamaican children with severe PCM upon admission (they also had evidence of infection), and at eight and 59 days post-admission. The plasma concentrations and synthesis rates for all three proteins were lower at admission than values at recovery. RBP and TTR concentrations and synthesis rates reverted to normal by eight days after the start of nutritional rehabilitation, whereas apoA1 was recovered more slowly [103]. The biochemical mechanisms underlying the effects of protein malnutrition and PCM on RBP and TTR have also been investigated. While TTR mRNA levels in livers of PCM rats are significantly reduced, those of RBP are not [104]. Thus, TTR synthesis may be impaired in protein malnutrition and PCM. It has been shown that hepatic retinyl palmitate hydrolase activity decreases during PCM and this may cause a local deficiency of retinol in hepatocytes, resulting in a secretory block of RBP [101]. Lyoumi et al. have recently demonstrated that protein malnutrition in rats can induce IL-6 and it is possible that the effect of protein malnutrition and PCM on circulating RBP and TTR levels shares a common biochemical mechanism with that of the APR [105].

Zinc

Zinc status is known to affect retinoid levels in the circulation [106, 107]. It has been proposed that zinc plays a regulatory role in the synthesis of RBP in the liver. Zinc deficient rats have been shown to have a reduced rate of hepatic RBP synthesis [107], lower levels of plasma RBP-ROH and lower hepatocyte concentrations of RBP when compared with pair-fed zinc-sufficient rats [108]. In humans, a linkage between retinoid and zinc status has not been conclusively demonstrated in either observational or cross-sectional studies or through intervention trials

[106]. Some studies of children have demonstrated significant relationships between serum zinc and retinol levels or night blindness; whereas others have failed to establish this relationship [106]. It has often been suggested that zinc-related night blindness might be attributed to a requirement for zinc by the 11-*cis*-retinol dehydrogenase of the visual cycle. However, the visual cycle dehydrogenases are members of the family of short-chain alcohol dehydrogenases, the entire family of which are not zinc-requiring enzymes [109, 110].

In biochemical terms, the actions of zinc in maintaining normal circulating levels of RBP-ROH are not well understood. Since many transcription factors require zinc, it is not difficult to understand how insufficient zinc status will influence protein synthesis, both generally and specifically. At present, zinc-dependent factors that are importantly and specifically involved in RBP synthesis and/or secretion or in RBP-ROH uptake and metabolism have not been identified. Thus, the effects of impaired zinc status on circulating retinoid levels may be secondary to other more general actions of zinc within the organism (e.g. zinc's role in gene transcription).

Steroid hormones

Both estrogen and dexamethasone have been reported to affect circulating RBP-ROH and/or tissue total retinol levels. It is known from early endocrinological studies that women taking oral contraceptives have elevated RBP-ROH levels [1]. In these early studies, the estrogen component of the oral contraceptive was proposed to be largely responsible for the increased RBP-ROH levels. This is verified by study of women using modern contraceptive formulations, which contain lower doses of estrogen than earlier formulations and which do not give rise to the elevations in RBP-ROH levels observed in the earlier studies (unpublished observations). RBP levels have also been found to show a bi-cyclic variation during the menstrual cycle [1]. Studies aimed at providing a molecular basis of these observations carried out in ovarectomized rats indicate that estrogen treatment results in a five-fold increase in kidney RBP mRNA levels [91]. Interestingly, no rise in liver RBP mRNA levels was observed [91]. Thus, estrogen may have a direct effect on RBP gene expression in the kidney, and this may account for the elevated levels of RBP-ROH associated with the use of contraceptives containing high doses of estrogens.

In rat studies, dexamethasone treatment at a dose of 0.5 mg/kg/day for seven days resulted in an elevation of serum RBP-ROH for both retinoid-sufficient and retinoid-deficient rats [111]. Hepatic and lung total retinol levels were significantly diminished in the dexamethasone treated retinoid-sufficient rats, possibly due to increased mobilization and/or utilization of tissue retinol stores. In this study, the retinoid-deficient rats had no appreciable hepatic or lung total retinol stores. Another study carried out in pregnant rhesus monkeys showed that dexamethasone treatment (0.1 to 15 mg/kg) at 132 days' gestation followed by cesarean section at day 135 resulted in fetal and maternal levels of RBP-ROH that were significantly elevated over those of placebo treated animals [112]. Additionally, there was a trend for fetal RBP levels to increase with increasing dexamethasone dose. Although the effect of dexamethasone on elevating circulating RBP-ROH levels is well established, the biochemical mechanisms underlying this response are not currently understood.

Genetic factors

There is only limited information available regarding genetic factors that influence circulating RBP-ROH levels. It is clear that circulating RBP-ROH levels are different across species and across different genetic strains within a species [113]. As described above, humans appear to maintain a homeostatic set point of circulating RBP-ROH levels. It would appear that this homeostatic set point is a property of the individual and, moreover, one that is influenced by genetic factors. However, the genetic factors that control or define this set point remain elusive.

Two independent families have been described in the literature that have abnormally low blood RBP-ROH levels [114–117]. In one family from Japan, several members reportedly have serum RBP-ROH levels that are approximately 50% that of normal [114–116]. These diminished RBP-ROH levels did not respond to oral administration of retinol or to a protein-rich diet [114–116]. RBP isolated from one affected member of this family demonstrated no differences in its molecular weight, isoelectric point, binding to TTR or immunological properties as compared to RBP isolated from unaffected family members [114–116]. At present it is not clear whether the diminished RBP-ROH levels measured in these individuals arises from some defect in the gene for RBP or if the defect exists in another gene that has an important influence on the normal physiology of RBP-ROH (i.e. on RBP synthesis, secretion, turnover or catabolism).

A recent report describing two sisters in Germany indicates that these women have no detectable RBP or retinol in their circulation [117]. Interestingly, they apparently harbor two distinct point mutations in their RBP genes, one on each allele of the gene [117]. At present, the mechanism whereby these two mutations contribute to the phenotype of undetectable RBP-ROH levels is not known. It is also unclear if the mutant RBP proteins are made but not secreted from the liver or if they are made and secreted, but then rapidly cleared from the circulation [117].

Regulation of RBP transcription and translation

The human gene for RBP has been investigated by Colantouni and colleagues [2, 118, 119]. These workers have shown that RBP gene expression can be upregulated in HepG2 hepatocytes upon exposure of the cells to either retinol or retinoic acid. Upregulation was both time- and dose-dependent and an increase in RBP mRNA levels was accompanied by a later increase in RBP protein in both the cells and culture medium. The region in the promoter of the RBP gene responsible for this upregulation was identified by band-shift assays and site-directed mutagenesis, and was mapped to two degenerate retinoic acid response elements separated by 30 nucleotides [118]. Included in this element is a GC-rich Sp1 consensus-like sequence. The entire element was required in order to confer retinoic acid inducibility to a heterologous promoter. In addition to retinoids, the transcription factor Sp1 or a closely related protein was shown to play an important role in regulating transcription of the RBP gene [118].

Recent investigations by Jessen and Satre indicate that the mouse gene for RBP is markedly upregulated by cAMP treatment in Hepa 1-6 hepatoma cells [120]. RBP mRNA levels were increased up to six-fold in a dose-dependent manner within 24 h after exposure to dibutyryl cAMP or forskolin. The increase in RBP mRNA levels was accompanied by an increase in RBP protein levels. This observation raises the question of whether there is a cAMP-response element in the mouse gene for RBP and if cAMP plays a role in regulating RBP gene expression.

Summary

Retinoid delivery to cells occurs through many diverse pathways and is influenced by a number of physiological factors. Ultimately, nutritional and health status are the most important factors which influence circulating retinoid levels. We have only a very limited understanding of how circulating levels of retinoids, whether retinol, retinyl ester or retinoic acid, are regulated and interact within the intact organism. This near lack of knowledge regarding how tissues and cells acquire retinoids needed for maintaining normal cellular actions is surprising considering the exquisite detail in which we now understand how retinoids act in regulating gene transcription. The presence of retinyl esters in chylomicrons was first reported in the 1950s, RBP was first identified in the late 1960s, and retinoic acid was identified as an endogenous component of plasma in the early 1980s. Retinoic acid receptors were first identified in the late 1980s and the retinoid X receptors in the early 1990s. Ironically, we have a much more detailed understanding of the molecular actions of the retinoid nuclear receptors than we do of retinoid delivery to tissues and cells. While important progress in the molecular biology of retinoid actions has been swift and informative, future studies of how retinoids are generated and transported in the body are required if this new molecular knowledge of retinoid actions is to be translated into improvements in the area of human health and disease.

References

1   Goodman DS (1984) Plasma retinol-binding protein. *In*: MB Sporn, AB Roberts, DS Goodman (eds): *The Retinoids*. Academic Press, Orlando, 41–88
2   Soprano DR, Blaner WS (1994) Plasma retinol-binding protein. *In*: MB Sporn, AB Roberts, DS Goodman (eds): *The Retinoids: Biology, Chemistry and Medicine*. Raven Press, New York, 257–282
3   Blaner WS, Olson JA (1994) Retinol and retinoic acid metabolism. *In*: MB Sporn, AB Roberts, DS Goodman (eds): *The Retinoids: Biology, Chemistry, and Medicine*. Raven Press, New York, 229–256
4   Krasinski SD, Cohn JS, Russell RM, Schaefer EJ (1990) Postprandial plasma vitamin A metabolism in humans: a reassessment of the use of plasma retinyl esters as markers for intestinally derived chylomicrons and their remnants. *Metabolism* 39: 357–365
5   Barua AB (1997) Retinoyl β-glucuronide: a biologically active form of vitamin A. *Nutr Rev* 55: 259–267
6   Cooper AD (1997) Hepatic uptake of chylomicron remnants. *J Lipid Res* 38: 2173–2192
7   Arnhold T, Tzimas G, Wittfoht W, Plonait S, Nau H (1996) Identification of 9-*cis*-retinoic acid and 9,13-di-*cis*-retinoic acid, and 14-hydroxy-4,14-retro-retinol in human plasma after liver consumption. *Life Sci* 59: PL169–PL177
8   Von Reinersdorff D, Bush E, Liberato DJ (1996) Plasma kinetics of vitamin A in humans after a single oral dose of [8,9,19-^{13}C]retinyl palmitate. *J Lipid Res* 37: 1875–1885
9   Eckhoff C, Collins MD, Nau H (1991) Human plasma all-*trans*, 13-*cis*- and 13-*cis*-4-oxoretinoic acid profiles during subchronic vitamin A supplementation: Comparison to retinol and retinyl ester plasma levels. *J Nutr* 121: 1016–1025
10 De Leenheer AP, Lambert WE, Claeys I (1982) All-*trans*-retinoic acid: Measurement of reference values in human serum by high performance liquid chromatography. *J Lipid Res* 23: 1362–1367
11 Eckhoff C, Nau H (1990) Identification and quantitation of all-*trans*- and 13-*cis*-retinoic acid and 13-*cis*-4-oxoretinoic acid in human plasma. *J Lipid Res* 31: 1445–1454
12 Cullum ME, Zile MH (1985) Metabolism of all-*trans*-retinoic acid and all-*trans*-retinyl acetate:

Demonstration of common physiological metabolites in rat small intestinal mucosa and circulation. *J Biol Chem* 260: 10 590–10 596

13 Napoli JL, Pramanik BC, Williams JB, Dawson MI, Hobbs PD (1985) Quantification of retinoic acid by gas-liquid chromatography-mass spectrometry: Total versus all-*trans*-retinoic acid in human plasma. *J Lipid Res* 26: 387–392

14 Tang GW, Russell RM (1990) 13-*cis* Retinoic acid is an endogenous compound in human serum. *J Lipid Res* 30: 175–182

15 Tang GW, Russell RM (1991) Formation of all-*trans*-retinoic acid and 13-*cis*-retinoic acid from all-*trans* retinyl palmitate in humans. *J Nutr Biochem* 2: 2100–2103

16 Folman Y, Russell RM, Tang GW, Wolf G (1989) Rabbits fed on β-carotene have higher serum levels of all-*trans*-retinoic acid than those receiving no β-carotene. *Brit J Nutr* 62: 195–201

17 Bhat PV, Poissant L, Lacroix A (1988) Properties of retinal-oxidizing enzyme activity in rat kidney. *Biochim Biophys Acta* 967: 211–217

18 Bhat PV, Poissant L, Falardeau P, Lacroix A (1988) Enzymatic oxidatation of all-*trans* retinal to retinoic acid in rat tissues. *Biochem Cell Biol* 66: 735–740

19 Noy N (1992) The ionization behavior of retinoic acid in aqueous environments and bound to serum albumin. *Biochim Biophys Acta* 1106: 152–158

20 Noy N (1992) The ionization behavior of retinoic acid in lipid bilayers and in membranes. *Biochim Biophys Acta* 1106: 159–164

21 Kurlandsky S, Gamble MV, Ramakrishnan R, Blaner WS (1995) Plasma delivery of retinoic acid to tissues in the rat. *J Biol Chem* 270: 17 850–17 857

22 Lewis KC, Green MH, Green JB, Zech LA (1990) Retinol metabolism in rats with low vitamin A status: a compartmental model. *J Lipid Res* 31: 1535–1548

23 Williams DL, Dawson PA, Newman TC, Rudel LL (1985) Apolipoprotein E synthesis in peripheral tissues of nonhuman primates. *J Biol Chem* 260: 2444–2451

24 Hamilton RL, Wong JS, Guo LSS, Krisans S, Havel RJ (1990) Apolipoprotein E localization in rat hepatocytes by immunogold labeling of cryothin sections. *J Lipid Res* 31: 1589–1603

25 Ji ZS, Fazio S, Lee YL, Mahley RW (1994) Secretion-capture role for apolipoprotein E in remnant lipoprotein metabolism involving cell surface heparan sulfate proteoglycans. *J Biol Chem* 269: 2764–2772

26 Ji ZS, Sanan DA, Mahley RW (1995) Intravenous heparinase inhibits remnant lipoprotein clearance from the plasma and uptake by the liver: *in vivo* role of heparan sulfate proteoglycans. *J Lipid Res* 36: 583–592

27 Ishibashi S, Herz J, Maeda N, Goldstein JL, Brown MS (1994) The two-receptor model of lipoprotein clearance: Tests of the hypothesis in "knockout" mice lacking the low density lipoprotein receptor, apolipoprotein E, or both proteins. *Proc Natl Acad Sci USA* 91: 4431–4435

28 Ishibashi S, Perrey S, Chen Z, Osuga J, Shimada M, Ohashi K, Harada K, Yazaki Y, Yamada N (1996) Role of the low density lipoprotein (LDL) receptor pathway in the metabolism of chylomicrons. *J Biol Chem* 37: 22 422–22 427

29 Mortimer BC, Beveridge DJ, Martins IJ, Redgrave TG (1995) Intracellular localization and metabolism of chylomicron remnants in the livers of low density lipoprotein receptor-deficient mice and apoE-deficient mice. *J Biol Chem* 270: 28 767–28 776

30 Goodman DS, Huang HS, Shiratori T (1965) Tissue distribution and metabolism of newly absorbed vitamin A in the rat. *J Lipid Res* 6: 390–396

31 van Bennekum AM, Kako Y, Weinstock PH, Harrison EH, Deckelbaum RJ, Goldberg IJ, Blaner WS (1999) Lipoprotein lipase expression level influences tissue clearance of chylomicron retinyl ester. *J Lipid Res* 40: 565–574

32 Vahlquist A, Peterson PA and Wibell L (1973) Metabolism of the vitamin A-transporting protein complex. *Eur J Clin Invest* 3: 352–362

33 Blaner WS (1989) Retinol-binding protein: the serumtransport protein for vitamin A. *Endocrine Rev* 10: 308–316.

34 Smith FR, Goodman DS, Arroyave G, Viteri F (1973) Serum vitamin A, retinol-binding protein, and prealbumin concentrations in protein-calorie malnutrition. II. Treatment including supplemental vitamin A. *Amer J Clin Nutr* 26: 982–987

35 Smith FR, Goodman DS, Zaklama MS, Gabr MK, El Maraghy S, Patwardhan VN (1973) Serum vitamin A, retinol-binding protein, and prealbumin concentrations in protein-calorie malnutrition. I. A functional defect in hepatic retinol release. *Amer J Clin Nutr* 26: 973–981

36 Ingenbleek Y, Van Den Schrieck HG, De Nayer P, De Visscher M (1975) The role of retinol-binding protein in protein-calorie malnutrition. *Metabolism* 24: 633–641

37 Blaner WS, Smith JE, Dell RB, Goodman DS (1985) Spatial distribution of retinol-binding protein and retinyl palmitate hydrolase activity in normal and vitamin A deficient rat livers. *J Nutr* 115: 856–864

38 Friedman SL, Wei S, Blaner WS (1993) Retinol release by activated rat hepatic lipocytes: regulation by Kupffer cell-conditioned medium and PDGF. *Amer J Physiol* 264(Pt 1): G947–G952

39 Suhara A, Kato M, Kanai M (1990) Ultrastructural localization of plasma retinol-binding protein in rat liver. *J Lipid Res* 31: 1669–1681

40 Gerlach TH, Zile MH (1991) Metabolism and secretion of retinol transport complex in acute renal failure.

J Lipid Res 32: 515–520
41 Gerlach TH, Zile MH (1990) Upregulation of serum retinol in experimental acute renal failure. *FASEB J* 4: 2511–2517
42 Green MH, Green JB, Lewis KC (1987) Variation in retinol utilization rate with vitamin A status in the rat. *J Nutr* 117: 694–703
43 Green MH, Uhl L, Green JB (1985) A multicompartmental model of vitamin A kinetics in rats with marginal liver vitamin A stores. *J Lipid Res* 26: 806–818
44 Kielson B, Underwood BA, Loerch JD (1979) Effects of retinoic acid on the mobilization of vitamin A from the liver in rats. *J Nutr* 109: 787–795
45 Lewis KC, Green MH, Underwood BA (1981) Vitamin A turnover in rats as influenced by vitamin A status. *J Nutr* 111: 1135–1144
46 Sundaresan PR (1977) Rate of metabolism of retinol in retinoic acid-maintained rats after a single dose of radioactive retinol. *J Nutr* 107: 70–78
47 Underwood BA, Loerch JD, Lewis KC (1979) Effects of dietary vitamin A deficiency, retinoic acid and protein quantity and quality on serially obtained plasma and liver levels of vitamin A in rats. *J Nutr* 109: 796–806
48 Centanni M, Maiani G, Vermiglio F, Canettieri G, Sanna AL, Moretti F, Trimarchi F, Andreoli M (1998) Combined impairment of nutritional parameters and thyroid homeostasis in mildly iodine-deficient children. *Thyroid* 8: 155–159
49 Centanni M, Maiani G, Parkes AB, N'Diaye AM, Ferro-Luzzi A, Lazarus JH (1995) Thyroid homeostasis and retinol circulating complex relationships in a severe iodine-deficient area of Senegal. *J Endocrinol Invest* 18: 608–612
50 Jaconi S, Rose K, Hughes GJ, Saurat JH, Siegenthaler G (1995) Characterization of two post-translationally processed forms of human serum retinol-binding prote*In*: altered ratios in chronic renal failure. *J Lipid Res* 36: 1247–1253
51 Basu TK, Basualdo C (1997) Vitamin A homeostasis and diabetes mellitus. *Nutrition* 13: 804–806
52 Muto Y, Smith JE, Milch PO, Goodman DS (1972) Regulation of retinol-binding protein metabolism by vitamin A status in the rat. *J Biol Chem* 247: 2542–2550
53 Venkataswamy G, Glover J, Cobby M, Pirie A (1977) Retinol-binding protein in serum of xeropthalmic, malnourished children before and after treatment at a nutrition center. *Amer J Clin Nutr* 30: 1968–1973
54 Dixon JL, Goodman DS (1987) Studies on the metabolism of retinol-binding protein by primary hepatocytes from retinol-deficient rats. *J Cell Physiol* 130: 14–20
55 Bellovino D, Morimoto T, Tosetti F, Gaetani S (1996) Retinol binding protein and transthyretin are secreted as a complex formed in the endoplasmic reticulum in HepG2 human hepatocarcinoma cells. *Exp Cell Res* 222: 77–83
56 Melhus H, Nilsson T, Peterson PA, Rask L (1991) Retinol-binding protein and transthyretin expressed in HeLa cells form a complex in the endoplasmic reticulum in both the absence and the presence of retinol. *Exp Cell Res* 197: 119–124
57 Kaji EH, Lodish HF (1993) *In vitro* unfolding of retinol-binding protein by dithiothreitol. Endoplasmic reticulum-associated factors. *J Biol Chem* 268: 22 195–22 202
58 Kaji EH, Lodish HF (1993) Unfolding of newly made retinol-binding protein by dithiothreitol. Sensitivity to retinoids. *J Biol Chem* 268: 22 188–22 194
59 Zhou M, Wu X, Huang L, Ginsberg HN (1995) Apolipoprotein B100, an inefficiently translocated secretory protein, is bound to the cytosolic chaperone, heat shock protein 70. *J Biol Chem* 270: 25 220–25 224
60 Bendich A, Langseth L (1989) Safety of vitamin A. *Amer J Clin Nutr* 49: 358–371
61 Goodman DS (1984) Vitamin A and retinoids in health and disease. *New Engl J Med* 310: 1023–1031
62 Armstrong RB, Ashenfelter KO, Eckhoff C, Levin AA, Shapiro SS (1994) General and reproductive toxicology of retinoids. *In*: MB Sporn, AB Roberts, DS Goodman (eds): *The Retinoids. Biology, Chemistry and Medicine.* Raven Press, New York, 545–572
63 Navab M, Smith JE, Goodman DS (1977) Rat plasma prealbumin. Metabolic studies on effects of vitamin A status and on tissue distribution. *J Biol Chem* 252: 5107–5114
64 Natarajan V, Holven KB, Reppe S, Blomhoff R, Moskaug JO (1996) The C-terminal RNLL sequence of the plasma retinol-binding protein is not responsible for its intracellular retention. *Biochem Biophys Res Commun* 221: 374–379
65 Episkopou V, Maeda S, Nishiguchi S, Shimada K, Gaitanaris GA, Gottesman ME, Robertson EJ (1993) Disruption of the transthyretin gene results in mice with depressed levels of plasma retinol and thyroid hormone. *Proc Natl Acad Sci USA* 90: 2375–2379
66 Wei S, Episkopou V, Piantedosi R, Maeda S, Shimada K, Gottesman ME, Blaner WS (1995) Studies on the metabolism of retinol and retinol-binding protein in transthyretin-deficient mice produced by homologous recombination. *J Biol Chem* 270: 866–870
67 Berni R, Malpeli G, Folli C, Murrell JR, Liepnieks JJ, Benson MD (1994) The Ile-84→Ser amino acid substitution in transthyretin interferes with the interaction with plasma retinol-binding protein. *J Biol Chem* 269: 23 395–23 398
68 Waits RP, Yamada T, Uemichi T, Benson MD (1995) Low plasma concentrations of retinol-binding protein in individuals with mutations affecting position 84 of the transthyretin molecule. *Clin Chem* 41: 1288–1291

69 Davis JT, Ong DE (1992) Synthesis and secretion of retinol-binding protein by cultured rat Sertoli cells. *Biol Reprod* 47: 528–533

70 Zovich DC, Orologa A, Okuno M, Wong Yen Kong L, Talmage DA, Piantedosi R, Goodman DS, Blaner WS (1992) Differentiation-dependent expression of retinoid-binding proteins in BFC-1β adipocytes. *J Biol Chem* 267: 13 884–13 889

71 Zanotti G, D'Acunto MR, Malpeli G, Folli C, Berni R (1995) Crystal structure of the transthyretin-retinoic-acid complex. *Eur J Biochem* 234: 563–569

72 Smith TJ, Davis FB, Deziel MR, Davis PJ, Ramsden DB, Schoenl M (1994) Retinoic acid inhibition of thyroxine binding to human transthyretin. *Biochim Biophys Acta* 1199: 76–89

73 Brouwer A, van den Berg KJ (1986) Binding of a metabolite of 3,4,3',4'-tetrachlorobiphenyl to transthyretin reduces serum vitamin A transport by inhibiting the formation of the protein complex carying both retinol and thyroxine. *Toxicol Appl Pharmacol* 85: 301–312

74 Barua AB, Duitsman PK, Kostic D, Baura M, Olson JA (1997) Reduction of serum retinol levels following a single oral dose of all-*trans* retinoic acid in humans. *Int J Vitam Nutr Res* 67: 423–426

75 Shankar S, De Luca LM (1988) Retinoic acid supplementation of vitamin A-deficient diet inhibits retinoid loss from hamster liver and serum pools. *J Nutr* 118: 675–680

76 Butler JC, Havens PL, Sowell AL, Huff DL, Peterson DE, Day SE, Chusid MJ, Bennin RA, Circo R, Davis JP (1993) Measles severity and serum retinol (vitamin A) concentration among children in the United States. *Pediatrics* 91: 1176–1181

77 Coutsoudis A, Coovadia HM, Broughton M, Salisbury RT, Elson I (1991) Micronutrient utilisation during measles treated with vitamin A or placebo. *Int J Vitam Nutr Res* 61: 199–204

78 Reddy V (1981) Fat-soluble vitamin deficiencies in children in relation to protein energy malnutrition and environmental stress. *Prog Clin Biol Res* 77: 109–117

79 Velasquez-Melendez G, Okani ET, Kiertsman B, Roncada MJ (1995) Vitamin A status in children with pneumonia. *Eur J Clin Nutr* 49: 379–384

80 Willumsen JF, Simmank K, Filteau SM, Wagstaff LA, Tomkins AM (1997) Toxic damage to the respiratory epithelium induces acute phase changes in vitamin A metabolism without depleting retinol stores of South African children. *J Nutr* 127: 1339–1343

81 Rosales FJ, Ross AC (1998) A low molar ratio of retinol binding protein to transthyretin indicates vitamin A deficiency during inflammation: Studies in rats and a posteriori analysis of vitamin A-supplemented children with measles. *J Nutr* 128: 1681–1687

82 Thurnham DI, Singkamani R (1991) The acute phase response and vitamin A status in malaria. *Trans Roy Soc Trop Med Hyg* 85: 194–199

83 Filteau SM, Morris SS, Raynes JG, Arthur P, Ross DA, Kirkwood BR, Tomkins AM, Gyapong JO (1995) Vitamin A supplementation, morbidity, and serum acute-phase proteins in young Ghanian children. *Amer J Clin Nutr* 62: 434–438

84 Hussey GD, Klein M (1990) A randomized, controlled trial of vitamin A in children with severe measles. *New Engl J Med* 323: 160–164

85 Felding P, Fex G (1985) Rates of synthesis of prealbumin and retinol-binding protein during acute inflammation in the rat. *Acta Physiol Scand* 123: 477–483

86 Stephensen CG, Alvarez JO, Kohatsu J, Hardmeier R, Kennedy JI Jr, Gammon RB Jr (1994) Vitamin A is excreted in the urine during acute infection. *Amer J Clin Nutr* 60: 388–392

87 Kanda Y, Yamamoto N, Yoshino Y (1990) Utilization of vitamin A in rats with inflammation. *Biochim Biophys Acta* 1034: 337–341

88 Schreiber G, Tsykin A, Aldred AR, Thomas T, Fung WP, Dickson PW, Cole T, Birch H, DeJong FA, Milland J (1989) The acute phase response in the rodent. *Ann N Y Acad Sci* 557: 61–86

89 Rosales FJ, Ritter SJ, Zolfaghari R, Smith JE, Ross AC (1996) Effects of acute inflammation on plasma retinol, retinol-binding protein, and its mRNA in the liver and kidneys of vitamin A-sufficient rats. *J Lipid Res* 37: 962–971

90 Rosales FJ, Ross AC (1998) Acute inflammation induces hyporetinemia and modifies the plasma and tissue response to vitamin A supplementation in marginally vitamin A-deficient rats. *J Nutr* 128: 960–966

91 Whitman MM, Harnish DC, Soprano KJ, Soprano DR (1990) Retinol-binding protein mRNA is induced by estrogen in the kidney but not the liver. *J Lipid Res* 31: 1483–1490

92 Banks RE, Forbes MA, Storr M, Higginson J, Thompson D, Raynes J, Illingworth JM, Perren TJ, Selby PJ, Whicher JT (1995) The acute phase protein response in patients receiving subcutaneous IL-6. *Clin Exp Immunol* 102: 217–223

93 Bartalena L, Farsetti A, Flink IL, Robbins J (1992) Effects of interleukin-6 on the expression of thyroid hormone-binding protein genes in cultured human hepatoblastoma-derived (Hep G2) cells. *Mol Endocrinol* 6: 935–942

94 Kopf M, Baumann H, Freer G, Freudenberg M, Lamers M, Kishimoto T, Ainkernagel R, Bluethmann H, Kohler G (1994) Impaired immune and acute-phase responses in interleukin-6-deficient mice. *Nature* 368: 339–342

95 Fattori E, Cappelletti M, Costa P, Sellitto C, Cantoni L, Carelli M, Faggioni R, Fantuzzi G, Ghezzi P, Poli V (1994) Defective inflammatory response in interleukin 6-deficient mice. *J Exp Med* 180: 1243–1250

96 Morrone G, Poli V, Hassan JH, Sorrentino V (1992) Effect of TGF β on liver genes expression. Antagonistic effect of TGF β on IL-6-stimulated genes in Hep 3B cells. *FEBS Lett* 301: 1–4
97 Meraihi Z, Lutz O, Scheftel J, Frey A, Ferezou J, Bach AC (1991) Decreased lipolytic activity in tissues during infectious and inflammatory stress. *Nutrition* 7: 93–97
98 Shetty PS, Jung RT, Watrasiewica KE, James WPT (1979) Rapid-turnover transport proteins: an index of subclinical protein-energy malnutrition. *Lancet* 2: 230–232
99 Large S, Neal G, Glover J, Thanangkul O, Olson RE (1980) The early changes in retinol-binding protein and prealbumin concentrations in plasma of protein-energy malnourished children after treatment with retinol and an improved diet. *Brit J Nutr* 43: 393–402
100 Underwood BA (1980) Effect of protein quantity and quality on plasma response to an oral dose of vitamin A as an indicator of hepatic vitamin A reserves in rats. *J Nutr* 110: 1635–1640
101 Tsin ATC, Chambers JP, Garcia MH, Flores JM (1986) Decreased hepatic retinyl palmitate hydrolase activity in protein-deficient rats. *Biochim Biophys Acta* 878: 20–24
102 Sommer A, Tarwotjo I (1982) Protein deficiency and treatment of xerophthalmia. *Arch Ophthalmol* 100: 785–787
103 Morlese JF, Forrester T, Del Rosario M, Frazer M, Jahoor F (1998) Repletion of the plasma pool of nutrient transport proteins occurs at different rates during the nutritional rehabilitation of severely malnourished children. *J Nutr* 128: 214–219
104 Perozzi G, Mengheri E, Faraonio R, Gaetani S (1989) Expression of liver-specific genes coding for plasma proteins in protein deficiency. *FEBS Lett* 257: 215–218
105 Lyoumi S, Tamion F, Petit J, Dechelotte P, Dauguet C, Scotte M, Hiron M, Leplingard A, Salier JP, Daveau M et al (1998) Induction and modulation of acute-phase response by protein malnutrition in rats: comparative effect of systemic and localized inflammation on interleukin-6 and acute-phase protein synthesis. *J Nutr* 128: 166–174
106 Christian P, West KP (1998) Interactions between zinc and vitamin A: an update. *Amer J Clin Nutr* 68: 435S–441S
107 Kimball SR, Chen SJ, Risica R, Jefferson LS, Leure-dePree AE (1995) Effect of zinc on protein synthesis and expression of specific mRNAs in rat liver. *Metabolism* 44: 126–133
108 Shingwekar AG, Mohanram M, Reddy V (1979) Effect of zinc supplementation on plasma levels of vitamin A and retinol-binding protein in malnourished children. *Clin Chim Acta* 93: 97–100
109 Duester G (1996) Involvement of alcohol dehydrogenase, short-chain dehydrogenase/reductase, aldehyde dehydrogenase, and cytochrome P450 in the control of retinoid signalling by activation of retinoic acid synthesis. *Biochemistry* 35: 12 221–12 227
110 Persson B, Krook M, Jornvall H (1995) Short-chain dehydrogenases/reductases. *Adv Exp Med Biol* 372: 383–395
111 Georgieff MK, Radmer WJ, Sowell AL, Yeager PR, Blaner WS, Gunter EW, Johnson DE (1991) The effect of glucocorticoids on serum, liver and lung vitamin A and retinyl ester concentrations. *J Pediat Gastroenterol Nutr* 13: 376–382
112 Hustead VA, Zachman RD (1986) The effect of antenatal dexamethasone on maternal and fetal retinol-binding protein. *Amer J Obstet Gynecol* 154: 203–205
113 Vitamin A (1998) IARC Handbooks of Cancer Prevention, IARC Press, Lyon
114 Matsuo T, Matsuo N, Shiraga F, Koide N (1988) Keratomalacia in a child with familial hypo-retinol-binding proteinemia. *Jpn J Ophthalmol* 32: 249–254
115 Matsuo T, Matsuo N (1988) Characterization of retinol-binding protein in familial hypo-retinol-binding proteinemia. *Jpn J Ophthalmol* 32: 379–384
116 Matsuo T, Noji S, Taniguchi S, Matsuo N (1990) No major defect detected in the gene of familial hypo-retinol-binding proteinemia. *Jpn J Ophthalmol* 34: 320–324
117 Frank J, Beck SC, Sellinger M, Wissinger M, Zernner E, Biesalski HK (1998) Biochemical vitamin A deficiency as a result of a mutation of the RBP gene. *FASEB J* 12: A352
118 Panariello L, Quadro L, Trematerra S, Colantuoni V (1996) Identification of a novel retinoic acid response element in the promoter region of the retinol-binding protein gene. *J Biol Chem* 271: 25 524–25 532
119 Mourey MS, Quadro L, Panariello L, Colantuoni V (1994) Retinoids regulate expression of the retinol-binding protein gene in hepatoma cells in culture. *J Cell Physiol* 160: 596–602
120 Jessen KA, Satre MA (1998) Induction of mouse retinol binding protein gene expression by cyclic AMP in Hepa 1-6 cells. *Arch Biochem Biophys* 357: 126–130

Enzymology and biogenesis of retinoic acid

J.L. Napoli

Department of Nutritional Sciences, University of California, Berkeley, CA 94720, USA

Introduction

Use of retinoid binding proteins as substrates for identifying enzymes of retinoid metabolism has revealed compelling new candidates (e.g. lecitin:retinol acyltransferase [LRAT], intestinal microsomal retinal reductase, retinol dehydrogenase [RDH]), and has redirected attention away from historical candidates (intestinal cytosolic retinal reductase, alcohol dehydrogenase [ADH]), which, at any rate, had not been well characterized with respect to retinoid metabolism. The use by specific enzymes of retinol complexed with cellular retinol binding protein (CRBP) provides potential insight into the mechanisms of intracellular retinol access and of imposing specificity on retinoid metabolism. These data also suggest why relatively rapid depletion of retinol does not occur *in vivo via* enzymes generally associated with xenobiotic metabolism, such as myriad cytochrome P-450 s and ADHs, despite their activities with "free" retinol *in vitro*.

Overview of retinoic acid biosynthesis

The model in Figure 1 proposes an integrated relationship among enzymes that metabolize retinoids and retinoid binding-proteins that interact with these enzymes as substrates and effectors of activity. Retinol uptake from plasma results in formation of holo-CRBP, which prompts RE formation by the microsomal enzyme LRAT. The reaction does not require diffusion of retinol from holo-CRBP because LRAT recognizes the CRBP-retinol complex and accesses its retinol [1, 2]. Not only has this been verified by kinetics, but cross-linking data also indicate a physical interaction between holo-CRBP and LRAT. Treating microsomes with CRBP modified to effect covalent crosslinking, affinity labeling with all-*trans*-retinyl-α-bromoacetate, an irreversible agent specific for microsomal LRAT, and cDNA cloning have all identified a 25 kDa monomer as LRAT [3–5].

Depletion of plasma retinol and/or intracellular retinol metabolism results in compensatory elevation of intracellular retinol through mobilization of retinyl ester (RE). Mechanisms appear to include inhibition of LRAT and stimulation of RE hydrolysis by apo-CRBP [6–8]. Observation and partial purification of membrane-associated, bile salt-independent retinyl ester hydrolase (REH) provided insight into intracellular RE hydrolysis. This REH lacks activity for cholesterol esters, has reduced activity for triacylglycerol, and distributes widely in retinoid target tissues [9–11]. To date, two probable physiologically significant, bile salt-*in*dependent

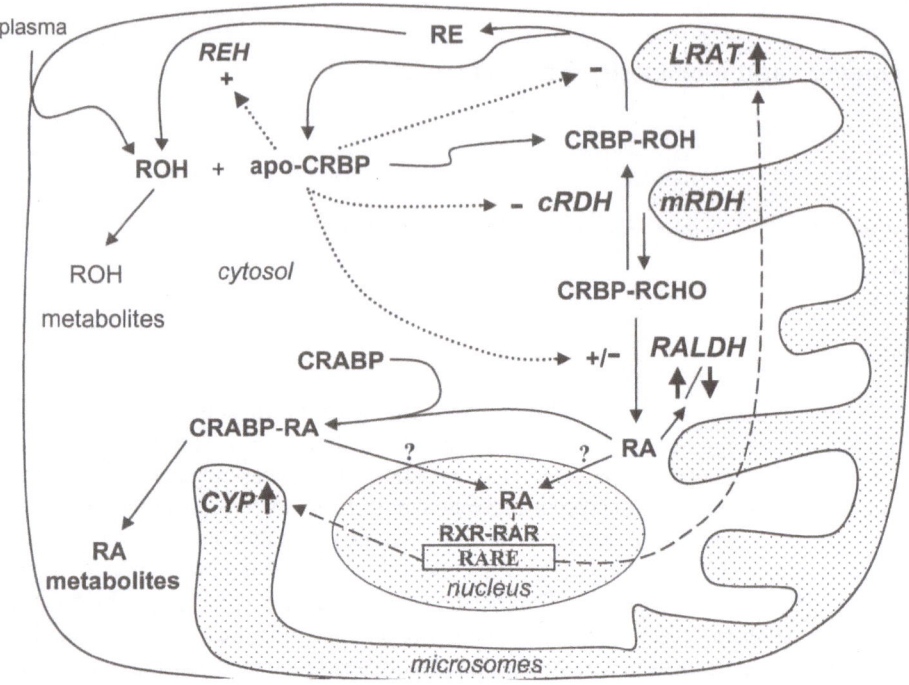

Figure 1. A model of proposed integrated relationships among enzymes that metabolize retinoids and retinoid binding proteins.

REHs have been purified and identified as ES-10 and ES-2 (also known as serum esterase) [12]. Future work should sort out the specific functions of ES-2, ES-10 and other candidate REHs (e.g. ES-4) in uptake of retinyl esters from chylomicron remnants and mobilization of RE from intracellular stores.

Holo-CRBP also supports RA biosynthesis by serving as substrate for production of the intermediate retinal. Just as in RE biosynthesis with LRAT, retinal biosynthesis catalyzed by RDH does not require diffusion of retinol from holo-CRBP. RDH isozymes access retinol bound in the holo-CRBP complex [5, 13–15]. Retinal dehydrogenase (RALDH) isozymes complete RA biosynthesis by irreversibly converting retinal into RA, in a step that may also involve CRBP. At least two RALDH isozymes (types 1 and 2) recognize retinal in the presence of CRBP [16–18].

Most retinol added to cultured mammalian cells undergoes conversion into RE, less than 5% initially undergoes conversion into RA [19–21]. The primary determinants of the relative flux of retinol into RE and RA appear to be the activities of the two irreversible enzymes LRAT and RALDH. By removing retinol and retinal, respectively, LRAT and RALDH would affect the equilibrium between retinol and retinal catalyzed by the readily reversible RDH.

RA itself affects the rates of its synthesis directly and indirectly by inducing the transcription of CRBP and LRAT [22, 23], altering RALDH mRNA levels [24], and inducing its own metabolism [19, 25, 26]. Such auto-regulation has been confirmed through cumulative experiments done with a variety of cultured mammalian cells [19–21, 25–30].

RA metabolism diminishes its biopotency, at least in some cultured cells and *in vivo;* conversely, inhibition of RA metabolism increases its biopotency [25, 31–33]. RA has been identified as the major retinoid in the nuclei of F9 embryonal carcinoma cells induced to differentiate with RA [25]. Possibly, however, a specific metabolite of RA, such as 4-oxo-RA, could serve as an activated retinoid in certain cell types or situations, as suggested for xenopus [34], but this has not been established. Although the metabolism of RA proceeds *in vitro* in the absence of binding protein, the CRABP-RA complex also acts as substrate for RA metabolism [35, 36]. Therefore, cells that express cellular retinoic acid binding protein (CRABP) continue to metabolize RA, while experiencing greatly decreased concentrations of "free" RA. CRABP does not have the same impact on all RA metabolites. 18-OH-RA was not isolated from incubation medium bound to CRABP, but appreciable amounts 4-OH-RA and 4-oxo-RA were. Unlike RA, 4-OH-RA and 4-oxo-RA have vastly decreased rates of metabolism when bound to CRABP, in contrast to their "free" states. The *effects* are not understood of these CRABP *affects* of reducing "free" RA, while permitting RA metabolism and shielding some metabolites from turnover. The function(s) of CRABPII also is (are) not understood. Possible multiple functions of CRABP and/or CRABPII may include: retaining RA in cells that make it (CRABPII), or concentrating RA in cells that do not (CRABPI); transporting RA to the nucleus while allowing metabolism of any "excess"; protecting metabolites (e.g. 4-OH-RA, 4-oxo-RA) from metabolism (to enhance their concentrations for gene activation in select circumstances or loci?).

Retinol dehydrogenases (RDH)

It would seem impetuous to focus on any one of the several enzymes that convert "free" retinol into retinal (Tab. 1) without additional discriminating data. The last column of Table 1 provides such discriminating data: some dehydrogenases recognize the major physiological form of retinol as substrate, CRBP-bound retinol (holo-CRBP). Evidence for RDH interaction with holo-CRBP stems from kinetic and selectivity experiments similar to those that support an interaction between holo-CRBP and LRAT [12, 13]. Detailed crosslinking data have also been generated with both microsomes and partially purified microsomal RDH [5]. Crosslinking of RDH and CRBP correlated well with their enzyme-substrate reaction patterns; cofactor was required and apo-CRBP was a weak inhibitor. Use of microsomes, which contain many dehydrogenases, confirmed the specificity of interaction, because only two bands crosslinked with holo-CRBP, the 34 kDa RDH and the 25 kDa LRAT. Moreover, mutation of CRBP residue L35 to change its size (L35A) or charge (L35R), had no significant impact on the K_d value for retinol or the K_m value with microsomal RDH, but diminished the V_m of the reaction by 25–30% [37]. This represents a remarkable effect for a single residue (especially because other mutants either had no effect or increased the V_m), and probably reflects the L35 location near the putative portal of retinol access. These data show that the affinity (K_d) of CRBP for ligand alone does not determine the efficiency of holo-CRBP as substrate, and provide the first evidence of a function for an exterior CRBP residue.

Table 1. Dehydrogenases known to recognize retinol and/or CRBP-retinol as substrate(s)

activity	family	subcellular fraction	cofactor	active with	
				retinol	CRBP-retinol
[a]RDH	SDR	microsomes	$NADP^+/NAD^+$	+	+
[a]unnamed	SDR	microsomes	NAD^+	+	[b]NA
[a]RDH	SDR?	cytosol	$NADP^+$	+	+
[a]RDH	SDR?	cytosol	NAD^+	+	+
[c]Class I	ADH	cytosol	NAD^+	+	NA
[c]Class II	ADH	cytosol	NAD^+	+	NA
[c]Class IV	ADH	cytosol	NAD^+	+	NA

[a]Multiple isozymes constitute microsomal and perhaps cytosolic RDHs. Some prefer $NADP^+$ *in vitro* but also recognize NAD^+; the cofactor these use *in vivo* remains uncertain.
[b]NA, no activity with the complex CRBP-retinol as substrate.
[c]Each of these represents a single isoform, except Class I which has three isoforms in human, but only one in rat, housemouse and deermouse.

The microsomal RDH candidates appear to be more consequential than the cytosolic ones (at least on first consideration) because microsomal RDHs account for 80 to 90% of the holo-CRBP-recognizing RDH units in the presence of apo-CRBP. Also, apo-CRBP in modest concentrations has little effect on microsomal RDHs, but inhibits cytosolic RDH potently [15].

Ten cDNA clones have been published that encode microsomal members of the short-chain dehydrogenase/reductase (SDR) superfamily, which catalyze retinoid metabolism [38–51]. Table 2 lists their names as they have appeared and new designations proposed by the HUGO/GDB mouse nomenclature committee, along with amino acid similarity comparisons. These *Rdh* genes strongly express mRNA intensely in adult liver, with several also showing strong expression in various other adult tissues (e.g. mammary gland, retina) and in the mouse embryo (*Rdh1, Rdh2, Rdh4*) [123, 125]. Indeed, detectable expression occurs for all but *Rdh3* in most adult tissues (e.g. see the human multiple tissue blot exposed to an *Rdh5* probe [51]).

In situ hybridization has colocalized RDH1 and RDH2 with CRBP throughout adult rat liver, kidney, testis and lung and in mouse embryo from d.p.c. 8 to 15 (unpublished observations of Y. Zhai and J. Napoli, [52]). Immunohistochemical techniques localized RDH (antibodies raised to an RDH epitope in all but retSDR1), CRBP and RALDH (antibodies not specific for RALDH1 or RALDH2) to the same loci of presomitic (d.p.c. 8) and to somite pair 1-3, 8-12 and 25-30 rat embryos [53]. No expression of RDH occurred in areas devoid of CRBP expression. These data support the model by showing that the substrate (CRBP) and enzymes involved in RA biosynthesis occur in the same cellular loci *in vivo*.

To the extent tested, each RDH shows a unique pattern of catalysis with various retinoids and, except retSDR1, androgens (Tab. 3). Recombinant RDH1 and RDH2 recognize holo-CRBP as substrate, but have not been tested with other retinoids [38, 39]. RDH4, 5, and 6 recognize 9-*cis*-retinol with much higher efficiencies than all-*trans*-retinol, and could serve extra-occularly as 9-*cis*-retinol dehydrogenases. RDH5, 6 and 7 show equivalent or better activity

Table 2. Comparisons of amino acid identities/similarities among SDRs that catalyze retinoid metabolism

HUGO/GDB proposed /common name (ref.)	RDH1 rat	RDH2 rat	RDH3 rat	RDH4 mouse	RDH5 human	RDH6 mouse	RDH7 mouse	RDHx human	RDHy human	RDHz human
rRDH1/RoDH1 [38]	100/100	81/85	97/97	53/62	52/62	80/84	83/87	63/71	13/30	72/77
rRDH2/RoDH2 [39]		100/100	82/86	52/60	52/60	85/87	79/84	61/69	13/30	71/77
rRDH3/RoDH3 [40]			100/100	52/62	52/62	80/84	83/86	64/72	13/30	73/77
mRDH4 [41]				100/100	87/90	52/60	51/60	47/56	27/37	51/61
hRDH5/11-*cis*-RDH [42–45]					100/100	52/61	52/61	48/56	27/36	53/62
mRDH6/CRAD1 [46]						100/100	82/87	61/69	17/35	74/81
mRDH7/CRAD2 [47]							100/100	63/70	17/35	73/78
hRDHx [48]								100/100	17/35	62/70
hRDHy/retSDR1 [49]									100/100	17/35
hRDHz/RoDHIV [50]										100/100
total amino acids	317	317	317	318	318	317	316	317	302	317

Note: hRDHx,y,z have been given letter designations because they have not been assigned numbers by the nomenclature committee. Also, it is not clear whether x and z represent interspecies homologs of rodent RDHs or unique forms.

Table 3. Substrates recognized by RDH isozymes[a]

RDH	retinoids	steroids	Ref.
1	holo-CRBP	3α-adiol > androsterone >> testosterone no activity with dihydrotestosterone, estradiol	38
2	holo-CRBP	-	39
3	-	-	-
4	9-*cis*-retinol >> all-*trans*-retinol	-	41
5	11-*cis*-retinol ≅ 9-*cis*-retinol >> all-*trans*-retinol	androsterone ≅ 3α-adiol no activity with testosterone, dihydrotestosterone, estradiol, corticosterone	51
6	11-*cis*-retinol ≅ 9-*cis*-retinol >> all-*trans*-retinol	3α-adiol > androsterone >> testosterone no activity with dihydrotestosterone, estradiol, corticosterone	46
7	11-*cis*-retinol >> all-*trans*-retinol >	3α-adiol > androsterone >> testosterone >> 9-*cis*-retinoldihydrotestosterone no activity with estradiol, corticosterone	47
x	-	3α-adiol > androsterone >> dihydrotestosterone >>> hydrotestosterone	48
y	all-*trans*-retinol	little or no activity with steroids	49
z	all-*trans*-retinol, 13-*cis*-retinol	3α-adiol > androsterone > dihydrotestosterone	50

[a]Substrates not listed have not been tested with enzymes expressed from the cDNA.

with 11-*cis*-retinol compared to 9-*cis*-retinol [42–47]. RDH4 has not been tested with 11-*cis*-retinol [41]. The intense expression of RDH5 in the retinal pigment epithelium and its activity with 11-*cis*-retinol suggest a function in the visual cycle, as does the characteristics of retSDR1 [49]. Eye also expresses both *Rdh*6 and *7*, but less intensely than RDH5 and retSDR1 [46, 47]. Both of these could also serve in the eye as 11-*cis*-retinol oxidoreductases. Since 11-*cis*-retinol has not been detected extra-occularly, it seems unlikely that the RDHs with dual 11-*cis*-/9-*cis*-retinol would function outside of the eye as 11-*cis*-retinol oxidoreductases. Therefore, if they function outside of the eye as RDH, they might function as 9-*cis*-retinoid oxidoreductases.

Retinoid-androgen interactions

RA and its isomers inhibit prostate epithelial cell growth [54]. Inhibition of RA metabolism raises plasma RA and inhibits relapse in the rat Dunning prostate cancer model [55]. RA decreases the concentrations of dihydrotestosterone, 3α-adiol and androsterone in serum, seems to cause a metabolic deviation away from the 5α-reductase path in liver [56], and causes a three-fold decrease in androgen receptor binding capacity [57]. Conversely, androgens affect the actions of retinoids by decreasing the mRNA of RARα approximately five-fold in prostate epithelia and 15 to 20-fold in seminal vesicles, while increasing it two-fold in kidney [58].

Prostate epithelial cells convert testosterone into dihydrotestosterone (5α-reductase pathway) as their predominant androgen, whereas prostate stromal cells convert dihydrotestosterone into 3α-adiol and then into androsterone, two weak androgens [59]. Two enzymes catalyze the reactions that inactivate dihydrotestosterone: 3α-hydroxysteroid dehydrogenases (members of the aldo-keto reductase superfamily) and 17β-hydroxysteroid dehydrogenases [59, 61]. In constrast, 3α-hydroxysteroid dehydrogenases that function oxidatively would convert 3α-adiol and androsterone into dihydrotestosterone and androstanedione (a potential dihydroxytestosterone precursor), respectively. Several RDHs (RDH1, 6 and 7) do function with 3α-adiol (5α-androstan-3α,17β-diol) and androsterone as the most efficient 3α-hydroxysteroid dehydrogenases known. These RDHs show little or no activity as 17β-hydroxysteroid dehydrogenases with testosterone, dihydrotestosterone and estradiol and do not function as 11β-hydroxysteroid dehydrogenases with corticosterone. RDH isozymes seem also to function *in vivo*, since despite low affinity for the androgen receptor, 3α-adiol stimulates prostate growth *in vivo* and in organ culture, consistent with conversion into dihydrotestosterone [62–64].

3α-Adiol may not represent solely an androgen catabolite or precursor. Mice with a null allele in the 5α-reductase type I gene enter prolonged labor on approximately day 21–22 and fail to deliver [65]. 3α-Adiol increases the incidence of normal parturition by greater than tree-fold in these mutants. Dihydrotestosterone is less effective, suggesting that 3α-adiol serves as a parturition initiation agent in mice. If so, RDH could affect the onset of parturition through affecting 3α-adiol concentration.

Retinal dehydrogenases (RALDH)

Candidate RALDHs that have been purified and/or cloned include mouse AHD2 and ADH7 isozymes [66], human ALDH1 [67], rat RALDH1 [16, 18, 24, 68], rat and mouse RALDH2 [17, 69] and chick and bovine retina enzymes [70, 71]. These candidates have low $K_{0.5}$ values for retinal with relatively high V_m values; no other candidates have emerged with similar efficiencies ($V_m/K_{0.5}$ values) for converting retinal into RA (Tab. 4). With the exception of AHD7, which hasn't been cloned or sequenced, these enzymes share close amino acid sequence similarity and, with the possible exception of RALDH2, are Class I aldehyde dehydrogenases.

The most studied enzymes, RALDH1 and RALDH2, produce RA from CRBP-retinal, as well as from "free" retinal [17, 18]. Both, however, catalyze conversion of retinal into RA more efficiently (greater $V/K_{0.5}$ and lower $K_{0.5}$ values) when CRBP presents the retinal (Tab. 5). This use of CRBP-retinal reinforces the observation of a chromatographic fraction (PI) prepared from rat liver cytosol that synthesized RA from CRBP-retinol and microsomes in the absence of exogenous retinal [16]. These data suggest that the cytosolic RALDH1 and RALDH2 contribute a segment of an "RA generating system" that includes holo-CRBP and microsomal RDH.

Both RALDH1 and RALDH2 also catalyze the conversion of 9-*cis*-retinal into 9-*cis*-RA [72, 73]. Therefore, they could also serve as partners with the "*cis*-retinoid" RDH in a pathway that generates 9-*cis*-RA from 9-*cis*-retinol. Alternatively, they could catalyze synthesis of 9-*cis*-RA from 9-*cis*-retinal produced by central cleavage of 9-*cis*-β-carotene [74].

Table 4. Candidate retinal dehydrogenases

dehydrogenase	% aa similarity		$^aK_{0.5}$
	1	2	μM
rat RALDH1 (P1, rat kidney)	100	85	1.7
rat/mouse RALDH2 (V2)	85	100	0.8
mouse AHD-2	99	85	0.6
mouse AHD-7	-	-	0.7
human ALDH1	94	87	0.3
chick retina	90	87	-
bovine retina	94	86	9

[a]Free retinal concentration that produced half-maximal velocity.

Table 5. Comparison of RALDH1 and RALDH2

substrate	$K_{0.5}$ μM (relative $V/K_{0.5}$)[a]	
	RALDH1	RALDH2
CRBP-retinal	0.8 (20)	0.2 (100)
all-*trans*-retinal	1.4 (12)	0.7 (48)
9-*cis*-retinal	5.2 (9)	0.5 (12)[c]
13-*cis*-retinal	low	0.2 (137)[c]
11-*cis*-retinal	-	0.3 (69)[c]
acetaldehyde	127 (3)	645 (<1)
tissue	mRNA[b] (retinoid effect)	
liver	34 (induction)	4 (none)
kidney	100 (induction)	3 (none)
lung	88 (induction)	7 (none)
testis	8 (repression)	100 (induction)
brain	22 (induction)	6 (none)
embryo	—	++

[a]Data have been normalized to RALDH2 data with CRBP-retinal at 37 °C.
  Data obtained at 25 °C have been multiplied ten-fold to compensate for reaction temperature.
[b]Values have been normalized independently to the tissue of highest expression for each.
[c]X. Wang and J.L. Napoli, unpublished observations.

Apo-CRBP inhibits RALDH1 with an IC_{50} of approximately 1.4 μM, but does not affect RALDH2 [17, 18]. RA also affects the mRNA levels of the two differently and affects the mRNA of each differently in different tissues (Tab. 5). These differences, along with the unique cellular localizations of each in liver and testis, for example suggest different roles for each in generating RA.

Both RALDH1 and RALDH2 inefficiently catalyze acetaldehyde metabolism *in vitro*. Nevertheless, one wonders whether the acetaldehyde produced during chronic alcohol intake competes with retinal metabolism, perhaps contributing to fetal alcohol syndrome. This simple postulate, however, has complex caveats. Not the least of which includes the imprudence of ascribing a specific mechanism to a highly reactive non-specific agent.

The three-dimensional structure of RALDH2 has been solved by X-ray crystallography [75, 76]. RALDH2 seems unique relative to other aldehyde dehydrogenases insomuch as it has unusually mobile catalytic machinery and substrate access channel, which may reflect retinal binding. This perhaps explains the ability of RALDH2 to efficiently catalyze RA formation, while also recognizing structurally simpler medium-chain aldehydes.

References

1   Ong DE, MacDonald PN, Gubitosi AM (1988) Esterification of retinol in rat liver: possible participation by cellular retinol-binding protein and cellular retinol binding-protein II. *J Biol Chem* 263: 5789–5796
2   Yost RW, Harrison EH, Ross AC (1988) Esterification by rat liver microsomes of retinol bound to cellular retinol-binding protein. *J Biol Chem* 263: 18 693–18 701
3   Shi YQ, Furuyoshi S, Hubacek I, Rando RR (1993) Affinity labeling of lecithin retinol acyltransferase. *Biochemistry* 32: 3077–3080
4   Ruiz A, Winston A, Lim Y-H, Gilbert BA, Rando RR, Bok D (1999) Molecular and biochemical characterization of lecithin retinol acyltransferase (LRAT). *J Biol Chem* 274: 3834–3841
5   Boerman MHEM, Napoli JL (1995) Characterization of a microsomal retinol dehydrogenase: a short-chain alcohol dehydrogenase with integral and peripheral membrane forms that interacts with holo-CRBP (type I). *Biochemistry* 34: 7027–7037
6   Herr F, Ong DE (1992) Differential interaction of lecith*In*:retinol acyltransfersase with cellular retinoid-binding proteins. *Biochemistry* 31: 6748–6755
7   Ottonello S, Petrucco S, Maraini G (1987) Vitamin A uptake from retinol-binding protein in a cell-free system from pigment epithelial cells of bovine retina: retinol transfer from plasma retinol-binding protein to cytoplasmic retinol-binding protein with retinyl-ester formation as the intermediate step. *J Biol Chem* 262: 3975–3981
8   Boerman MHEM, Napoli JL (1991) Cholate-independent retinyl ester hydrolysis: stimulation by apo-cellular retinol binding protein. *J Biol Chem* 266: 22 273–22 278
9   Napoli JL, Pacia EB, Salerno GJ (1989) Cholate-independent hydrolysis of all-*trans*-retinyl palmitate by rat tissues: solubilization of multiple kidney microsomal hydrolases. *Arch Biochem Biophys* 274: 192–199
10 Harrison EH, Gad M (1989) Hydrolysis of retinyl palmitate by enzymes of rat pancreas and liver: differentiation of bile salt-dependent and bile salt-independent, neutral retinyl ester hydrolases in rat liver. *J Biol Chem* 264: 17 142–17 147
11 Gad MZ, Harrison EH (1991) Neutral and acid retinyl ester hydrolases associated with rat liver microsomes: relationships to microsomal cholesteryl ester hydrolases. *J Lipid Res* 32: 685–694
12 Sun G, Alexon SEH, Harrison EH (1997) Purification and characterization of a neutral, bile salt-independent retinyl ester hydrolase from rat liver microsomes: relationship to rat carboxylesterase ES-2. *J Biol Chem* 272: 24 488–24 493
13 Posch KC, Boerman MHEM, Burns RD, Napoli JL (1991) Holo-cellular retinol binding protein as a substrate for microsomal retinal synthesis. *Biochemistry* 30: 6224–6230
14 Ottonello S, Scita G, Mantovani G, Cavazzini D, Rossi GL (1993) Retinol bound to cellular retinol binding-protein is a substrate for cytosolic retinoic acid synthesis. *J Biol Chem* 268: 27 133–27 142
15 Boerman MHEM, Napoli JL (1996) Cellular retinol-binding protein-supported retinoic acid synthesis: relative roles of microsomes and cytosol. *J Biol Chem* 271: 5610–5616
16 Posch KC, Burns RD, Napoli JL (1992) Biosynthesis of all-*trans*-retinoic acid from retinal: recognition of retinal bound to cellular retinol binding protein (type I) as substrate by a purified cytosolic dehydrogenase. *J Biol Chem* 267: 19 676–19 682
17 Wang X, Penzes P, Napoli JL (1996) Cloning of a cDNA encoding an aldehyde dehydrogenase and its expression in *Escherichia coli*: recognition of retinal as substrate and role in retinoic acid synthesis. *J Biol Chem* 271: 16 288–16 293
18 Penzes P, Wang X, Napoli JL (1997) Enzymatic characteristics of retinal dehydrogenase type I expressed in *Escherichia coli*. *Biochim Biophys Acta* 1342: 175–181

19 Napoli JL (1986) Retinol metabolism in LLC-PK₁ cells: characterization of retinoic acid synthesis by an established mammalian cell line. *J Biol Chem* 261: 13 592–13 597

20 Sigenthaler G, Saurat J-H, Ponec M (1990) Retinol and retinoic acid metabolism: relationship to the state of differentiation of cultured human keratinocytes. *Biochem J* 268: 371–378

21 Randolph RK, Simon M (1993) Characterization of retinol metabolism in cultured human epidermal keratinocytes. *J Biol Chem* 268: 9198–9205

22 Haq RU, Chytil F (1988) Retinoic acid rapidly induces lung cellular retinol-binding protein mRNA levels in retinol deficient rats. *Biochem Biophys Res Commun* 156: 712–716

23 Matsuura T, Ross AC (1993) Regulation of hepatic lecith*In*:retinol acyltransferase activity by retinoic acid. *Arch Biochem Biophys* 301: 221–227

24 Penzes P, Wang X, Napoli JL (1997) Cloning of a cDNA encoding retinal dehydrogenase isozyme type and its expression in *Escherichia coli*. *Gene* 191: 167–172

25 Williams JB, Napoli JL (1985) Metabolism of retinoic acid and retinol during differentiation of F9 embryonal carcinoma cells. *Proc Natl Acad Sci. USA* 82: 4658–4662

26 Kurlandsky SB, Duell EA, Kang S, Voorhees JJ, Fisher GJ (1996) Autoregulation of retinoic acid biosynthesis through regulation of retinol esterification in human keratinocytes. *J Biol Chem* 271: 15 346–15 352

27 Sigenthaler G, Saurat J-H, Ponec M (1990) Retinol and retinoic acid metabolism: relationship to the state of differentiation of cultured human keratinocytes. *Biochem J* 268: 371–378

28 Randolph RK, Simon M (1993) Characterization of retinol metabolism in cultured human epidermal keratinocytes. *J Biol Chem* 268: 9198–9205

29 Shingleton JL, Skinner MK, Ong DE (1989) Characteristics of retinol accumulated from serum retinol-binding protein by cultured Sertoli cells. *Biochemistry* 28: 9641–9647

30 Bishop PD, Griswald MD (1987) Uptake and metabolism of retinol in cultured Sertoli cells: evidence for a kinetic model. *Biochemistry* 26: 7511–7518

31 van Wauwe JP, Coene M-C, Goossens J, van Nyen J, Cools W, Lauwers W (1988) Ketoconazole inhibits the *in vivo* and *in vitro* metabolism of all-*trans*-retinoic acid. *J Pharmacol Exp Ther* 245: 718–722

32 van Wauwe JP, Coene MC, Goossens J, Cools W, Monbaliu J (1990) Effects of cytochrome P-450 inhibitors on the *in vivo* metabolism of all-*trans*-retinoic acid in the rats. *J Pharmacol Exp Ther* 252: 365–369

33 van der Leede BM, van den Brink CE, Pijnappel WWM, Sonneveld E, van der Saag PT, van der Burg B (1997) Autoinduction of retinoic acid metabolism to polar derivatives with decreased biological activity in retinoic acid-sensitive, but not in retinoic acid-resistant human breast cancer cells. *J Biol Chem* 272: 17 921–17 928

34 Pijnappel WWM, Hendricks HFJ, Folkers GE, van den Brink CE, Dekker EJ, Edelenbosch C, van der Saag PJ, Durston AJ (1993) The retinoid ligand 4-*oxo*-retinoic acid is a highly active modulator of positional specification. *Nature* 366: 340–344

35 Fiorella PD, Napoli JL (1991) Expression of cellular retinoic acid binding protein in *Escherichia coli*: characterization and evidence that holo-CRABP is a substrate in retinoic acid metabolism. *J Biol Chem* 266: 16 572–16 579

36 Fiorella PD, Napoli JL (1994) Microsomal retinoic acid metabolism: effects of cellular retinoic acid-binding protein (type I) and C18-hydroxylation as an initial step. *J Biol Chem* 269: 10 538–10 544

37 Penzes P, Napoli JL (1999) Holo-cellular retinol-binding protein: distinction of ligand-binding affinity from efficiency as substrate in retinal biosynthesis. *Biochemistry* 38: 2088–2093

38 Chai X, Boerman MHEM, Zhai Y, Napoli JL (1995) Cloning of a cDNA for liver microsomal retinol dehydrogenase: a tissue-specific, short-chain alcohol dehydrogenase. *J Biol Chem* 270: 3900–3904

39 Chai X, Zhai Y, Popescu G, Napoli JL (1995) Cloning of a cDNA for a second retinol dehydrogenase, type II: expression of its mRNA relative to type I. *J Biol Chem* 270: 28 408–28 412

40 Chai X, Zhai Y, Napoli JL (1996) Cloning of a cDNA encoding a retinol dehydrogenase isozyme type III. *Gene* 169: 219–222

41 Romert A, Tuvendal P, Simon A, Dencker L, Eriksson U (1998) The identification of a 9-*cis*-retinol dehydrogenase in the mouse embryo reveals a pathway for synthesis of 9-*cis* retinoic acid. *Proc Natl Acad Sci. USA* 95: 4404–4409

42 Simon A, Hellman U, Wernstedt C, Eriksson U (1995) The retinal pigment epithelial-specific 11-*cis* retinol dehydrogenase belongs to the family of short chain alcohol dehydrogenases. *J Biol Chem* 270: 1107–1112

43 Simon A, Lagercrantz J, Bajalica-Lagercrantz S, Eriksson U (1996) Primary structure of human 11-*cis* retinol dehydrogenase and organization and chromosomal localization of the corresponding gene. *Genomics* 36: 424–430

44 Driessen CAGG, Janssen BPM, Winkens HJ, van Vugt AHM, de Leeuw TLM, Janssen JJM (1995) Cloning and expression of a cDNA encoding bovine retinal pigment epithelial 11-*cis* retinol dehydrogenase. *Invest. Ophthalmol. Vis. Sci.* 36: 1988–1996

45 Mertz JR, Shang E, Pianedosi R, Wei S, Wolgemuth DJ, Blaner WS (1997) Identification and characterization of a stereospecific human enzyme that catalyzes 9-*cis*-retinol oxidation: a possible role in 9-*cis*-retinoic acid formation. *J Biol Chem* 272: 11 744–11 749

46 Chai X, Zhai Y, Napoli JL (1997) cDNA cloning and characterization of a *cis*-retinol/3α-hydroxysterol short-chain dehydrogenase. *J Biol Chem* 272: 33125–33131

47 Su J, Chai X, Kahn B, Napoli JL (1998) cDNA cloning, tissue distribution, and substrate characterization of

a *cis*-retinol/3α-hydroxysterol short-chain dehydrogenase isozyme. *J Biol Chem* 273: 17 910–17 916
48 Biswas MG, Russell DW (1997) Expression cloning and characterization of oxidative 17β- and 3α-hydroxysteroid dehydrogenases from rat and human prostate. *J Biol Chem* 272: 15 959–15 966
49 Haeseleer F, Huang J, Lebioda L, Saari JC, Palczewski K (1998) Molecular characterization of a novel short-chain dehydrogenase/reductase. *J Biol Chem* 273: 21 790–21 799
50 Gough WH, VanOoteghem S, Sint T, Kedishvili N Y (1998) cDNA cloning and characterization of a new human microsomal NAD+-dependent dehydrogenase that oxidizes all-*trans*-retinol and 3α-hydroxysteroids. *J Biol Chem* 273: 19 778–19 785
51 Wang J, Chai X, Eriksson U, Napoli JL (1999) Activity of human Rdh5 with steroids and retinoids and expression of its mRNA in extra-ocular human tissue. *Biochem J* 338: 23–27
52 Zhai Y, Higgins D, Napoli JL (1997) Coexpression of the mRNAs encoding retinol dehydrogenase isozymes and cellular retinol-binding protein. *J Cell Physiol* 173: 36–43
53 Båvik CO, Ward SJ, Ong DE (1997) Identification of a mechanism to localize generation of retinoic acid in rat embryos. *Mech Develop* 69: 155–167
54 Young CY, Murtha PE, Andrews PE, Lindzey JK, Tindall DJ (1994) Antagonism of androgen action in prostate tumor cells by retinoic acid. *Prostate* 25: 39–45
55 De Coster R, Wouters W, Bruynseels J (1996) P450-dependent enzymes as targets for prostate cancer therapy. *J Steroid Biochem Mol Biol* 56: 133–143
56 Boudou P, Chivot M, Vexiau P, Soliman H, Villette JM, Belanger A, Fiet J (1994) Evidence for decreased androgen 5α-reductase in skin and liver of men with severe acne after 13-*cis*-retinoic acid treatment. *J Clin Endocrinol Metab* 78: 1064–1069
57 Boudou P, Soliman H, Chivot M, Villette JM, Vexiau P, Belanger A, Fiet J (1995) Effect of oral isotretinoin treatment on skin androgen receptor levels in male acneic patients. *J Clin Endocrinol Metab* 80: 1158–1161
58 Huang HF, Li MT, Von Hagen S, Zhang YF, Irwin RJ (1997) Androgen modulation of the messenger ribonucleic acid of retinoic acid receptors in the prostate seminal vesicles and kidney in the rat. *Endocrinology* 138: 553–559
59 Russell DW, Wilson JD (1994) Steroid 5α-reductase: two genes/two enzymes. *Annu Rev Biochem* 63: 25–61
60 Taurog JD, Moore RJ, Wilson JD (1975) Partial characterization of the cytosol 3α-hydroxysteroid:NAD(P)+ oxidoreductase of rat ventral prostate. *Biochemistry* 14: 810–817
61 Lin H-K, Jez JM, Schlegel BP, Peehl DM, Pachter JA, Penning TM (1997) Expression and characterization of recombinant type 2 3α-hydroxysteroid dehydrogenase (HSD) from human prostate: demonstration of bifunctional 3α/17β-HSD activity and cellular distribution. *Mol Endocrinol* 11: 1971–1984
62 Bruchovsky N (1971) Comparison of the metabolites formed in rat prostate following the *in vivo* administration of seven natural androgens. *Endocrinology* 89: 1212–1222
63 Moore RJ, Wilson JD (1993) The effect of androgenic homones on the reduced nicotinamide adenine dinucleotide phosphate δ-4-3-ketosteroid-5α-oxidoreductase of rat ventral prostate. *Endocrinology* 93: 581–592
64 Schultz FM, Wilson JD (1974) Virulization of the Wolffian duct in the rat fetus by various androgens. *Endocrinology* 94: 979–986
65 Mahendroo MS, Cala KM, Russell DW (1996) 5α-Reduced androgens play a key role in murine parturition. *Mol Endocrinol* 10: 380–392
66 Lee MO, Manthey CL, Sladek NE (1991) Identification of mouse liver aldehyde dehydrogenases that catalyze the oxidation of retinaldehyde to retinoic acid. *Biochem Pharmacol* 42: 1279–1285
67 Dockham PA, Le e MO, Sladek NE (1992) Identification of human liver aldehyde dehydrogenases that catalyze the oxidation of aldophosphamide and retinaldehyde. *Biochem Pharmacol* 43: 2453–2469
68 Bhat PV, Labrecque J, Boutin JM, Lacroix A, oshida A (1995) Cloning of a cDNA encoding rat aldehyde dehydrogenase with high activity for retinal oxidation. *Gene* 166: 303–306
69 Zhao D, McCaffery P, Ivins KJ, Neve RL, Hogan P, Dräger U (1996) Molecular identification of a major retinoic-acid-synthesizing enzyme, a retinaldehyde-specific dehydrogenase. *Eur J Biochem* 240: 5–22
70 Saari JC, Champer RJ, Asson-Batres M, Garwin GG, Huang J, Crabb JW, Milam AH (1995) Characterization and localization of an aldehyde dehydrogenase to amacrine cells of bovine retina. *Visual Neurosci* 12: 263–272
71 Godbout R, Packer M, Poppema S, Dabragh L (1996) Localization of cytosolic aldehyde dehydrogenase in the developing chick retina: *in situ* hybridization and immunochemical analyses. *Develop Dyn* 205: 319–331
72 El Akawi Z, Napoli JL (1994) Rat liver cytosolic retinal dehydrogenase: comparison of 13-*cis*-9-*cis*-, and all-*trans*-retinal as substrates and effects of cellular retinoid-binding proteins and retinoic acid on activity. *Biochemistry* 33: 1938–1943
73 Labrecque J, Dumas F, Lacroix A, Bhat PV (1995) A novel isoenzyme of aldehyde dehydrogenase specifically involved in the biosynthesis of 9-*cis* and all-*trans* retinoic acid. *Biochem J* 305: 681–684
74 Nagao A, Olson JA (1994) Enzymatic formation of 9-*cis*, 13-*cis*, and all-*trans*-retinals from isomers of β-carotene. *FASEB J* 8: 968–973
75 Lamb AL, Wang X, Napoli JL, Newcomer ME (1998) Purification, crystallization and preliminary X-ray diffraction studies of retinal dehydrogenase type II. *Acta Crystallogr* 54: 639–642
76 Lamb A, Newcomer M (1999) The structure of retinal dehydrogenase type II at 2.7 Å resolution: implications for retinal specificity. *Biochemistry* 38; *in press*

Requirements and safety of vitamin A in humans

J.A. Olson

Department of Biochemistry, Biophysics and Molecular Biology, 3252 Molecular Biology Building, Iowa State University, Ames, IA 50011, USA

Introduction

Night blindness and some eye disorders, which were well recognized in ancient Egypt, were treated by the topical application of juice squeezed from cooked liver or by prescribing liver in the diet [1]. This medical lore was lost over the centuries, and night blindness, called "this curious and obscure disease" by R.J. Hicks [2], a surgeon serving with the Confederate Army in the American Civil War, plagued armies throughout the world in the Nineteenth Century. The active principle of liver in treating eye disease was vitamin A, which was identified as a necessary fat-soluble factor for rat growth in 1913 and was structurally elucidated in 1930. The biological conversion of β-carotene into vitamin A was shown the same year. These early studies on vitamin A have been well reviewed [1, 3].

Chemistry and nomenclature

The parent compound in the vitamin A group is called all-*trans* retinol (Fig. 1A) [4]. Its aldehyde and acid forms are retinal (Fig. 1B) and retinoic acid (Fig. 1C). The active form of vitamin A in vision is 11-*cis* retinal (Fig. 1D), and a therapeutically useful form (accutane, isotretinoin) is 13-*cis* retinoic acid (Fig. 1E). Retinyl palmitate (Fig. 1F) is a major storage form, and retinoyl β-glucuronide is a biologically active, relatively non-toxic water-soluble metabolite (Fig. 1G). A synthetic aromatic analog (etretin, acitretin), shows therapeutic usefulness (Fig. 1H). Finally, β-carotene, a major provitamin A carotenoid, is shown in Figure 1I.

In a nutritional sense, the vitamin A family includes all naturally occurring compounds with the biological activity of retinol [4]. Because provitamin A carotenoids are nutritionally active, they are included in the vitamin A family. Only 50 of approximately 600 carotenoids found in nature, however, are converted into vitamin A. Common provitamin A carotenoids in foods are β-carotene, α-carotene, γ-carotene and β-cryptoxanthin. The all-*trans* isomer of each of these carotenoids is more active nutritionally than any of the *cis* isomers.

Dietary sources and equivalencies

Common dietary sources of preformed vitamin A are liver; various dairy products, such as milk, cheese, butter, and ice cream; and fish, such as herring, sardines, and tuna. The richest sources

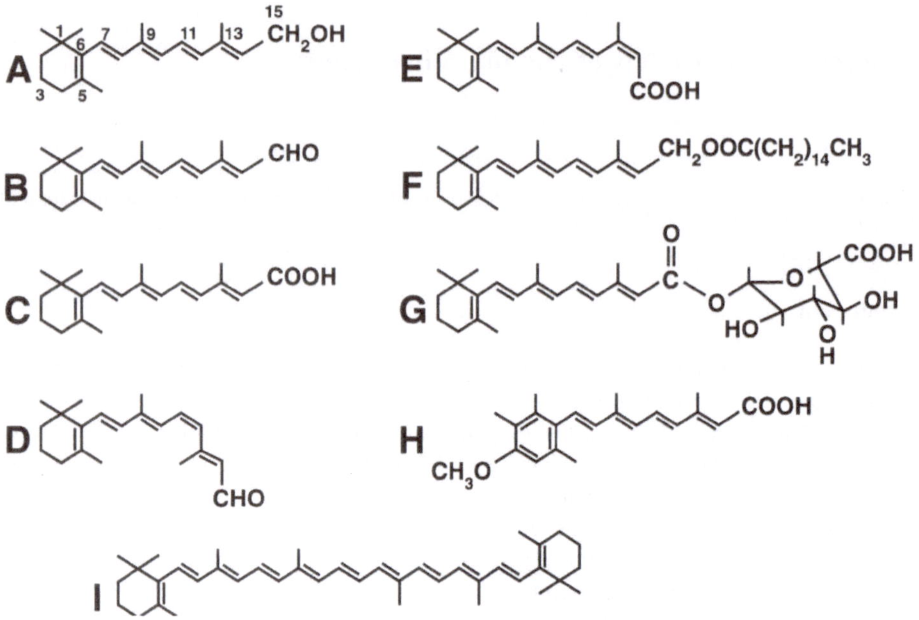

Figure 1. Formulas of major retinoids and of β-carotene. A, all-*trans* retinol; B, all-*trans* retinal; C, all-*trans* retinoic acid; D, 11-*cis* retinal; E, 13-*cis* retinoic acid; F, all-*trans* retinyl palmitate; G, all-*trans* retinoyl β-glucuronide; H, the trimethyl methoxyphenol analog of all-*trans* retinoic acid (etretin, acitretin); I, all-*trans* β-carotene. From [5], p. 110, reprinted with the permission of the International Life Sciences Institute, Washington DC 20036-4810, Copyright 1996.

of preformed vitamin A, although rarely ingested now in Europe or in the United States, are liver oils of the shark; of marine fish, such as cod and halibut; and of marine mammals, such as the polar bear [5–7].

Common dietary sources of provitamin A carotenoids are carrots, yellow squash, dark-green leafy vegetables, corn, tomatoes, papaya, and oranges. Red palm oil, which contains a high concentration of α-carotene and β-carotene, is used for frying foods in parts of Africa and Brazil, but to a much lesser extent elsewhere. The color of fruits and vegetables, however, is not necessarily an indicator of its concentration of provitamin A; tomatoes, for example, are particularly rich in lycopene, which is nutritionally inactive, and the green color of leafy vegetables is due to chlorophyll, which masks the yellow color of the carotenoids. The bioavailability of carotenoids in foods, but not of preformed vitamin A, can vary greatly [8]. Nonetheless, as a rough general guide, most national and international committees have accepted the convention that 6 μg of all-*trans* β-carotene or 12 μg of other all-*trans* provitamin A carotenoids in food are equivalent nutritionally to 1 μg all-*trans* retinol. Thus, dietary intakes of vitamin A are expressed in terms of μg retinol equivalents (REQ) by using the above relationships.

At an earlier time, International Units (IU) were used to express the amount of vitamin A and carotenoids in foods. Because one IU of vitamin A (0.3 µg all-*trans* retinol) in food is not nutritionally equivalent to one IU of all-*trans* β-carotene (0.6 µg) in food, the IU designation in food composition tables has resulted in confusion. As a partial way of resolving the ambiguity, the IU of retinol has been designated IU_a and that of β-carotene IU_b, where $1\ IU_a = 3\ IU_b$ [9]. The IU designation continues to be used, however, for supplements. In the labeling of components of foods in the United States, the percentage of the so-called "Daily Value" is given, an amount based on the recommended dietary allowance and an assumed energy intake of 2000 kilocalories (~8400 kilojoules).

When the concentration of a solution of a given retinoid or carotenoid is known, the best way to express the value is by using molar units, e.g. a plasma concentration of 2 µmol/l is equivalent to 57.2 µg/dl. In such chemical calculations, molecular weights are used; 286 g/mol for vitamin A and 537 g/mole for β-carotene.

Dietary intakes

Mean and median intakes of vitamin A by men in the United Kingdom are approximately 1800 and 1100 µg REQ, respectively [10], and in the United States are approximately 1400 and 900 µg REQ, respectively [11]. Mean intakes for women and young children in the United Kingdom and the United States are 70–90% of that for men. Mean intakes in other European nations also appear to be adequate [12]. The considerably higher values of means relative to medians indicate a skewing of the distribution curve towards lower intakes. In both European countries and the United States, 60–80% of the REQ are derived from preformed vitamin A. In southern Asia and Africa, however, carotenoids provide the major portion of dietary vitamin A. Thus, the bioavailability issue becomes of greater significance in such regions as opposed to in Europe or in most of North America [8].

In addition to dietary sources, commercial supplements of vitamin A as pills are available in most countries. In most instances, the multivitamin pill contains 5000 International Units (IU_a), which is 1.50 and 1.87 times the RDA of preformed vitamin A for men (3333 IU_a) and women (2667 IU_a), respectively, in the United States [11]. Vitamin A supplements containing 10 000 IU_a or even larger amounts per pill are also available. Thus, some individuals are almost certainly ingesting large, unneeded, and possibly toxic supplements of vitamin A.

Vitamin A inadequacy

Vitamin A inadequacy, together with those of iron and iodine, are major micronutrient problems in the world. Anemia caused by inadequate iron intake and iodine insufficiency are present in certain groups and geographical areas in Europe and the United States, although to a much lesser degree than in less-industrialized countries. Although clinical vitamin A deficiency is rare in otherwise healthy European and American children, significant numbers of children and pregnant women from socioeconomically disadvantaged American families show an inadequate vitamin A status as measured by the modified relative dose response (MRDR) test [13, 14]. Whether similar groups in Europe are at-risk has not yet been studied. Thus, complacency about the vitamin A nutriture of populations in industrialized countries is not warranted.

Assessment of vitamin A status

Status categories

Vitamin A status can best be defined in terms of the total body reserves of the vitamin, which are metabolized in healthy men with a mean half-life of 120–150 days [15, 16]. Thus, only 0.5% of the total body reserves are catabolized each day. This slow turnover explains the observation that very long periods are required to induce vitamin A deficiency in well-nourished adult subjects fed a vitamin A-free diet [16]. In reference to total body reserves, vitamin A status can further be classified into five categories: deficient, marginal, satisfactory, excessive, and toxic.

The *deficient state* is characterized by the presence of clinical signs, which appear most strikingly and diagnostically in the conjunctiva and cornea of the eye. Bitot's spots in preschool-age children are the most used diagnostic indicator, although other clinical signs have also been employed.

The *marginal state*, which is also termed preclinical deficiency, is characterized by an inadequate concentration (<0.07 µmol/g wet weight) of vitamin A in the liver, which is the major storage organ for vitamin A, and an increased susceptibility to severe infections. Clinical signs are not present in the marginal state. Marginal status can be measured by several relatively new procedures: the relative dose-response test (RDR), the modified relative dose-response test (MRDR), frequency analysis of serum retinol concentrations before and after supplementation, conjunctival impression cytology, vision restoration time, and the pupillary response test [17–21]. Of the biochemical indicators, the modified relative dose-response test is proving to be a highly reliable procedure as performed by different laboratories in many different countries [18–21].

The *satisfactory state* implies the absence of clinical signs, full physiological functions that are dependent either directly or indirectly on vitamin A, and an adequate total body reserve to meet stresses of various kinds and to provide adequate vitamin A during periods of low dietary intake. Mean total body contents of vitamin A that fulfill all functions of the vitamin and provide a three month reserve on a low vitamin A intake for a 76 kg male and for a 62 kg female are 0.18 mmol and 0.14 mmol, respectively. These values are derived from a satisfactory liver vitamin A concentration of 0.07 µmol vitamin A/g in both sexes [10, 12, 15].

The *excessive state* is characterized by high intakes of preformed vitamin A, usually as supplements, by enhanced total body reserves of vitamin A and by increased steady-state concentrations of retinyl ester in the plasma. No gross signs of toxicity are present. Finally, the *toxic state* is a clinical condition that will be discussed later.

Status indicators

The most commonly used biochemical indicator of vitamin A status in the past has been the measurement of serum retinol concentrations. Although the distribution of such values in a population can be instructive, serum retinol values are homeostatically controlled over a wide range of total body reserves, are depressed during fever and infections, and are adversely affected by inadequate intakes of other nutrients, such as protein and zinc. Thus, serum retinol concentrations are not very sensitive indicators of vitamin A status, unless they are very low (<0.35 µmol/l, i.e.<10 µg/dl) or are clearly adequate (>1.05 µmol/l in children, >1.40 µmol/l in adults). As a result, the response tests, RDR and MRDR, are increasingly used, which combine serum retinol concentrations with the individual's response to a small oral dose of a retinoid.

Mean serum retinol values in men and women in the United Kingdom are 2.3 and 1.8 µmol/l, respectively [10], and in the United States are 2.1 and 1.7 µmol/l, respectively [11]. Mean serum values for three to five year-old-old children of both sexes in the United States are 1.2 µmol/l, whereas values <0.35 µmol/l are extremely rare. Thus, in Europe and the United States, clinically apparent vitamin A deficiency currently is very rare, if existent at all, in otherwise healthy children.

The most general method for assessing vitamin A status is the isotope dilution technique, in which a dose of deuterated vitamin A is allowed to equilibrate with endogenous reserves [18, 22]. This procedure has now been effectively used in several different countries [23–25].

Breast milk vitamin A concentrations are of particular utility in assessing the vitamin A status of lactating women and infants, two groups at high-risk of being in a deficient or marginal state [19]. Because vitamin A is largely in the form of lipid-soluble retinyl ester in breast milk, the ratio of vitamin A to fat in breast milk seems to be a better indicator of maternal vitamin A status than vitamin A concentrations alone [26]. The latter can vary greatly in milk samples taken at different times during feeding and during the day.

Night blindness in young children as assessed by interview of the mother, tear fluid vitamin A concentrations, and yet other indicators have been used successfully under appropriate conditions [17–19]. The approximate relationships between the dietary intake of vitamin A, its liver reserves, and the useful ranges for selected indicators of vitamin A status are given in Figure 2.

Requirements and recommended intakes

The average daily amount of vitamin A that should be ingested by healthy individuals varies with age, body mass, metabolic activity, and special conditions, e.g. pregnancy and lactation [5–7, 10–12, 27]. The operational end point must also be defined; namely, whether the objective is just to prevent deficiency or also to provide for a suitable body reserve. Finally, the variance in the requirement of a given nutrient in a population group must be considered. In some cases, recommendations have also considered the possible beneficial actions of given nutrients against chronic diseases. In general, however, recommended dietary intakes for population groups have tended to decrease as more information has become available.

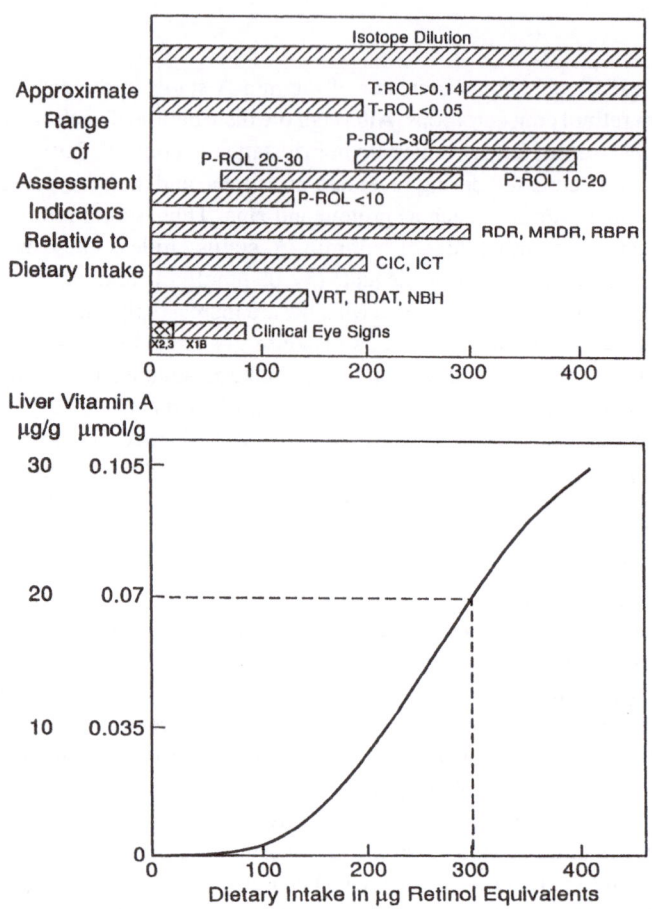

Figure 2. An approximate relationship between the average daily dietary intake of vitamin A and provitamin A carotenoids, expressed as μg retinol equivalents, and the vitamin A concentration in the liver of a preschool child. Above the figure, the approximate ranges for the responses of various indicators of vitamin A status, relative to dietary intake, are given. Thus, clinical signs appear when the average daily intake of vitamin A is very low, which corresponds to negligible liver reserves. Abbreviations for the indicators are: X2, corneal xerosis; X3, corneal ulceration and keratomalacia; X1B, Bitot's spots with conjunctival xerosis; VRT, vision restoration time; RDAT, rapid dark-adaptation test time; NBH, night blindness by history; CIC, conjunctival impression cytology test; ICT, impression cytology with transfer test; RDR, relative dose response test; MRDR, modified relative dose response test; RBPR, retinol-binding protein response test; P-ROL, plasma (or serum) retinol concentration; and T-ROL, tear fluid retinol concentration, followed by values expressed in μg retinol/dl. Vitamin A equivalencies are: 1 μmol retinol = 286 μg retinol; 1 μg retinol = 1 μg retinol equivalents = 0.0035 μmol retinol. From [5], p 114, with the permission of the International Life Sciences Institute, Washington DC 20036-4810, Copyright 1996.

Several global recommendations

Five sets of recommended dietary intakes (RDI) for vitamin A are presented in Table 1: RDI values for Italy [28]; recommended dietary allowances (RDA) of the Food and Nutrition Board,

Table 1. Recommended dietary intakes of vitamin A in retinol equivalents[a]

Group	RDI (Italy)[b]	RDA (USA)[c]	RDI (FAO/WHO)[d] Basal	RDI (FAO/WHO)[d] Safe	DRV (UK)[e] RI (CEC)[f] Lower reference nutrient intake / Lowest threshold intake[f]	DRV (UK)[e] RI (CEC)[f] Estimated average requirement / Average requirement[f]	DRV (UK)[e] RI (CEC)[f] Reference nutrient intake / Population reference intake[f]
Infants							
0–0.5 y	375	375	180	350	150	250	350
0.5–1 y	375	375	180	350	150	250	350
Children							
1–2 y	375	400	200	400	200	300	400
2–6 y	400	500	200	400	200	300	400
6–10 y	500	700	250	400	250	350	500
Males							
11–14 y	700	1000	300	500	250	400	600
14–70+ y	700	1000	300	600	300	500	700
Females							
11–70+ y	600	800	270	500	250	400	600
Pregnancy	+200	+0	+100	+100	+100	+100	+100
Lactation							
0–6 months	+400	+500	+180	+350	+350	+350	+350
>6 months	+400	+400	+180	+350	+350	+350	+350

[a] A retinol equivalent is defined as 1 μg retinol, which is considered equal to 6 μg β-carotene or 12 μg of mixed pro-vitamin A carotenoids.
[b] [28].
[c] [111].
[d] [29].
[e] [10].
[f] [12].

National Research Council, U.S. National Academy of Sciences [11]; recommended dietary intakes (RDI) of the Food and Agriculture Organization and the World Health Organization [29]; dietary reference values (DRV) for the United Kingdom [10]; and reference intakes (RI) of the Commission of the European Communities (EC)[12]. Because the values at different tiers are essentially the same for the DRV and the RI, they are given together in Table 1. The listed age categories in the table for different sets are largely, but not entirely, concordant.

Systems

Three different systems have been employed; single values in the RDI (Italy) and the RDA (USA), a two-tier system in the RDI of the FAO/WHO, and a three-tier system in the DRV (UK) and RI (CEC). The single RDI and RDA values, the higher tier in the FAO/WHO system, termed "safe" intake, and the top tier in the UK and CEC systems, termed "reference nutrient intake (RNI)" or "population reference intake (PRI)", are roughly equivalent conceptually. All presume the presence of an adequate total body reserve of vitamin A for most individuals in society but also include, in the UK system, the requirements of "those few members of the community with particularly high needs" [10].

Tier definitions

The lowest tiers in the FAO/WHO system and in the DRV and CEC systems show similar values, but are defined differently. The FAO/WHO defines basal requirement as the "(average daily) amount needed to prevent clinically demonstrable impairment of function" [29]. The lower reference nutrient intake (LRNI) in the UK system is rather defined as "the (average daily) amount of the nutrient that is enough for only the few people in a group who have low needs" [10]. Finally, the lowest threshold intake (LTI) in the CEC system is defined as an intake "below which almost all individuals will be unable to maintain metabolic integrity" [12].

Definition of requirements

The estimated average requirement (EAR) in the UK system and the average requirement (AR) in the CEC system are the (average daily) amount of a nutrient that meets the needs of 50% of the analyzed group of people. In the UK and CEC systems, the LRNI (or LTI) is two standard deviations below the EAR (or AR), and the RNI (or PRI) is two standard deviations above the EAR. Assigned values for the LRNI (LTI) in the UK and CEC systems in large part denote *inadequate intakes* for most people, whereas the basal requirement in the FAO/WHO classification represents a *minimally adequate intake* for most people. These confounding differences may well be due to different reference standards for weight that are used between the FAO/WHO report and the European reports. The reference weights for adult men and women, respectively, in the FAO/WHO report are 65 and 55 kg, in the UK report are 74 and 60 kg, in the CEC report are 75 and 62 kg, and in the US report are 76 and 62 kg. Thus, since the needs for vitamin A in adults may well be partially dependent on weight, a given intake of vitamin A may be minimally adequate for reference adults with 10–13% lower weight but not for their heavier European counterparts. Although the reference nutrient intakes for infants and children are similar in most sets of recommendations, the RDA for adults in the United States is generous relative to those of many other countries.

Current approaches in the United States

The Food and Nutrition Board of the U.S. National Academy of Sciences has recently devised a new nomenclature for human nutrient needs [30]. Dietary Reference Intakes (DRI) refer to a set of four nutrient-based reference values, termed the estimated average requirement (EAR), the recommended dietary allowance (RDA), adequate intake (AI), and the tolerable upper intake level (UL). The EAR is defined in the same way as in the CEC and UK recommendations. The RDA is now somewhat more rigorously defined to be the EAR + 2 SD (standard deviations), i.e. equivalent to the RNI and PRI in the UK and CEC systems.

The AI is used when insufficient scientific evidence is available to calculate an EAR or a RDA. AI values consider not only nutritional data but also contemporary concepts of disease risk. AI values are different from the EAR and RDA and should not be applied too rigidly. No equivalent value exists in any other system of nutritional recommendations.

The UL is the maximum level of daily intake that is unlikely to pose risks of adverse health effects to most individuals in a given age/sex category. UL does not imply a possible health benefit from intakes at that level [30]. Although the EAR, RDA and UL are relatively straightforward recommendations that are in keeping with global use, the AI value, which is less rigorously defined, may have limited utility. New recommendations for vitamin A in the United States have not yet been defined.

Health and genetic effects

Recommended nutrient intakes are established for healthy individuals. Sickness, and particularly febrile conditions and lipid malabsorption, can markedly increase needs. Genetic defects in the handling of vitamin A can also significantly affect requirements. Thus, the inability to convert β-carotene into vitamin A and genetic defects in the intestinal absorption and plasma transport of vitamin A [31] should increase dietary needs. On the opposite tack, signs of hypervitaminosis A appear in some individuals who ingest only moderate amounts of vitamin A, a condition termed "vitamin A intolerance" [32]. These latter instances, although rare, also seem to have a genetic basis.

Toxicity

When ingested in large doses, vitamin A can be toxic [5–7, 27, 33–36]. The three categories of toxicity are acute, chronic, and teratogenic.

Acute toxicity

Acute toxicity is produced by one or several closely spaced, very large doses of vitamin A, usually >100 times the recommended intake (RDI or RDA) in adults and >20 times the RDI in children. Early signs of acute toxicity include nausea, vomiting, headache, vertigo, blurred vision, muscular incoordination, and, in infants, bulging of the fontanelle. These signs are usually transient and disappear within a few days. When the dose to primates is very large, a second phase, characterized by drowsiness, malaise, inappetence, physical inactivity, itching, skin exfoliation, and recurrent vomiting, follows during the next week [37]. When lethal doses are given to monkeys, the terminal phase includes coma, convulsions, respiratory abnormalities, and then death by respiratory failure or convulsions within 1–16 days [37]. For young monkeys the LD_{50} value (i.e. the single intramuscular dose that kills half of the treated animals) is 0.59 mmol retinol (560 000 IU_a)/kg body weight. In humans, the only comparable case is that of a one month-old male infant weighing 2.25 kg who died after receiving 1.05 mmole (1 000 000 IU_a) of vitamin A during an eleven day period, or a total dose of 0.46 mmole (~440 000 IU_a)/kg [38]. Nonetheless, after acute dosing with smaller but still toxic amounts, recovery usually is complete within a few weeks.

Chronic toxicity

Chronic toxicity, which is much more common than acute toxicity, is induced by the recurrent ingestion over a period of weeks to years of excessive doses of vitamin A that are usually ≥ 10 times the recommended intake (RDI or RDA). Toxic signs commonly include headache, alopecia, cracking of the lips, dry and itchy skin, hepatomegaly, bone and joint pain, as well as many other complaints [5–7, 33–36]. Most cases of chronic hypervitaminosis have been reported in children with daily intakes of 12 000–600 000 IU_a (2000–60 000 $IU_a \cdot kg^{-1} \cdot d^{-1}$) and, in adults, with daily intakes of 50 000–1000 000 IU_a (700–15 000 $IU_a \cdot kg^{-1} \cdot d^{-1}$) [34]. As already indicated, signs of chronic toxicity were noted in a very few children and adults who were presumably ingesting much smaller amounts of vitamin A daily; i.e. 6,000–53 000 IU_a, or 200–800 $IU_a \cdot kg^{-1} \cdot d^{-1}$) [32]. These cases of vitamin A intolerance seem to have a genetic basis, although the metabolic defect has not been defined [32]. After terminating dosing, most patients recover fully from chronic toxicity. Permanent damage to liver, bone, and vision as well as chronic muscular and skeletal pain, however, results in some instances.

Teratogenic effects

The most serious teratogenic effects of vitamin A include fetal resorption, abortion, birth defects, and permanent learning disabilities in the progeny [33–36]. Permanent learning disabilities in animals occur at lower doses than those that cause gross abnormalities [39]. Generally, the drugs accutane (13-*cis* retinoic acid) and etretinate, an aromatic analog of the ethyl ester of all-*trans* retinoic acid, have been most implicated in producing human terata [33, 40]. In comparison with all-*trans* retinoic acid, both all-*trans* retinol and 13-*cis* retinoic acid are less toxic in several pregnant animal models [39]. Because of differences in the sensitivity of various species to teratogenic doses of retinoids, extrapolation of the toxicity of given doses among species must be done with caution [41]. Interestingly, high oral doses of all-*trans* retinoyl β-glucuronide and of some retinoylamides are much less teratogenic, if at all, in rodent models [42–44].

The dose of vitamin A that can cause fetal defects in pregnant women depends on the gestational age, the isomer given, intestinal absorption efficiency, and metabolic and genetic factors. Although an epidemiological survey, based on telephone interviews, suggested that oral daily supplements of preformed vitamin A > 10 000 IU_a (3 mg retinol) could be teratogenic [45], at least a two-fold higher daily dose of vitamin A is needed to increase teratogenic risk in most studies [46]. Indeed, exposures of women to daily oral doses of up to 30 000 IU_a (9 mg retinol) did not significantly increase the blood levels of all-*trans* retinoic acid above the physiological range [46].

Recommendations of safe intakes

Because of concern about birth defects that might be caused by excessive intakes of vitamin A, the International Vitamin A Consultative Group (IVACG) [47] recommends maximal average daily intakes of 650 µg REQ for pregnant women. IVACG approves the use of daily supplements of 10 000 IU_a for pregnant women to prevent deficiency-induced fetal abnormalities only in regions of the world where vitamin A deficiency is common. The Teratology Society recommends that women of child-bearing age limit their total daily intake of preformed vitamin A, including food and supplements, to ≤8000 IU_a and that supplements, if taken, be in the form of provitamin A carotenoids [48]. The Council for Responsible Nutrition, a trade association of the nutrient-supplement industry, reports a no observed adverse effect level (NOAEL) of 10 000 IU_a (3000 µg retinol) and a lowest observed adverse effect level (LOAEL) of 21 600 IU_a (6,500 µg retinol) for adults, including pregnant women [49]. Thus, several groups with different orientations suggest maximal daily intakes of ≤10 000 IU_a of preformed vitamin A during pregnancy.

Supplements are not needed by healthy persons ingesting a balanced diet. In this regard, the American Institute of Nutrition, the American Society for Clinical Nutrition, and the American Dietetic Association issued a formal joint statement that supplements of vitamins and minerals were not needed by well-nourished, healthy individuals except for some specific exceptions, e.g. folate and iron for pregnant women and vitamin K for the newborn [50].

Carotenoid effects

Carotenoids in foods are not known to be toxic in healthy persons, even when ingested in large amounts [51]. Nonetheless, hypercarotenosis, a benign condition characterized by a jaundice-like yellowing of the skin and high plasma carotenoid concentrations can result when large amounts of carotene-rich foods (i.e. tomato juice, carrot juice, or daily β-carotene supplements (>30 mg)) are ingested [5, 51, 52]. When the liver is damaged by the ingestion of alcohol or the lung by heavy smoking, however, supplements of β-carotene can show adverse effects [53, 54]. Furthermore, canthaxanthin retinopathy can occur in patients with photosensitivity disorders who are treated therapeutically with large daily doses (50–100 mg) of this 4,4'-diketo derivative of β-carotene for long periods [55]. Canthaxanthin, however, is not a common dietary constituent. Furthermore, after cessation of intake, the crystalline canthaxanthin inclusion bodies in the retina slowly disappear [55].

As an action independent of their conversion into vitamin A, both nutritionally active and nutritionally inactive carotenoids (e.g. zeaxanthin, lutein and lycopene) may have protective effects in reducing oxidative stress and some forms of chronic disease [56–60]. The extent to which carotenoid ingestion affects the onset of chronic disease in humans, however, is still unclear.

Conclusion

Vitamin A is an essential nutrient for humans. Dietary vitamin A consists of preformed vitamin A, present mainly in foods of animal origin, and provitamin A carotenoids, present primarily in colored fruits and vegetables. The bioavailability of dietary preformed vitamin A in healthy humans is good (>50%), whereas that of provitamin A carotenoids varies, depending on a variety of factors. Currently 6 μg of all-*trans* β-carotene are considered to be equivalent, on average, to 1 μg all-*trans* retinol, although higher ratios have been reported. Median intakes of total dietary vitamin A in industrial countries are generally adequate, although some children and pregnant women from socioeconomically disadvantaged segments of society show inadequate vitamin A status. Useful techniques in assessing the vitamin A statuses of individuals include isotope-dilution methods, response tests, conjunctival histological examinations, and dark-adaptation procedures. A frequency pattern of serum retinol values can also be useful in assessing vitamin A nutriture in populations. Recommended intakes of vitamin A, in the past expressed as a single value for a given population group, have recently been expanded to include two or three tiers of recommendations. Each tier is defined in similar, but not in identical, ways by different expert groups. At higher intakes, vitamin A is toxic. Three categories of toxicity exist: acute, chronic, and teratogenic. Dietary carotenoids tend to show little, if any, toxicity in humans. Supplements of β-carotene, however, when taken in large doses over significant periods, can show adverse effects on livers damaged by alcohol and on lungs damaged by smoking. Large doses of canthaxanthin can induce a reversible retinopathy.

Acknowledgment
This work was supported in part by grants from the NIH (DK 39733) and the USDA (NRICGP 97-37200-0490). This is paper J-18203 of the Iowa Agriculture and Home Economics Experiment Station, Ames, IA (Project 3335), and supported by Hatch Act and State of Iowa funds.

References

1   Wolf G (1996) A history of vitamin A and retinoids. *FASEB J* 10: 1102–1107
2   Hicks RJ (1867) Night blindness in the Confederate army. *Richmond Med J* 3: 34–38
3   Moore T (1957) *Vitamin A. Part I. Historical introduction.* Elsevier, Amsterdam, 3–29
4   American Institute of Nutrition (1990) Nomenclature policy: generic descriptors and trivial names for vitamins and related compounds. *J Nutr* 120: 12–19
5   Olson JA (1996) Vitamin A. *In*: EE Ziegler, LJ Filer Jr (eds): *Present knowledge of nutrition*, 7th ed. ILSI Press, Washington, 109–119
6   Ross AC (1998) Vitamin A and retinoids. *In*: ME Shils, JA Olson, M Shike, AC Ross (eds): *Modern nutrition in health and disease*, 9th ed., Williams and Wilkins, Baltimore, 305–328
7   Olson JA (1998) Fat-soluble vitamins: Vitamin A. *In*: J Garrow, WPT James, A Ralph (eds): *Human nutrition and dietetics*, 10th ed. Churchill Livingstone, Edinburgh, 211–220
8   Castenmiller JJM, West CE (1998) Bioavailability and bioconversion of carotenoids. *Annu Rev Nutr* 18: 19–38
9   Olson JA (1994) Vitamin A, retinoids and carotenoids. *In*: Shils ME, Olson JA, Shike M (eds): *Modern nutrition in health and disease*, 8th ed. Lea and Febiger, Philadelphia, 287–307
10 Department of Health (1991) Dietary reference values for food energy and nutrients for the United Kingdom. Report No. 41 on health and social subjects. HMSO, London
11 Food, Nutrition Board National Research Council (1989) *Recommended dietary allowances*, 10th ed. National Academy Press, Washington,
12 Commission of the European Communities (1993) *Nutrient and energy intakes for the European Community*. Report of the Scientific Committee for Food, thirty-first series. Official publication of the European Communities, Luxembourg
13 Duitsman PK, Cook LR, Tanumihardjo SA, Olson JA (1995) Vitamin A inadequacy in socioeconomically disadvantaged pregnant Iowan women as assessed by the modified relative dose response (MRDR) test. *Nutr Res* 15: 1263–1276
14 Spannaus-Martin DJ, Cook LR, Tanumihardjo SA, Duitsman PK, Olson JA (1997) Vitamin A and vitamin E statuses of preschool children of socioeconomically disadvantaged families living in the Midwestern United States. *Eur J Clin Nutr* 51: 864–869
15 Olson JA (1987) Recommended dietary intakes (RDI) of vitamin A in humans. *Amer J Clin Nutr* 45: 704–716
16 Sauberlich HE, Hodges RE, Wallace DL, Kolder H, Canham JE, Hood J, Raica Jr, N, Lowry LK (1974) Vitamin A metabolism and requirements in the human studied with the use of labelled retinol. *Vitamins and Hormones* (USA) 32: 251–275
17 Sommer A, West Jr, KP (1996) *Vitamin A deficiency – health, survival and vision*. Oxford University Press, Oxford, 1–437
18 Underwood BA, Olson JA (eds) (1993) *A brief guide to current methods of assessing vitamin A status*. International Vitamin A Consultative Group. ILSI, Washington, 1–37
19 World Health Organization (1996) *Indicators for assessing vitamin A deficiency and their application in monitoring and evaluating intervention programs*. Micronutrient series 96-10. World Health Organization, Geneva, 1–66
20 van den Berg H (1996) Vitamin A intake and status. *Eur J Clin Nutr* 50(suppl 3):S7–S12
21 Tanumihardjo SA, Cheng J-C, Permaesih D, Muherdiyantiningsih Rustan E, Muhilal Karyadi D, Olson JA (1996) Refinement of the modified relative dose response test as a method for assessing vitamin A status in a field setting: experience with Indonesian children. *Am J Clin Nutr* 64: 966–971
22 Furr HC, Amedee-Manesme O, Clifford AJ, Bergen HRIII, Jones AD, Anderson DP, Olson JA (1989) Vitamin A concentrations in liver determined by isotope dilution assay with tetra-deuterated vitamin A and by biopsy in generally healthy adult humans. *Amer J Clin Nutr* 49: 713–716
23 Haskell MJ, Handelman GJ, Peerson JM, Jones AD, Rabbi MA, Awal MA, Wahed MA, Mahalanabis D, Brown KH (1997) Assessment of vitamin A status by the deuterated-retinol-dilution technique and comparison with hepatic vitamin A concentration in Bangladeshi surgical patients. *Amer J Clin Nutr* 66: 67–74
24 Haskell MJ, Islam MA, Handelman GJ, Peerson JM, Jones AD, Wahed MA, Mahalanabis D, Brown KH (1998) Plasma kinetics of an oral dose of [²H₄]retinyl acetate in human subjects with estimated low or high total body stores of vitamin A. *Amer J Clin Nutr* 68: 90–95
25 Ribaya-Mercado JD, Mazariegos M, Tang G, Romero-Abel ME, Mena I, Solomons NW, Russell RM (1999) Assessment of total body stores of vitamin A in Guatemalan elderly by the deuterated-retinol-dilution method. *Amer J Clin Nutr* 69: 278–284

26 Stoltzfus RJ, Habicht J-P, Rasmussen KM, Hakimi M (1993) Evaluation of indicators for use in vitamin A intervention trials targeted at women. *Int J Epidemiol* 22: 1111–1118
27 Gerster H (1997) Vitamin A functions, dietary requirements and safety in humans. *Int J Vitam Nutr Res* 67: 71–90
28 Mariani-Costantini A (ed.) (1989) *Guidelines for a healthy Italian diet*. National Institute of Nutrition Press, Rome, 14–15
29 Food, Agriculture Organization/World Health Organization (1989) *Requirements of vitamin A, iron, folate, and vitamin B_{12}*. Report of a joint FAO/WHO Expert Committee. FAO Food and Nutrition Series 23, Rome
30 Food and Nutrition Board, Institute of Medicine (USA) (1997) Dietary reference intakes. *Nutr Rev* 55: 319–326
31 McLaren DS (1993) Fat-soluble vitamins: vitamin A. *In*: JS Garrow, WPT James (eds): *Human nutrition and dietetics*, 9th ed. Churchill Livingstone, Edinburgh, 208–217
32 Olson JA (1989) Upper limits of vitamin A in infant formulas, with some comments on vitamin K. *Amer J Clin Nutr* 119: 1820–1824
33 Armstrong RB, Ashenfelter KO, Eckhoff C, Levin AA, Shapiro SS (1994) General and reproductive toxicology of retinoids. *In*: MB Sporn, AB Roberts, DS Goodman (eds): *The retinoids: biology, chemistry, and medicine*, 2nd ed. Raven Press, New York, 545–572
34 Bauernfeind JC (1980) *The safe use of vitamin A*. International Vitamin A Consultative Group. Nutrition Foundation, Washington, 1–44
35 Bendich A, Langseth L (1989) Safety of vitamin A. *Amer J Clin Nutr* 49: 358–371
36 Hathcock JN, Hattan DG, Jenkins MV, McDonald JT, Sundaresan PR, Wilkening VL (1990) Evaluation of vitamin A toxicity. *Amer J Clin Nutr* 52: 183–202
37 Macapinlac MP, Olson JA (1981) A lethal hypervitaminosis A syndrome in young monkeys following a single intramuscular dose of a water-miscible preparation containing vitamins A, D_2 and E. *Int J Vitam Nutr Res* 51: 331–341
38 Bush ME, Dahms BB (1984) Fatal hypervitaminosis in a neonate. *Arch Pathol Lab Med* 108: 838–842
39 Adams J (1993) Structure-activity and dose-response relationships in the neural and behavioral teratogenesis of retinoids. *Neurotoxicol Teratol* 15: 193–202
40 Lammer EJ, Chen DT, Hoar RM, Agnish ND, Banks PJ, Braun JT, Curry CJ, Bernhoff PM, Grix AW, Lott IT et al (1985) Retinoic acid embryopathy. *New Engl J Med* 313: 837–841
41 Howard WB, Willhite CC (1986) Toxicity of retinoids in humans and animals. *J Toxicol-Toxin Rev* 5: 55–94
42 Gunning DB, Barua AB, Olson JA (1993) Comparative teratogenicity and metabolism of all-*trans* retinoic acid, all-*trans* retinyl β-glucose, and all-*trans* retinyl β-glucuronide in pregnant Sprague-Dawley rats. *Teratology* 47: 29–36
43 Barua AB (1997) Retinoyl β-glucuronide: a biologically active form of vitamin A. *Nutr Rev* 55: 259–267
44 Formelli F, Barua AB, Olson JA (1996) Bioactivities of N-(4-hydroxyphenyl)retinamide and retinoyl β-glucuronide. *FASEB J* 10: 1014–1024
45 Rothman KJ, Moore LL, Singer ML, Nguyen UDT, Mannino S, Milinksy A (1995) Teratogenicitiy of high vitamin A intake. *New Engl J Med* 333: 1369–1373
46 Miller RK, Hendrickx AG, Mills JL, Hummler H, Wiegand U-W (1998) Periconceptual vitamin A use: How much is teratogenic? *Reprod Toxicol* 12: 75–88
47 Underwood BA (1986) *The safe use of vitamin A by women during the reproductive years*. International Vitamin A Consultative Group, ILSI-Nutrition Foundation, Washington
48 Teratology Society (1987) Position paper: Recommendations for vitamin A use during pregnancy. *Teratology* 35: 269–275
49 Hathcock JN (1997) *Vitamin and mineral safety: A summary review*. Council for Responsible Nutrition Press, Washington
50 American Institute of Nutrition American Society for Clinincal Nutrition American Dietetic Association (1987) Joint Statement on vitamin and mineral supplements. *J Nutr* 117: 1649
51 Olson JA (1998) Carotenoids. *In*: ME Shils, JA Olson, M Shike, AC Ross (eds): *Modern nutrition in health and disease*, 9th ed. Williams and Wilkins, Baltimore, 525–542
52 Micozzi MS, Brown ED, Taylor PR, Wolfe E (1988) Carotenoids in men with elevated carotenoid intake from foods and β-carotene supplements. *Am J Clin Nutr* 48: 1061–1064
53 Leo MA, Aleynik SI, Aleynik MK, Lieber CS (1997) β-Carotene beadlets potentiate hepatotoxicity of alcohol. *Amer J Clin Nutr* 66: 1461–1469
54 Omenn GS, Goodman GE, Thornquist MD, Balmes J, Cullen MR, Glass A, Keogh JP, Meyskens FL, Valaris B, Williams JH et al (1996) Effects of a combination of β carotene and vitamin A on lung cancer and cardiovascular disease. *New Engl J Med* 334: 1150–1155
55 Weber U, Georz G, Baseler H, Michaelis L (1992) Canthaxanthin retinopathy: follow-up of over 6 years. *Klin Monatsbl Augenheilk* 201: 174–177
56 Bendich A, Olson JA (1989) Biological actions of carotenoids. *FASEB J* 3: 1927–1932
57 Canfield LM, Krinsky NI, Olson JA (eds) (1993) *Carotenoids in human health*, vol. 691. The New York Academy of Sciences, New York
58 Mayne ST (1996) Carotenoids 5: β carotene, carotenoids, and disease prevention in humans. *FASEB J* 10:

690–701
59 Weisburger JH (ed) (1998) International symposium: Lycopene and tomato products in disease prevention. *Proc Soc Exp Biol Med* 218: 93–143
60 Snodderly DM (1995) Evidence for protection against age-related macular degeneration by carotenoids and antioxidant vitamins. *Amer J Clin Nutr* 62: 1448S–1461S

Vitamin A and retinoids: an update of biological aspects and clinical applications
M.A. Livrea (ed.)
© 2000 Birkhäuser Verlag Basel/Switzerland

Current views on carotenoids: biology, epidemiology and trials

N.I. Krinsky[1] and S.T. Mayne[2]

[1] Department of Biochemistry, School of Medicine and USDA Human Nutrition Research Center on Aging, Tufts University, Boston, MA 02111, USA
[2] Department of Epidemiology and Public Health, Yale University School of Medicine and Yale Cancer Center, New Haven, CT 06520, USA

Introduction

There have been well over 600 carotenoids described in nature with the vast majority having colors ranging from pale yellow to deep red [1]. We find them not only in our fruits and vegetables, but in the flowers and foliage decorating our environment. In addition, many animal species contain visible coloration due to the ingestion of carotenoid-containing plant material. The pinkish tone of many fish species, some of the color in the skin and plumage of birds, and even the yellow color in the fat associated with meat products represents the transfer of carotenoids from plants to animals. The human is not exempt from this process, for we absorb numerous carotenoids from our diet, and then deposit them in our tissues [2]. Figure 1 represents the structures of the most common carotenoids found in humans, with relative levels depending on regional differences in diet. It should be noted that some of the naturally occurring carotenoids are found in the *cis*-configuration, which can significantly alter their physical properties and may impact on their biological properties. An example of this phenomenon relates to β-carotene. In our usual plant sources of β-carotene, the predominant isomeric form is the all-*trans*-β-carotene. In some algal sources, such as *Dunaliella*, there is a 50:50 mixture of the all-*trans*-β-carotene and 9-*cis*-β-carotene [3]. The latter isomer is much more soluble, presumably due to an inability to form tightly structured crystals. This change in solubility has been suggested as one reason for differences in the biological properties of these two isomers [4].

The best known biological action of carotenoids is the ability of some of these compounds to be metabolized in animals to form vitamin A, or retinol [5]. The carotenoids that can undergo this metabolism are referred to as provitamin A carotenoids, and are indicated as such in Figure 1. It has been estimated that in the United States, 30–40% of our daily intake of vitamin A comes from the metabolism of provitamin A carotenoids [6]. This percentage will vary from country to country, depending on the availability of fruits and vegetables in diets, as well as the availability of pre-formed vitamin A, which is found in meat, liver and dairy products.

The metabolism of a provitamin A carotenoid such as β-carotene is depicted in Figure 2. The major pathway involves cleavage by a 15,15'-central dioxygenase that results in the formation of retinal, which can either be reduced to retinol or oxidized to retinoic acid (see Napoli, this volume). This process occurs in the cytoplasm of many cell types, and produces two

MAJOR HUMAN SERUM CAROTENOIDS

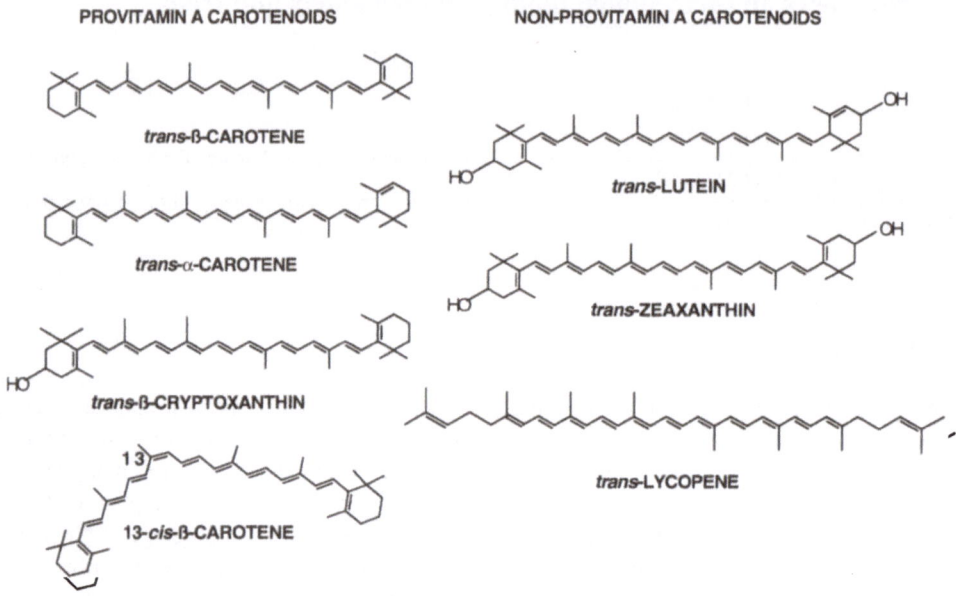

Figure 1. Structures of the provitamin A and non-provitamin A carotenoids present in human serum.

ß-CAROTENE METABOLISM

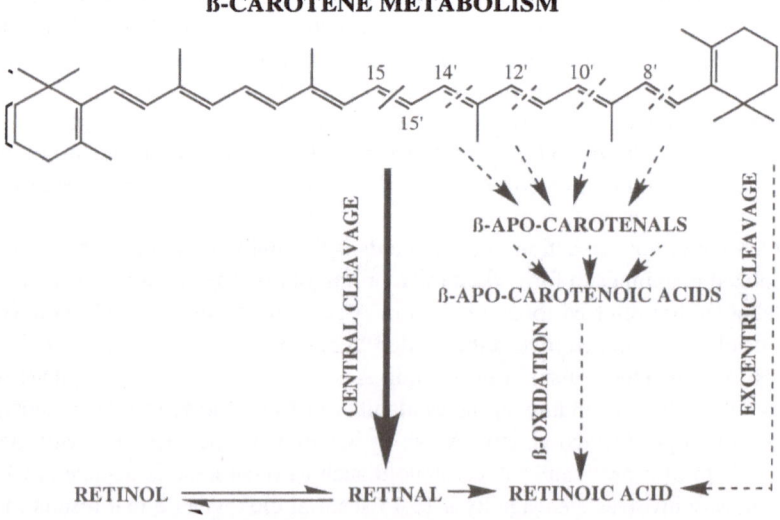

Figure 2. Metabolism of β-carotene to retinal via central cleavage and to retinoic acid via excentric cleavage.

molecules of retinal from each molecule of β-carotene [7, 8]. In addition, there is a minor pathway associated with mitochondrial oxidation of β-carotene breakdown products and produces retinoic acid without requiring the intermediacy of retinal [9]. This process has been referred to as an excentric cleavage mechanism, and the overall reactions appear to be analogous to mitochondrial fatty acid β-oxidation [10].

Antioxidant role of carotenoids

In the thirty years since Foote and his associates demonstrated that singlet excited oxygen ($^1O_2$) could be inactivated, or quenched, by β-carotene at a near diffusion limited rate [11], there have been numerous articles describing the mechanism, kinetics, and efficacy of quenching $^1O_2$ by various carotenoids [12, 13]. With respect to dietary carotenoids, lycopene appears to be a somewhat better quencher than β-carotene, but the range between the most effective and least effective dietary carotenoid is only an order of magnitude [14]. $^1O_2$ is usually formed through photochemical reactions involving the absorption of light by a sensitizer molecule (S) which converts it to the singlet sensitizer (^1S). Through a process known as intersystem crossing (ISC), ^1S is then converted to a meta-stable triplet state (^3S). The ^3S can react with ground state oxygen, which exists in a triplet configuration ($^3O_2$), in an energy transfer reaction, to form ground state sensitizer (S) and $^1O_2$. These reactions are depicted below:

$$S + light \longrightarrow {}^1S$$

$$^1S \underline{\quad} ISC \longrightarrow {}^3S$$

$$^3S + {}^3O_2 \longrightarrow S + {}^1O_2$$

It is this $^1O_2$ that is a highly reactive species, capable of oxidizing nucleic acids, various amino acids in proteins, and unsaturated fatty acids. Fortunately for plants, carotenoids (CAR) are the most effective quenchers of $^1O_2$ found in nature, *via* the following reactions:

$$^1O_2 + CAR \longrightarrow {}^3O_2 + {}^3CAR$$

$$^3CAR \longrightarrow CAR + heat$$

Again, because of the long conjugated polyene nature of these molecules, they lose the excess energy in the excited state (^3CAR) *via* vibrational and rotational interactions with the solvent system, ultimately re-forming the ground state carotenoid (CAR), ready to begin another cycle of $^1O_2$ quenching. It has been estimated that each carotenoid molecule can quench 1000 $^1O_2$ molecules before it reacts chemically, and form products, thus serving as very effective antioxidants [15].

In vitro *evidence*

It is beyond the scope of this chapter to do more than touch on this area of carotenoid research. Suffice it to say that in addition to $^1O_2$ quenching, carotenoids also interfere with radical-initiated reactions, particularly with those that result in lipid peroxidation. Several reviews on this topic have appeared [12, 16, 17] and newer results suggest that this protection is unlike that seen with classical dietary antioxidants such as α-tocopherol. Burton and Ingold first pointed out the unique antioxidant effects of β-carotene [18] and recent studies of the *in vitro* antioxidant activities of β-carotene and other carotenoids demonstrate excellent activity, in some cases showing a more powerful antioxidant action than α-tocopherol [19].

Low-density lipoprotein (LDL) oxidation continues to interest investigators as a means of assessing whether carotenoids can act as either *in vitro* or *in vivo* antioxidants. LDL can be loaded with carotenoids above and beyond what is normally present by either feeding large doses of carotenoids to volunteers and isolating their LDL, or by adding the carotenoids directly to serum or to isolated LDL. Several studies have involved feeding a mixture of antioxidants and reporting a decrease in the susceptibility of LDL to oxidative stress, but it is not clear if one or more of the compounds is responsible for this effect. In recent studies, where β-carotene or other carotenoids were fed or added directly to LDL, the results are conflicting, with several groups reporting a protective effect [20–22], while others find either no effect [23–25] or a prooxidant effect [26, 22]. It would appear that the final story of the role of carotenoids in LDL is not yet resolved.

In vivo *evidence*

The studies referred to above use LDL in an *in vitro* assay, but the ultimate antioxidant test is whether these compounds work in animals or humans. There are very few examples of an *in vivo* antioxidant action of a carotenoid in humans. The best recent examples come from studies with children suffering from cystic fibrosis who are known to have relatively low levels of carotenoids in their serum. These patients are supplemented with large amounts of vitamin E, but still show significantly elevated levels of malondialdehyde in their serum. Two groups have supplemented patients with this disease with β-carotene, and in each case, they reported a decrease in the malondialdehyde levels [27, 20] as well as increased resistance of LDL to oxidant stress [27].

Pro-oxidant role of carotenoids

Burton and Ingold suggested that β-carotene acts as a prooxidant at 100% oxygen, but their observation merely represents a decrease in antioxidant activity at higher oxygen tensions during radical-initiated lipid peroxidation [18]. Since then, several studies have reported that at 100% oxygen tension, β-carotene actually acts as a prooxidant whereas at ambient oxygen tension (20%), it continues to act as an antioxidant. Palozza and her associates demonstrated that β-carotene is an effective antioxidant in protecting both rat liver microsomes [28] and mouse thymocytes [29] against radical-induced lipid peroxidation under air, but at 100% oxygen, the carotenoid acts as a prooxidant. In another study, it was demonstrated that dietary β-carotene actually enhanced tissue lipid peroxidation in mice treated with methyl mercuric chloride [30]. Truscott proposed a theoretical basis for antioxidant and prooxidant effects of β-carotene under low and high oxygen tensions [31], and it will be interesting to see if this can be demonstrated experimentally.

Relationship between observational epidemiology and a health-related role for carotenoids

Observational studies of carotenoids and chronic disease have examined the association between carotenoid intake and disease incidence, or between blood or tissue levels of carotenoids, and disease incidence. One type of observational study is prospective, with the carotenoid intake and/or status measured years before the diagnosis of disease. Another type of observational study is retrospective, wherein individuals with disease and comparable controls are asked to recall usual dietary intake prior to the development of the disease (cases) or prior to interview (controls). Retrospective studies of carotenoids and disease commonly rely on recalled dietary intake rather than blood carotenoid levels to assess exposure, as altered blood carotenoid status could easily be a consequence, rather than a cause, of the disease, particularly for diseases like cancer.

Several reviews of the literature concerning health effects of carotenoids have been published recently [32–34] with considerable attention given to the observational studies. Upon examination of this literature, it is apparent that persons who consume higher levels of carotenoids or who have higher blood levels, as compared to persons with lower intake or blood levels, are at lower risk for a number of chronic diseases including a variety of different cancers [35], cardiovascular disease [36], cataract [37], age-related macular degeneration [38], and even all-cause mortality [39]. Perhaps the most striking finding to emerge from the observational studies of carotenoids and chronic disease is the strong and consistent inverse association between β-carotene-rich foods and lung cancer risk. That is, persons who consume greater quantities of β-carotene from fruits and vegetables are far less likely to develop lung cancer than persons who consume lesser quantities of β-carotene from fruits and vegetables, as reviewed extensively elsewhere [40]. For example, intake of carotenoid-rich fruits and vegetables was associated with reduced lung cancer risk in eight of eight prospective studies and 18 of 20 retrospective studies reviewed [40]. The reduced risk is observed in men and women, in various countries, in current smokers, ex-smokers and never-smokers, and for all histological types of lung cancer.

Carotenoids, and β-carotene, are inversely associated with risk of several cancers in addition to lung cancer, most notably upper aerodigestive tract cancers including oral cavity, pharynx, larynx, and esophageal cancers [34]. More recent epidemiological studies have also used newer carotenoid databases [41] in order to estimate intake of several different carotenoids. Results suggest that several different carotenoids such as α-carotene, β-carotene and lutein/zeaxanthin are inversely associated with risk of lung [42, 43] and other cancers [34].

Also, lycopene, a carotenoid concentrated in tomatoes and tomato products, has been inversely associated with risk of prostate cancer in some epidemiological studies [44].

While the epidemiological data demonstrating an association between lower carotenoid intakes and higher disease risk is quite compelling, the interpretation is difficult in that carotenoids are arguably the best marker for fruit and vegetable intake. Fruits and vegetables contain numerous substances, including vitamin C and other phytochemicals such as polyphenols, that might play a role in the prevention of cancers or other diseases [45]. Thus, beneficial effects observed could be attributable to factors other than carotenoids in fruits and vegetables, or even could be due to health behaviors that correlate with increased fruit and vegetable consumption, such as lesser consumption of foods of animal origin, more exercise, or other unknown behavioral correlates. This problem of confounding is insurmountable in observational epidemiological studies; therefore, these studies comprise only one aspect of the totality of the evidence concerning health effects of carotenoids in humans.

Clinical trials with carotenoids

Basis for trials

The consistency of the literature from observational studies of carotenoids and risk of certain cancers, particularly lung cancer, combined with evidence of a consistent cancer preventive effect in several animal carcinogenesis model systems, most notably hamster buccal pouch and mouse skin tumor models [34], led to a large number of cancer prevention trials involving β-carotene in humans. The focus on β-carotene and not on other carotenoids was largely a practical consideration, in that β-carotene was the only carotenoid that was commercially available in large quantities for which human data supporting safety existed. As chemoprevention trials are generally done in disease-free populations, strong evidence supporting a lack of acute and chronic toxicity is critical prior to launching population trials, and supplemental β-carotene had been used for years in the treatment of the photosensitivity disorder erythropoietic protoporphyria without any evidence of toxicity [46].

Endpoints of clinical trials

Some of the first trials involving β-carotene used an intermediate endpoint as the outcome for the trial. Trials of intermediate endpoints generally require smaller sample sizes and interventions of shorter duration than trials with incident cancer as the outcome of interest. However, an important limitation to the use of intermediate markers in chemoprevention trials is that it has not been demonstrated for any of the intermediate endpoints that modulation of the endpoint by an agent is predictive of modulation of invasive malignancy. Therefore, trials that use cancer incidence or mortality as the primary endpoint have also been conducted, the results of which are detailed below.

Completed clinical trials

A large number of clinical intervention studies have examined the effect of supplemental β-carotene on intermediate cancer endpoints, as shown in Table 1. The results of these trials indicate that supplemental β-carotene consistently results in regression of oral precancerous lesions (oral leukoplakia, oral dysplasia), and a decreased frequency of micronucleated buccal mucosal cells. While many of the trials of oral precancerous endpoints were not placebo-controlled, and thus somewhat difficult to interpret because spontaneous regression can occur, those that were placebo-controlled nonetheless demonstrated significant benefit to β-carotene relative to placebo. From Table 1 it appears that the chemopreventive efficacy of β-carotene varies by site, with evidence for efficacy in the oral cavity and possibly esophagus, mixed evidence in cervix and lung, and convincing evidence of a lack of efficacy in the prevention of recurrent colorectal polyps.

Completed trials of β-carotene using cancer incidence or mortality as the outcome are shown in Table 2. Given the large sample size and resulting cost of these trials, only six total trials have been completed, with mixed results. The α-Tocopherol β-Carotene (ATBC) trial involved 29 133 males aged 50–69 years old from Finland [47], who were heavy cigarette smokers at entry (average one pack/day for 36 years). Participants were randomized to receive either supplemental β-carotene (20 mg/day), α-tocopherol (50 mg/day), the combination, or placebo for five to eight years. Unexpectedly, participants receiving β-carotene (alone or in combination with α-tocopherol) had a statistically significant 18% increase in lung cancer incidence (relative risk [RR] = 1.18; 95% confidence interval [CI] = 1.03–1.36) and 8% increase in total mortality (RR = 1.08, 95% CI = 1.01–1.16) relative to participants receiving placebo. Supplemental β-carotene did not significantly increase the incidence of other major cancers occurring in this population.

The finding of an increased incidence of lung cancer in β-carotene supplemented smokers was apparently replicated in another major trial. The Carotene and Retinol Efficacy Trial (CARET) was a multi-center lung cancer prevention trial of supplemental β-carotene (30 mg/day) plus retinol (25 000 IU/day) versus placebo in asbestos workers and smokers [48]. CARET was terminated nearly two years early in January 1996, because interim analyses of the data indicated that should the trial have continued for its planned duration, it was highly unlikely that the intervention would have been found to be beneficial, given the results as of late 1995. Furthermore, the interim results indicated that the supplemented group was developing more lung cancer, not less, consistent with the results of the Finnish trial. Overall, lung cancer incidence was increased by 28% in the supplemented subjects (RR = 1.28; 95% CI = 1.04–1.57) and total mortality was also increased (RR = 1.17, 95% CI = 1.03–1.33). The increase in lung cancer following supplementation with β-carotene and retinol was observed for current but not former smokers.

The Physicians' Health Study (PHS) of supplemental β-carotene versus placebo in 22 071 male U.S. physicians reported no significant effect—positive or negative—of 12 years of supplementation of β-carotene (50 mg every other day) on total cancer, lung cancer, or cardiovascular disease [49]. With regard to lung cancer, there was no indication of excess lung cancer in the β-carotene supplemented individuals, even among smokers. The relative risk for lung can-

Table 1. Completed β-carotene trials: precancerous endpoints

Tumor site	Total n	β-Carotene dose	Outcome	Author (ref)
ORAL CAVITY				
Oral leukoplakia	24	30 mg/day	71% Overall response*	Garewal [54]
Oral leukoplakia	50	60 mg/day	52% Overall response	Garewal [55]
Oral leukoplakia	18	90 mg/day	44% Overall response	Toma [56]
Oral leukoplakia	111	180 mg/week +100 000 IU Vit. A/week	15% Complete response 28% Complete response	Stich† [57]
Oral leukoplakia/ esophagitis	384	40 mg/day +100 000 IU Vit. A +80 mg Vit. E/week	Odds ratio (OR) = 0.62 for leukoplakia (CI = 0.39–0.98)	Zaridze† [58]
Oral dysplasias	18	120 mg/day	44% Overall response	Malaker [59]
Oral leukoplakia	79	30 mg/day +1000 mg Vit. C + 800 IU vit. E/day	49% Overall response	Kaugars [60]
Oral micronuclei	43	180 mg/week	66% decrease in micronuclei vs. 1% in placebo arm	Stich† [61]
Oral micronuclei	54	180 mg/week	60% decrease in micronuclei vs. 0.5% in placebo arm	Stich† [62]
Oral leukoplakia	112	180 mg/week +100 000 IU Vit. A/week	71% decrease in micronuclei 71% decrease in micronuclei vs. 8% increase in placebo arm	Stich† [57]
LUNG				
Sputum atypia	755	50 mg/day +25 000 IU Vit. A every other day	OR = 1.24 atypia (CI = 0.78–1.96)	McLarty† [52]
Sputum micronuclei	114	20 mg/day	27% decrease micronuclei vs. placebo (CI = 9–41% decrease)	Van Poppel [63]
ESOPHAGUS				
Oral Leukoplakia/ Esophagitis	291	40 mg/day +100 000 IU Vit. A +80 mg Vit. E/week	35% Dec. in risk of progression (esophagitis)	Zaridze† [58]
COLON				
Adenomatous polyps	751	25 mg/day ± 1 g Vit. C and 400 mg Vit. E/day	RR = 1.01 recurrence (CI = 0.85–1.20)	Greenberg† [64]

Table 1. (continued)

Adenomatous polyps	390	20 mg/day ± wheat bran ± low fat diet	OR = 1.4 recurrence (CI = 0.8–2.3) 2 yrs.	MacLennan[†] [65]
Adenomatous polyps	257	15 mg/day	29% recurrence versus 24% in placebo	Kikendall[†] [66]
CERVIX				
Cervical dysplasia	278	10 mg/day	OR = 0.68 for regression (CI = 0.28–1.60)	de Vet[†] [67]
Cervical dysplasia	69	30 mg/day	OR = 1.53 for persistent cervical intraepithelial neoplasia p = 0.55	Romney[†] [68]
Cervical dysplasia	30	30 mg/day	70% response	Meyskens [69]

*Complete response plus partial response
[†]Placebo-controlled studies

Table 2. Completed β-carotene cancer prevention clinical trials: cancer incidence or mortality as primary endpoints

Tumor Site	Total n	β-Carotene dose	Relative risk (95% CI)*	Author [ref]
Lung	29 133	20 mg/day	1.18 (1.03–1.36)	ATBC [47]
Lung	18 314	30 mg/day +25 000 IU Vit. A/day	1.28 (1.04–1.57)	CARET [48]
Esophagus/stomach	29 584	15 mg/day +30 mg Vit. E +50 µg Se/day	0.79 (0.64–0.99) Stomach; 0.96 (0.78–1.18) Esophagus	Blot [51]
Esophagus/stomach	3 318	15 mg/day +Multivit/ Multimineral	1.18 (0.76–1.85) Stomach; 0.84 (0.54–1.29) Esophagus	Li [70]
Skin (non-melanoma)	1 805	50 mg/day	1.05 (0.91–1.22)	Greenberg [50]
Total cancer	22 071	50 mg every other day	0.98 (0.91–1.06)	PHS [49]

* Risk for site-specific cancer incidence (ATBC, CARET, Greenberg, PHS) or mortality (Blot, Li)

cer in current smokers randomized to β-carotene was 0.90 (95% CI = 0.58–1.40). Among non-smokers, the relative risk was 0.78 (95% CI 0.34–1.79). The apparent lack of an effect of long-term supplementation of β-carotene on lung cancer incidence, even in baseline smokers who took the supplements for up to 12 years, is noteworthy and will be discussed further below.

In contrast to the lack of effect [50, 49] and adverse effects [47, 48] reported for β-carotene in cancer prevention are the findings from the Linxian (China) cancer prevention trial [51]. Nearly 30 000 men and women aged 40–69 took part in the study, which tested the efficacy of four different nutrient combinations at inhibiting the development of esophageal and gastric cancers. After a five year intervention period, those who were given the combination of β-carotene, vitamin E and selenium had a 13% reduction in total cancer deaths (RR = 0.87; 95% CI = 0.75–1.00), a 4% reduction in esophageal cancer deaths (RR = 0.96; 95% CI = 0.78–1.18), and a 21% reduction in gastric cancer deaths (RR = 0.79; 95% CI = 0.64–0.99). None of the other nutrient combinations reduced gastric or esophageal cancer deaths significantly in this trial. The finding that vitamin supplements reduced cancer deaths in this population provides compelling data supporting the concept of cancer prevention via nutrients; however, the applicability of these results for populations with adequate nutritional status and for other tumor sites may be limited. Also, it is unclear which nutrient or combination (β-carotene, vitamin E, or selenium) was responsible for the observed protection.

A logical question to raise at this juncture is whether or not the results of the intermediate marker trials of β-carotene (Tab. 1) are concordant with those using cancer as the primary outcome (Tab. 2). McLarty et al. [52] reported an insignificant increase in sputum atypia with the combination of β-carotene plus retinol, a result that is consistent with the increased lung cancer incidence seen in CARET, which used the same combination intervention. Notably absent from Table 2, however, are trials aimed at the prevention of oral cancer, given the promising results of β-carotene in oral premalignant lesions. Two trials, one in the U.S. [53] and one in Italy, are aimed at the prevention of oral cancers with β-carotene. The results of these trials, which are expected shortly, should contribute to our understanding of whether chemopreventive efficacy in intermediate endpoint studies predicts efficacy in the prevention of human cancer.

Interpretation of trials

To date, the only carotenoid that has been studied in trials aimed at the prevention of chronic disease is β-carotene. The general lack of efficacy reported in the major trials, and the possible suggestion of harm with regard to lung cancer in cigarette smokers, is clearly at odds with the observational data on carotenoids from foods, specifically with regard to lung cancer. There are many explanations for why this might occur, for example, β-carotene could simply be a marker for something else found in fruits and vegetables that is protective against cancer, or alternatively, β-carotene might be protective at a certain exposure range but harmful at extremely high exposures. As described in detail elsewhere [35], the median plasma β-carotene concentrations achieved in participants in the ATBC and CARET trials were 300 and 210 μg/dl, respectively. In contrast, the levels associated with a reduced risk of lung cancer in observational studies are in the order of about 20 μg/dl. The Physicians' Health Study and the Linxian trials had intermediate levels (120 and 85 µg/dl, respectively), with no evidence of harm and possible evidence of benefit. Thus, it is possible that β-carotene, like many chemical substances, can have adverse effects at very high doses, but null or beneficial effects at lower doses. There are currently a number of large trials ongoing around the world using β-carotene alone or in combination with other agents, including the Women's Antioxidant Cardiovascular Study in the U.S., the Heart Protection Study in the U.K., the SUVIMAX trial in France, and the Physicians' Health Study II in the U.S. The results of these trials, combined with those of the completed trials discussed previously, should help clarify the effects of various doses of β-carotene, alone or in combination with other agents, on the risk of chronic diseases in humans.

References

1 Britton G (1995) Structure and properties of carotenoids in relation to function. *FASEB J* 9: 1551–1558
2 Erdman JW Jr, Bierer TL, Gugger ET (1993) Absorption and transport of carotenoids. *In*: LM Canfield, NI Krinsky, JA Olson, (eds): *Carotenoids in Human Health*. New York Acad. Sci., New York, 691: 76–86
3 Ben-Amotz A, Lers A, Avron M (1988) Stereoisomers of β-carotene and phytoene in the alga, *Dunaliella bardawil*. *Plant Physiol* 86: 1286–1291
4 Ben-Amotz A, Mokady S, Edelstein S, Avron M (1989) Bioavailability of a natural isomer mixture as compared with synthetic all-*trans*-β-carotene in rats and chicks. *J Nutr* 119: 1013–1019
5 Moore T (1930) Vitamin A and carotene. vi. the conversion of carotene to vitamin A *in vivo*. *Biochem J* 24: 696–702
6 Anonymous (1995) Third Report on Nutrition Monitoring in the United States. *Life Sciences Research Office, FASEB, Interagency Board for Nutrition Monitoring and Related Research*. U.S. Government Printing Office, Washington, 2: VA-32
7 Devery J, Milborrow BV (1994) β-Carotene-15,15'-dioxygenase (*EC* 1.13.11.21) isolation reaction mechanism and an improved assay procedure. *Brit J Nutr* 72: 397–414
8 Nagao A, During A, Hoshino C, Terao J, Olson JA (1996) Stoichiometric conversion of all-*trans*-β-carotene to retinal by pig intestinal extract. *Arch Biochem Biophys* 328: 57–63
9 Wang X-D, Krinsky NI (1997) Identification and quantification of retinoic acid and other metabolites from β-carotene excentric cleavage in human intestine *in vitro* and ferret intestine *in vivo*. *Methods Enzymol* 282: 117–130
10 Wang X-D, Russell RM, Liu C, Stickel F, Smith D, Krinsky NI (1996) β-Oxidation in rabbit liver *in vitro* and in the perfused ferret liver contributes to retinoic acid biosynthesis from β-apo-carotenoic acids. *J Biol Chem* 271: 26 490–26 498
11 Foote CS, Denny RW (1968) Chemistry of singlet oxygen. VIII. Quenching by β-carotene. *J Amer Chem Soc* 90: 6233–6235
12 Krinsky NI (1989) Antioxidant functions of carotenoids. *Free Radical Biol Med* 7: 617–635
13 Sundquist AR, Briviba K, Sies H (1994) Singlet oxygen quenching by carotenoids. *Methods Enzymol* 234: 384–388
14 Di Mascio P, Kaiser S, Sies H (1989) Lycopene as the most efficient biological carotenoid singlet oxygen quencher. *Arch Biochem Biophys* 274: 532–538
15 Foote CS, Chang YC, Denny RW (1970) Chemistry of singlet oxygen. X. Carotenoid quenching parallels biological protection. *J Amer Chem Soc* 92: 5216–5218
16 Palozza P, Krinsky NI (1992) Antioxidant effects of carotenoids *in vitro* and *in vivo*: An overview. *Methods Enzymol* 213: 403–420
17 Krinsky NI (1998) The antioxidant and biological properties of the carotenoids. *Ann N Y Acad Sci* 854: 443–447
18 Burton GW, Ingold KU (1984) β-Carotene: an unusual type of lipid antioxidant. *Science* 224: 569–573
19 Miller NJ, Sampson J, Candeias LP, Bramley PM, Rice-Evans CA (1996) Antioxidant activities of carotenes and xanthophylls. *FEBS Lett* 384: 240–242
20 Lepage G, Champagne J, Ronco N, Lamarre A, Osberg I, Sokol RJ, Roy CC (1996) Supplementation with carotenoids corrects increased lipid peroxidation in children with cystic fibrosis. *Amer J Clin Nutr* 64: 87–93
21 Levy Y, Kaplan M, Ben-Amotz A, Aviram M (1996) Effect of dietary supplementation of β-carotene on human monocyte-macrophage-mediated oxidation of low density lipoprotein. *Isr J Med Sci* 32: 473–478
22 Dugas TR, Morel DW, Harrison EH (1998) Impact of LDL carotenoid and α-tocopherol content on LDL oxidation by endothelial cells in culture. *J Lipid Res* 39: 999–1007
23 Gaziano JM, Hatta A, Flynn M, Johnson EJ, Krinsky NI, Ridker PR, Hennekens CH, Frei B (1995)

Supplementation with β-carotene *in vivo* and *in vitro* does not inhibit low density lipoprotein (LDL) oxidation. *Atherosclerosis* 112: 187–195

24 Nenseter MS, Volden V, Berg T, Drevon CA, Ose L, Tonstad S (1995) No effect of β-carotene supplementation on the susceptibility of low density lipoprotein to *in vitro* oxidation among hypercholesterolemic, postmenopausal women. *Scand J Clin Lab Invest* 55: 477–485

25 Chopra M, McLoone U, O'Neill M, Williams N, Thurnham DI (1996) Fruit and vegetable supplementation —effect on *ex vivo* LDL oxidation in humans. *In*: JT Kumpulainen, JT Salonen (eds): *Natural Antioxidants and Food Quality in Atherosclerosis and Cancer Prevention*. The Royal Society of Chemistry, London, 150–155

26 Bowen HT, Omaye ST (1998) Oxidative changes associated with β-carotene and α-tocopherol enrichment of human low-density lipoproteins. *J Amer Coll Nutr* 17: 171–179

27 Winklhofer-Roob BM, Puhl H, Khoschsorur G, van't Hof MA, Esterbauer H, Shmerling DH (1995) Enhanced resistance to oxidation of low density lipoproteins and decreased lipid peroxide formation during β-carotene supplementation in cystic fibrosis. *Free Radical Biol Med* 18: 849–859

28 Palozza P, Calviello G, Bartoli GM (1995) Prooxidant activity of β-carotene under 100% oxygen pressure in rat liver microsomes. *Free Radical Biol Med* 19: 887–892

29 Palozza P, Luberto C, Calviello G, Ricci P, Bartoli GM (1997) Antioxidant and prooxidant role of β-carotene in murine normal and tumor thymocytes: effects of oxygen partial pressure. *Free Radical Biol Med* 22: 1065–1073

30 Andersen HR, Andersen O (1993) Effects of dietary α-tocopherol and β-carotene on lipid peroxidation induced by methyl mercuric chloride in mice. *Pharmacol Toxicol* 73: 192–201

31 Truscott TG (1996) β-Carotene and disease: a suggested pro-oxidant and anti-oxidant mechanism and speculations concerning its role in cigarette smoking. *J Photochem Photobiol B-Biol* 35: 233–235

32 Mayne ST (1996) β-carotene, carotenoids, and disease prevention in humans. *FASEB J* 10: 690–701

33 Rock CL (1997) Carotenoids: biology and treatment. *Pharmacol Ther* 75: 185–197

34 Anonymous (1998) *IARC Handbooks of Cancer Prevention, Vol. 2: Carotenoids*. IARC, Lyon

35 Mayne ST (1998) β-Carotene, carotenoids and cancer prevention. *Principles and Practice of Oncology Updates* 12: 1–15

36 Kohlmeier L, Hastings SB (1995) Epidemiologic evidence of a role of carotenoids in cardiovascular disease prevention. *Amer J Clin Nutr* 62 (Suppl): 1370S–1376S

37 Taylor A, Jacques PF, Epstein E (1995) Relations among aging, antioxidant status, and cataracts. *Amer J Clin Nutr* 62 (Suppl): 1439S–1447S

38 Snodderly DM (1995) Evidence for protection against age-related macular degeneration by carotenoids and antioxidant vitamins. *Amer J Clin Nutr* 62 (Suppl): 1448S–1461S

39 Greenberg ER, Baron JA, Karagas MR, Stukel TA, Nierenberg DW, Stevens MM, Mandel JS, Haile RW (1996) Mortality associated with low plasma concentration of β carotene and the effect of oral supplementation. *JAMA* 275: 699–703

40 Ziegler RG, Mayne ST, Swanson CA (1996) Nutrition and lung cancer. *Cancer Cause Control* 7: 157–177

41 Mangels AR, Holden JM, Beecher GR, Forman ME, Lanza E (1993) Carotenoid content of fruits and vegetables: An evaluation of analytic data. *J Amer Diet Assoc* 93: 284–296

42 Le Marchand L, Hankin JH, Kolonel LN, Beecher GR, Wilkens LR, Zhao LP (1993) Intake of specific carotenoids and lung cancer risk. *Cancer Epidem Biomarker Prev* 2: 183–187

43 Ziegler RG, Colavito EA, Hartge P, McAdams MJ, Schoenberg JB, Mason TJ, Fraumeni JF Jr (1996) Importance of α-carotene, β-carotene and other phytochemicals in the etiology of lung cancer. *J Nat Cancer Inst* 88: 612–615

44 Gerster H (1997) The potential role of lycopene for human health. *J Amer Coll Nutr* 16: 109–126

45 Dragsted LO, Strube M, Larsen JC (1993) Cancer-protective factors in fruits and vegetables; biochemical and biological background. *Pharmacol Toxicol (Suppl)* 72: S116–S135

46 Mathews-Roth MM (1997) Carotenoids and photoprotection. *Photochem Photobiol* 65S: 148S–151S

47 α-Tocopherol, β-Carotene Cancer Prevention Study Group (1994) The effect of vitamin E and β carotene on the incidence of lung cancer and other cancers in male smokers. *New Engl J Med* 330: 1029–1035

48 Omenn GS, Goodman GE, Thornquist MD, Balmes J, Cullen MR, Glass A, Keogh JP, Meyskens FL Jr, Valanis B, Williams JH Jr et al (1996) Effects of a combination of β carotene and vitamin A on lung cancer and cardiovascular disease. *New Engl J Med* 334: 1150–1155

49 Hennekens CH, Buring JE, Manson JE, Stampfer M, Rosner B, Cook NR, Belanger C, LaMotte F, Gaziano JM, Ridker PM et al (1996) Lack of effect of long-term supplementation with β-carotene on the incidence of malignant neoplasms and cardiovascular disease. *New Engl J Med* 334: 1145–1149

50 Greenberg ER, Baron JA, Stukel TA, Stevens MM, Mandel JS, Spencer SK, Elias PM, Lowe N, Nierenberg DW, Bayrd G et al (1990) A clinical trial of β carotene to prevent basal-cell and squamous-cell cancers of the skin. *New Engl J Med* 323: 789–795

51 Blot WJ, Li J-Y, Taylor PR, Guo W, Dawsey S, Wang G-Q, Yang CS, Zheng S-F, Gail M, Li G-Y et al (1993) Nutrition intervention trials in Linxian, China: supplementation with specific vitamin/mineral combinations, cancer incidence, and disease-specific mortality in the general population. *J Nat Cancer Inst* 85: 1483–1492

52 McLarty JW, Holiday DB, Girard WM, Yanagihara RH, Kummet TD, Greenberg SD (1995) β-Carotene, vita-

min A and lung cancer chemoprevention: results of an intermediate endpoint study. *Amer J Clin Nutr* 62 (Suppl): 1431S–1438S

53 Mayne ST, Zheng T, Janerich DT, Goodwin WJJ, Fallon BG, Cooper DL, Friedman CD (1992) A population-based trial of β-carotene chemoprevention of head and neck cancer. *In*: G Newell, WK Hong (eds): *The Biology and Prevention of Aerodigestive Tract Cancer*. Plenum Press, New York, 119–127

54 Garewal HS, Meyskens FL Jr, Killen D, Reeves D, Kiersch TA, Elletson H, Strosberg A, King D, Steinbronn K (1990) Response of oral leukoplakia to β-carotene. *J Clin Oncol* 8: 1715–1720

55 Garewal H, Meyskens F, Katz RV, Friedman S, Morse DE, Alberts D, Girodias K (1995) β-Carotene produces sustained remissions in oral leukoplakia: results of a 1 year randomized, controlled trial. *Proc Am Soc Clin Oncol* 14: 496

56 Toma S, Benso S, Albanese E, Palumbo R, Cantoni E, Nicolo G, Mangiante P (1992) Treatment of oral leukoplakia with β-carotene. *Oncology* 49: 77–81

57 Stich HF, Rosin MP, Hornby AP, Mathew B, Sankaranarayanan R, Nair MK (1988) Remission of oral leukoplakias and micronuclei in tobacco/betel quid chewers treated with β-carotene and with β-carotene plus vitamin A. *Int J Cancer* 42: 195–199

58 Zaridze D, Evstifeeva T, Boyle P (1993) Chemoprevention of oral leukoplakia and chronic esophagitis in an area of high incidence of oral and esophageal cancer. *Ann Epidemiol* 3: 225–234

59 Malaker K, Anderson BJ, Beecoft WA, Hodson DI (1991) Management of oral mucosal dysplasia with β-carotene retinoic acid: a pilot cross-over study. *Cancer Detect Prev* 15: 335–340

60 Kaugars GE, Silverman S, Lovas JGL, Brandt RB, Riley WT, Dao Q, Singh VS, Gallo J (1994) A clinical trial of antioxidant supplements in the treatment of oral leukoplakia. *Oral Surg Oral Med Oral Pathol* 78: 462–468

61 Stich HF, Rosin MP, Vallejera MO (1984) Reduction with vitamin A and β-carotene administration of proportion of micronucleated buccal mucosal cells in asian betel nut and tobacco chewers. *Lancet* I: 1204–1206

62 Stich HF, Hornby P, Dunn BP (1985) A Pilot β-Carotene intervention trial with inuits using smokeless tobacco. *Int J Cancer* 36: 321–327

63 van Poppel G, Kok FJ, Hermus RJJ (1992) β-Carotene supplementation in smokers reduces the frequency of micronuclei in sputum. *Brit J Cancer* 66: 1164–1168

64 Greenberg ER, Baron JA, Tosteson TD, Freeman DH Jr, Beck GJ, Bond JH, Colacchio TA, Coller JA, Frankl HD, Haile RW et al (1994) A clinical trial of antioxidant vitamins to prevent colorectal adenoma. *New Engl J Med* 331: 141–147

65 MacLennan R, Macrae F, Bain C, Battistutta D, Chapuis P, Gratten H, Lambert J, Newland RC, Ngu M, Russell A et al (1995) Randomized trial of intake of fat, fiber, and β carotene to prevent colorectal adenomas. *J Nat Cancer Inst* 87: 1760–1766

66 Kikendall JW, Mobarhan S, Nelson R, Burgess M, Bowen PE (1991) Oral β-carotene does not reduce the recurrence of colorectal adenomas. *Amer J Gastroenterol* 36: 1356

67 de Vet HCW, Knipschild PG, Willebrand D, Schouten HJA, Sturman F (1991) The effect of β-carotene on the regression and progression of cervical dysplasia: a clinical experiment. *J Clin Epidemiol* 44: 273–283

68 Romney SL, Ho GY, Palan PR, Basu J, Kadish AS, Klein S, Mikhail M, Hagan RJ, Chang CJ, Burk RD (1997) Effects of β-carotene and other factors on outcome of cervical dysplasia and human papillomavirus infection. *Gynecol Oncol* 65: 483–492

69 Meyskens FL Jr, Manetta A (1995) Prevention of cervical intraepithelial neoplasia and cervical cancer. *Amer J Clin Nutr* 62 (Suppl): 1417S–1419S

70 Li J-Y, Taylor PR, Li B, Dawsey S, Wang G-Q, Ershow AG, Guo W, Liu S-F, Yang CS, Shen Q et al (1993) Nutrition intervention trials in Linxian, China: multiple vitamin/mineral supplementation, cancer incidence, and disease-specific mortality among adults with esophageal dysplasia. *J Nat Cancer Inst* 85: 1492–1498

The role of vitamin A in visual transduction

R.K. Crouch and J.-X. Ma

Medical University of South Carolina, 96 Jonathan Lucas Street, P.O. Box 250617, Charleston SC 29425, USA

Introduction

The role of vitamin A aldehyde as the chromophore of vertebrate visual pigments was discovered by George Wald [1]. Vitamin A deprivation studies by John Dowling and his colleagues established the necessity of this essential compound to the visual process and the maintenance of the structural integrity of the photoreceptors [2]. Although it has been recognized that vitamin A is essential for vision for years, the transport and metabolism of vitamin A within the retina and pigment epithelium is far from being fully understood and there are obviously splendid opportunities for further discoveries in this area.

The native chromophore of vertebrate pigments is retinal and/or 3,4 dehydroretinal (found in certain aquatic species). In the invertebrate pigments, 3-hydroxyretinal [3, 4] and 4-hydroxyretinal [5] have also been found as the chromophore of visual pigments (Fig. 1). There has recently been an excellent review of the invertebrate pigments [6]. The discussion in this review will focus on the vertebrate systems.

Analogues of vitamin A have been used extensively for probing the transduction process (for a recent review see [7]). The visual pigment itself is particularly convenient for these types of studies *in vitro* as the native chromophore can be easily detached by the use of light and hydroxylamine. These studies were conducted on biochemical preparations, mainly isolated by detachment of the outer segments from bovine retina which consists mainly of rod photoreceptors. Recently, physiological studies in isolated photoreceptors have allowed the probing of the system in a more intact environment and, perhaps more importantly, have allowed the study of individual rods and cones. Retinal analogues can be easily incorporated into these photoreceptor (for a recent review, see [8]) and surprisingly, it is becoming evident that there are some fundamental differences in these two types of photoreceptors.

The use of site-directed mutagenesis to alter the amino acid composition of various proteins in the visual cycle has been very helpful. In rhodopsin, the key residues interacting with the chromophore have been identified by this method, as well as the sections of the protein critical for the role of this protein as a G-protein receptor [9]. Other proteins involved in the visual transduction process and in the metabolism of retinoids in the regeneration of the visual pigments are also being explored by this mechanism. The combined use of visual pigment mutation with retinal analogue probes (see e.g. [10]) has been limited by the quantities of proteins that can be expressed but is an obvious avenue of further study. The recent development of transgenic and knockout animal models in which various retinoid binding proteins have been mutat-

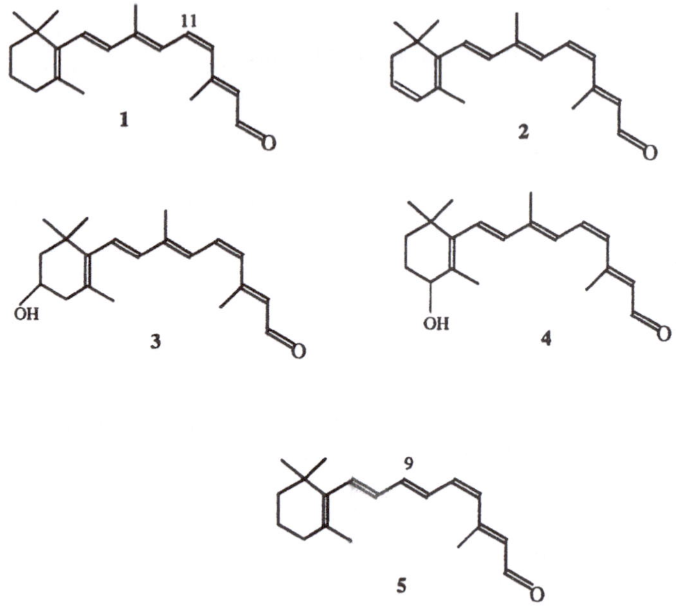

Figure 1. Structures of retinal and retinal analogues. 1. 11-*cis* retinal; 2. 11-*cis* 3,4-dehydroretinal; 3. 11-*cis* 3-hydroxyretinal; 4. 11-*cis* 4-hydroxyretinal; 5. 11-*cis* 9-desmethylretinal

ed or eliminated is a most exciting advance and no doubt will be the focus of attention in future years.

From the summary of findings to date, it will be obvious that both the role of retinal in initiating the visual process and the complicated metabolic pathway required for the production of the 11-*cis* aldehyde for pigment regeneration are far from being fully understood and indeed seem to become more intriguing as new discoveries are made. This discussion will focus on recent findings (see [11] for an earlier review) and is divided into (1) an examination of the binding site and role of the chromophore within visual pigments and (2) the process by which new 11-*cis* retinal is generated for the formation of pigments.

Structure and function of retinal in visual pigments

Binding site of the chromophore

In vertebrates, the bound chromophore of both rod and cone visual pigments is vitamin A in the aldehyde state, except in some aquatic species in which 3,4-dehydroretinal is employed. The isomeric form of the retinal is the high energy 11-*cis* isomer. The linkage to the protein is via a protonated Schiff base to a lysine deep in the seventh helix of the protein. On exposure to light, the critical 11-*cis* bond isomerizes to the more stable all-*trans* conformation, followed by deprotonation and hydrolysis of the Schiff base.

The interaction of the chromophore with the protein is known to control the wavelengths of the pigments and the increasing reports of the sequences of the various pigments is instructive in understanding this control of the wavelengths. Vertebrate visual pigments can be classified into four groups: long (L), two middle (M1 and M2), and short(S) wavelength pigments based on their absorbance spectra [12]. In most vertebrates there is a single rod pigment, known as the red rod. However, in some aquatic species, a green rod has also been detected [13], although there is no report to date of the cloning and sequencing of a green rod pigment. There are a variety of cone pigments that have been reported and the number and type appear to vary between species. Interestingly, Makino and Dodd [14] reported from spectroscopic measurements that there are two and sometimes three different opsins within a single cone photoreceptor. The presence of two visual pigments in a single photoreceptor cell was recently confirmed at the mRNA level in the butterfly [15].

Visual pigments all have a seven helical transmembrane structure with a lysine in the seventh helix for binding to the chromophore (see Fig. 2 for one example, the red rod pigment of

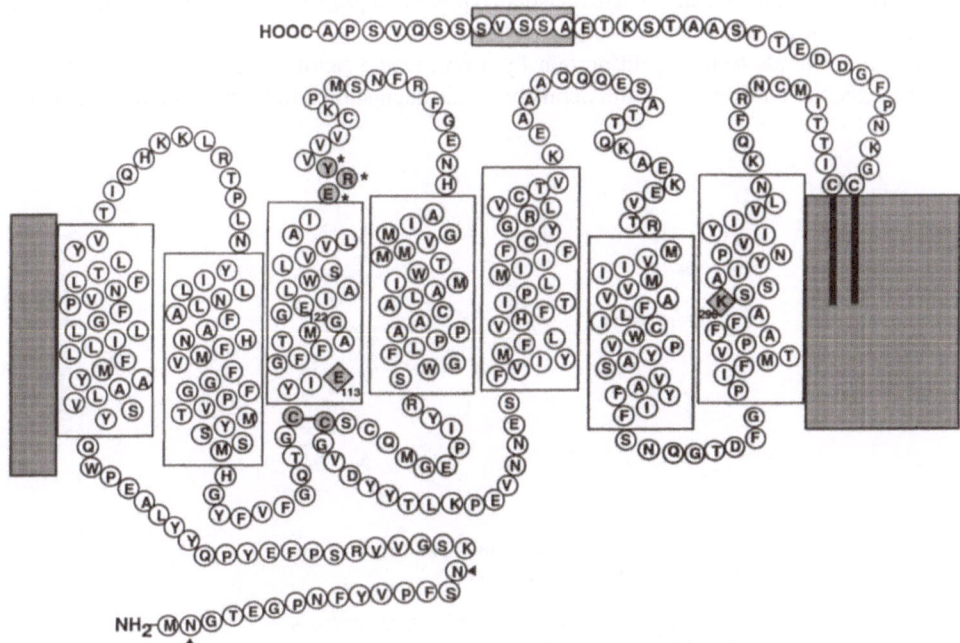

Figure 2. Predicted secondary structure of salamander red rod rhodopsin. The transmembrane helices (boxed) are defined based on Kyte-Doolittle hydropathy plots and similarity to bovine rhodopsin. Lys296 and Glu113 are indicated by diamonds. Putative glycosylation sites are indicated by filled triangles. The ERY motif is indicated by *. The shaded box in the carboxyl terminal region indicates the six extra amino acid residues which do not exist in mammalian rhodopsins.

the tiger salamander [16]). However, the presence of the protonated Schiff base has not been proven in all of the cone pigments, and is open to question in the shorter wavelength pigments. The pigments contain many conserved elements across photoreceptor types and across species. Examples are the presence in the C-terminus of a number of serines and threonines, which can serve as phosphorylation sites; two cysteines for the M1 and M2 pigments and a single cysteine for the S pigments, which are palmitoylated; a specific ERY (ERW or ERF for certain cones) motif at the boundary of helix 3 which is thought to define that boundary and have an important role in the determination of G-protein binding specificity; and glycosylation sites on the N-terminus of all these pigments.

The constraints of the binding site of rod rhodopsin have been extensively studied using retinal analogues [7] and from these studies the following generalizations can be made (Fig. 3): (1) all isomers (including di- and tri-*cis*) except for 13-*cis* and all-*trans* can be accommodated; (2) the methyl at C-9 on the polyene chain is critical to the absorption properties and activity of the protein; (3) the binding cavity has little tolerance for additional bulk in the region of the cyclohexyl ring, however, the ring itself is not essential for pigment formation as long as one (two gives increased stability) of the three ring methyl groups are present. Due to the lack of significant quantities of the cone opsins, these studies have been conducted almost exclusively with rod opsin, but the results to date on isolated photoreceptors have not indicated that the constraints for binding vary between various opsins.

The retinal binds to the opsin protein by a protonated Schiff base. Through site-directed mutagenesis, the counterion for the Schiff base has been shown to be the glutamate residue at

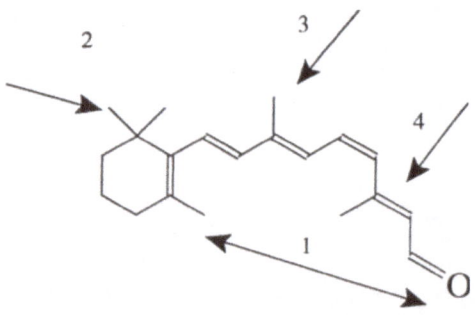

1. Distance between ring methyl and carbonyl is critical for pigment formation.

2. Ring methyls but not the full ring strucutre is required for pigment formation.

3. Removal 9-methyl group partially inactives the protein.

4. The 13-methyl is necessary for rapid pigment formation.

Figure 3. Constraints on retinal binding to opsins.

position 113 [17, 18]. This residue is conserved in cone pigments [19]. The position of the chromophore within the protein is still not determined, although progress is being made on obtaining crystals of rhodopsin that will be suitable for three-dimensional analysis [20]. The crosslinking studies by the Nakanishi group [21, 22] suggest that the ring portion of the chromophore is near helix F and that light isomerization moves the ring to a position near helix C.

Role of chromophore in control of pigment activity

The role of retinal in the visual pigment is to control the activity of the G-protein receptor. These rather beautifully designed receptors allow for a photon of light of a specific wavelength to isomerize the critical -11-*cis* double bond to the all-*trans* conformation, bringing the receptor from its inactive state to the active state. 11-*cis* Retinal acts as an inverse agonist, locking the receptor in an inactive conformation, which of course is critical in keeping the receptor "quiet" when not receiving light. The light-induced isomerization of the chromophore appears designed to break the glutamate 113/lysine 296 Schiff base salt bridge and change the conformation of the receptor to allow maximal interaction with the G-protein. In the all-*trans* conformation, the retinal is acting as an agonist activating the receptor (Fig. 4). At least two forms of the rod rhodopsin containing the all-*trans* conformation are proposed to have activity: (1) the metarhodopsin II (or R*) form which is very effective in activating the transduction process (for recent review, see [23]) and (2) a second form of metarhodopsin II, proposed to be phosphorylated and bound to arrestin, having an activity of 10^{-5} that of R* [24]. The apoprotein opsin with no ligand in the binding site also has some basal activity [25, 26], estimated at 10^{-7} that

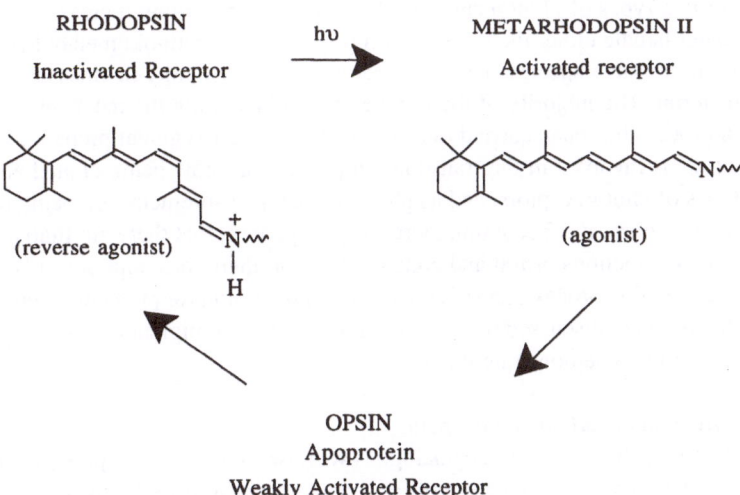

Figure 4. Activation states of rhodopsin. The conformation of retinal and its role as a ligand are presented.

of R* [27]. While this is decidedly a low level [28], it is quite meaningful on the physiological scale, as has been demonstrated using both isolated rods [27] and cones [29]. The retinal is thus critical to the control of the activity of these G-protein receptors.

A number of laboratories have also determined that combination of the apoprotein opsin with all-*trans* retinal forms a complex which has an activity resembling that of the photolyzed rhodopsin as measured by transducin activation [30], phosphorylation by rhodopsin kinase [31, 32], and arrestin binding [32]. The activity of this complex does not depend on the formation of a Schiff base [33, 34]. The role of such a complex in the physiological process is as yet unclear.

Mutation of certain amino acids results in rhodopsins that are also constitutively active, that is, the protein can activate transducin in the absence of light [35–37]. Rim and Oprian [38] have shown that rhodopsin mutants that are constitutively active are also phosphorylated by rhodopsin kinase. These results are in agreement with the results on the "active opsin complexes" discussed above. These mutants disrupt the salt bridge between lysine 296 and glutamate 113 [35–37] and therefore prevent the "inactive" conformation of the rhodopsin state. These studies have important clinical applications as it has been found that several of the mutations of rhodopsin found in individuals with retinal degenerative disorders result in constitutively active pigments [39].

Variation of rods and cones

As rods and cones have as their photosensitive moiety visual pigments composed of opsins of generally similar structures bound to the 11-*cis* form of vitamin A aldehyde, it might be reasonable to assume that the retinoid metabolism and general transduction process should be the same in these two types of photoreceptors. However, it is becoming increasingly obvious that the cones do not handle either the transduction process or the retinoid metabolism exactly like the rods within the same species. There has been a major difficulty in studying the cones due to lack of material. The majority of the biochemistry elucidating the mechanism of the visual transduction process has been carried out in bovine rod outer segment preparations. However, as various proteins involved in the transduction process are now being cloned and sequenced from both types of photoreceptors, and as physiological measurements are being made on both isolated rods and cones, it is becoming increasingly apparent that there are fundamental differences in (1) the interactions of rod and cone opsins with their chromophores, (2) the biochemistry of the transduction process, and (3) the retinoid processing for pigment regeneration. Some of these differences are discussed below in the regeneration of pigments. Three other examples are discussed below to demonstrate this point.

Analogue activation of rod and cone opsins
The development of the use of retinal analogues with isolated photoreceptors for physiological studies has proven to be a convenient tool for the comparison of rods and cones. A fundamental difference has been observed in the interaction of the retinal analogue β-ionone, which contains a truncated polyene chain (Fig. 5). In cones, it was determined that this analogue acts as

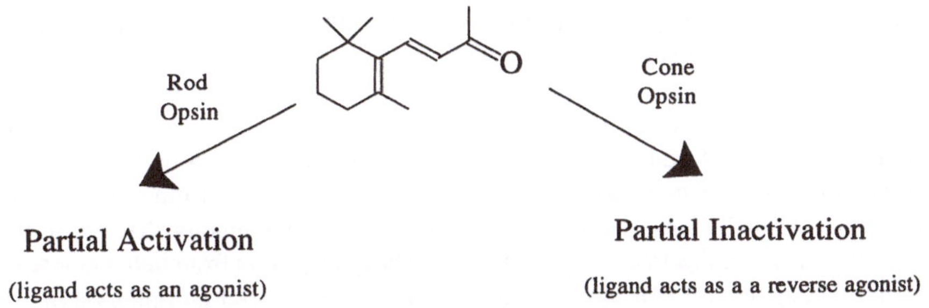

Figure 5. The role of β-ionone on the activation of rod and cone opsins.

an inverse agonist, desensitizing the receptor and reversing the bleach-induced guanylyl cyclase activity [29]. These results were in variance with the results from biochemical experiments (performed on outer segment preparations which were mostly the rod cells), which had shown both by transducin activation [40] and by phosphorylation by rhodopsin kinase [31] that this same analogue acts as an agonist, activating the receptor. In recent physiological experiments on isolated rods, it has been shown that β-ionone does indeed act as an agonist and that activation is obtained [41]. (Fig. 5). These results clearly demonstrate that the interactions of the rod and cone opsins with their chromophores are different. As the expressed cone pigments become more available for biochemical studies, it will be critical that these mechanisms are studied in detail.

Rod and cone transducins
In addition to the cone opsins, several other proteins specific to cones and involved in the transduction processes have been reported. This discussion will focus on the variations of the rod and cone transducins. Transducins are GTP-binding proteins coupled with visual pigments and consist of α, β and γ subunits. The rod transducin, composed of α_1, β_1 and γ_1 subunits, is known to interact specifically with red rod rhodopsin. The ERY motif at the boundary of the third transmembrane helix and the second intracellular loop is known to be responsible for specific interaction with rod transducin. Cone photoreceptor cells contain a cone-specific transducin with the subunits α_2, β_3 and γ_2 [42]. Later, another γ subunit, γ_8 was identified in cones and found to have the same distribution pattern as the β_3 subunit [43]. It is uncertain whether cone transducin is composed of α_2, β_3 and γ_2 or α_2, β_3 and γ_8. The cone transducin subunits share significant sequence homology with those of the rod transducin. Using double labeling with antibodies against cone transducin subunits and antibodies against red, green and blue opsins, it has been demonstrated that the α_2, β_3 and γ_2 subunits are expressed in all the blue, green and red-sensitive cone photoreceptors in the human retina [43, 44]. These findings suggest that all three cone opsins may interact with the same transducin although they have different G-protein interaction sequences. It is noteworthy that unlike other cone opsins, the UV cone opsin has a transducin interaction ERY motif identical to the red rod rhodopsin. Moreover, a single UV cone cell has been found to contain three different opsins [14]. Therefore, the interesting question to

be addressed is whether three opsins in the same UV cone cell share the same transducin or use different combinations of the subunits.

Rod and cone pigment phosphorylation

The visual pigments of both rods [45] and cones [46] are phosphorylated on light stimulation and this step is thought to be necessary to the quenching of the activated form of the visual pigments. The phosphorylation enzyme rhodopsin kinase has been immunolocalized to both rods and cones [47]. Recent studies from measurements of phosphorylation from intact mouse retina (a mainly rod retina) suggest that this is a complex process as several phosphorylation states are observed [48]. Combination *in vitro* of rod opsin with all-*trans* retinal mimics the meta II or activated state of the pigment and this form is phosphorylated by rhodopsin kinase [26]. This system has provided a convenient assay of the activity of retinal analogue pigments [26]. In the case of 9-desmethyl retinal, results on a rod outer segment preparation demonstrated that the opsin/9-des-all-*trans* complex is poorly phosphorylated [31, 49]. Data from isolated photoreceptors have shown that responses from rod photoreceptors regenerated with this analogue are reduced in magnitude and are prolonged which could be attributed to poor phosphorylation [50]. However, an interesting observation is that in cases of Oguchi disease, where mutations of rhodopsin kinase have been identified, the photopic vision (mediated by cones) is not affected whereas the rod-mediated scotopic vision is compromised [51]. These results suggest that cones either have a method for phosphorylation in addition to rhodopsin kinase or that phosphorylation is not critical to the inactivation of cone pigments.

Generation of 11-*cis* retinal

Retinal/retinol metabolism and transport

The regeneration of visual pigments depends on the availability of 11-*cis* retinal. The chromophore is detached from the opsin as the all-*trans* isomer in the alcohol oxidative state and therefore both an isomerization and oxidation are required for regeneration of the pigment. The overall scheme for the resupply of this retinoid is shown in Figure 6. This area has been recently reviewed [11] and thus this discussion will concentrate on the recent findings that have been made, with particular reference to some very intriguing knockout and transgenic animal models.

Detachment of the chromophore from the opsin

The chromophore dissociates from rhodopsin by reduction of the all-*trans* retinal to the alcohol form by the enzyme all-*trans*-retinol dehydrogenase. This reduction also results in the dissociation of arrestin [52], and the dephosphorylation of opsin. The all-*trans* retinol dehydrogenase has been purified [53] and found to react with all-*trans* retinal and 13-*cis* retinal but very slowly with 9- or 11-*cis* retinal. Removal of the 9-methyl group of the chromophore reduces its activity by 50% [31]. The enzyme also reduces all-*trans* retinal complexed with opsin, although

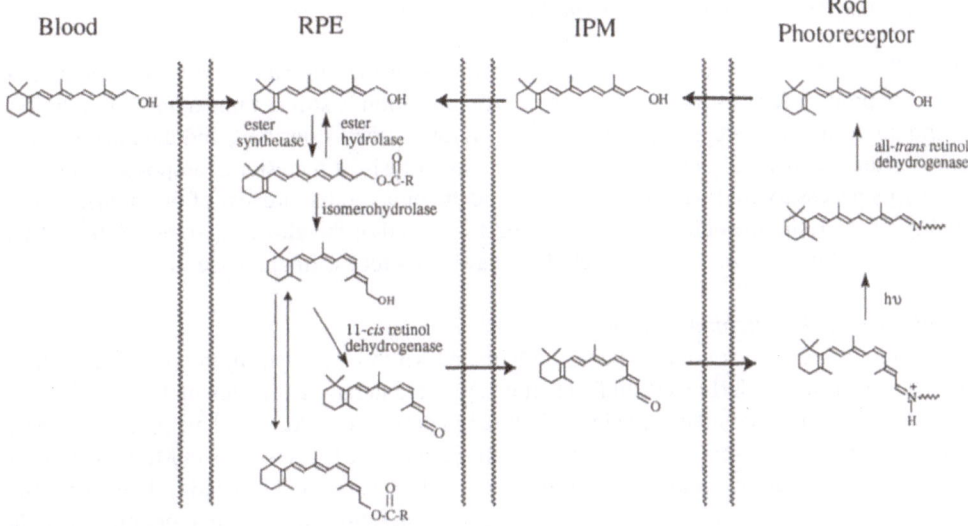

Figure 6. Pathway for retinal/retinol metabolism in the visual cycle.

the activity is lower than that obtained with photolyzed rhodopsin [31]. Interestingly, Saari et al. [54]. recently showed that in mice the rate of rhodopsin regeneration is limited by the reduction of all-*trans* retinal and therefore it is proposed that the accumulation of all-*trans* retinal may have physiological effects.

Transport of retinoids through the interphotoreceptor space

For the reformation of 11-*cis* retinal which occurs in the retinal pigment epithelium (RPE), it is required that the all-*trans* retinol, which is formed in the photoreceptor outer segments, be transported to the RPE and that the resulting 11-*cis* retinal be transported to the photoreceptors. The protein that has been implicated in this function is the interphotoreceptor retinoid-binding protein (IRBP) which both transports and protects retinoids in this process [55]. The number of retinoid binding sites on this protein has been of some controversy [56, 57] but it is apparent that the protein does have an affinity for both retinal and retinol. Experiments on eyecups [58] as well as isolated photoreceptors [59] have shown that IRBP can both deliver and accept retinoids from these tissues. However, it has been argued that the role of this protein is to buffer the retinoid rather than to act as a transport carrier [60]. Recently, Liou and his group [61] have developed a mouse model with targeted disruption of the IRBP gene. These animals demonstrate abnormalities of the photoreceptors but, surprisingly, retain some function by electroretinogramm (ERG) measurements. These results indicate that 11-*cis* retinal is still getting to the retina. Further studies are needed in order to determine the retinoid profile and rhodopsin concentrations in this important animal model.

Isomerization and oxidation in the retinal pigment epithelium

The RPE is the site of isomerization and oxidization of the all-*trans* retinol to 11-*cis* retinal [62]. The process for this non-photochemical isomerization is still not understood. Rando [63] has proposed that two steps occur: the all-*trans* retinol is first esterified, and then the all-*trans* retinyl ester is converted into 11-*cis* retinol. The hydrolysis of the ester is proposed to provide the energy necessary to drive the thermodynamically unfavorable process of the isomerization. Storage of the retinoid could occur both as the 11-*cis* and/or the all-*trans* isomer. A few of the steps involved in this process for which there have been recent findings are discussed below.

Cellular retinaldehyde-binding protein
There are a number of retinoid binding proteins known to be present in the pigment epithelium. One that is particularly well studied that is suggested to have a fundamental role in the vitamin A metabolism is cellular retinaldehyde-binding protein (CRALBP). This protein has been shown to carry both 11-*cis* retinal and 11-*cis* retinol in the RPE (as well as in Müller cells) and has been proposed to be a modulator of the storage of 11-*cis* retinol ester [64]. In experiments with RPE membranes, it has been shown that the isomerization step occurs more rapidly in the presence of CRALBP and lower concentrations of CRALBP are required for this effect than of bovine serum albumin [65]. Mutations of the protein have been linked with retinitis pigmentosa [66]. Recent studies on this protein probing the retinoid binding site have shown that the retinoid is not covalently bonded and have identified some of the amino acids in this site [67].

11-cis *ester hydrolase*
It is likely that the 11-*cis* retinol is stored as the ester and therefore it would be expected that an 11-*cis* retinyl ester hydrolase would be present in the RPE. 11-*cis* Retinyl esters have been shown to be present in the RPE, particularly in the dark [68]. Mata et al. [69] have observed an 11-*cis* ester hydrolase. This enzyme has been colocalized with 11-*cis* retinyl esters in the plasma membrane of the RPE [70]. Interestingly, these results suggest a compartmentalization of the retinoid esters, as the site of synthesis of these esters is in the endoplasmic reticulum utilizing the enzyme lecithin:retinol acyltransferase (LRAT) [71].

RPE65
The protein RPE65 has been identified as a protein abundantly expressed in the RPE [72]. Mutations in the protein have been shown to result in at least two forms of early-onset retinal dystrophies in humans [73–75] and congenital stationary night blindness in the dog [76]. The RPE65-deficient mouse model has been generated and studies on this animal have demonstrated that the protein is essential for production of 11-*cis* retinal [77]. These animals exhibit many of the characteristics of a vitamin A-deprived animal. Rhodopsin is not present in the 20 week animal, the rod outer segments are disorganized and lipid droplets are observed in the RPE. ERG results on the knockout mouse indicate that the rods are not functioning properly. Therefore, although the exact role of this protein remains to be elucidated, it is evidently essential for the normal production of 11-*cis* retinal and regeneration of visual pigment. Interestingly, the pro-

tein has been localized in cone outer segments in the salamander [78], another demonstration that the retinoid processing and pigment regeneration differs for these two photoreceptors.

Pigment regeneration

There is strong evidence that the regeneration of pigment in rods and cones may involve different mechanisms. The rate of pigment regeneration is much faster in cones than in rods [59, 79]. Using isolated retinas it has been shown that pigment is regenerated in rods only if the RPE is present. However, cone pigments are reported to regenerate in the absence of the RPE [80]. In isolated rods, pigment regeneration occurs on the addition of 11-*cis* retinal to the outer segments but no regeneration is observed if the inner segment is exposed to the retinal. Interestingly, the sensitivity of the cone is restored, indicating that pigment is regenerated on exposure of the inner segment to 11-*cis* retinal. These results suggest that retinal can move from the cone inner segment to the cone outer segment [79]. Both 11-*cis* retinal and retinol have been reported in the Müller cells (glial cells) of the retina [81] and the formation of 11-*cis* retinol from all-*trans* retinol has been observed in cultures of Müller cells obtained from the cone-rich retina of the chicken [82]. A further intriguing observation is that isolated salamander cones are sensitized when these cells are exposed to 11-*cis* retinol [59]. No sensitization was obtained in similar experiments on rod cells. Finally, the results on the RPE65 knockout animal described above [77], in which retinoid metabolism is altered but cone function is not compromised, indicate that the cones have a source of 11-*cis* retinal not available to the rods. These results all suggest that there are differences in the regeneration machinery of these two types of photoreceptors.

Summary

From the above discussion, it is clear that there is abundant evidence to show that retinal is a key factor in the control of the activity of the visual pigments. The processes by which 11-*cis* retinal is formed and the pigments are regenerated are not yet understood. Insights into the deactivation of the pigments and detachment of the chromophores likewise require further study. Finally, the recent findings that demonstrate that there are fundamental differences in the interaction of rods and cone pigments with their chromophore present another area to explore. Evidence is accumulating as to the relevance of these processes to human clinical disorders. With the powerful new animal models being produced and the evolving sophisticated molecular, analytical and physiological methodologies, there is every likelihood of rapid progress toward answering these questions over the next few years.

Acknowledgments
Supported by NIH grants EY04939 (RKC) and EY12231 (J-XM), NSF grants MCB96-00772 and EPS96-30167 (J-xM), and by an unrestricted grant from Research to Prevent Blindness, Inc.

References

1   Wald G (1939) On the distribution of vitamins A_1 and A_2. *J Gen Physiol* 22: 391–415
2   Dowling JE (1960) The chemistry of dark adaptation. *Nature* 168: 114–118
3   Vogt K (1983) Is the fly visual pigment a rhodopsin? *Z Naturforsch Sect C Bioscience* 38: 329–333
4   Shimazaki Y, Eguchi E (1995) Light-dependent metabolic pathway of 3-hydroretinoids in the eye of a butterfly *Papilio xuthus. J Comp Physiol* 176: 661–671
5   Matsui S, Seidou M, Uchiyama I, Sekiya N, Hiraki K, Yoshihara K, Kito Y (1988) 4-Hydroxyretinal, a new visual pigment chromophore found in the bioluminescent squid *Watasenia scintillans. Biochim Biophy Acta* 966: 370–374
6   Gardner W, Towner P (1995) Invertebrate visual pigments. *Photochem Photobiol* 62: 1–16
7   Nakanishi K, Crouch RK (1996) Application of artificial pigments to structure determination and study of photoinduced transformations of retinal proteins. *Isr J Chem* 35: 253–272
8   Corson DW, Crouch RK (1996) Physiological activity of retinoids in natural and artificial visual pigments. *Photochem Photobiol* 63: 595–600
9   Sakmar TP (1998) Rhodops*In*: a prototypical G protein-coupled receptor. *Progr Nucl Acid Res Mol Biol* 59: 1–34
10 Han M, Groesbeek M, Smith SO, Sakmar TP (1998) Role of the C9 methyl group in rhodopsin activation: characterization of mutant opsins with the artificial chromophore 11-*cis*-9-demethylretinal. *Biochemistry* 37: 538–45
11 Crouch RK, Chader GJ, Wiggert B, Pepperberg D (1996) Retinoids and the visual process. *Photochem Photobiol* 64: 613–621
12 Yoshizawa T (1994) Molecular basis for color vision. *Biophys Chem* 50: 17–24
13 Makino-Tasaka M, Suzuki T (1984) The green rod pigment of the bullfrog, *Rana catesbeiana. Vision Res* 24: 309–22
14 Makino CL, Dodd RL (1996) Multiple visual pigments in a photoreceptor of the salamander retina. *J Gen Physiol* 108: 27–34
15 Kitamoto J, Sakamoto K, Ozaki K, Mishina Y, Arikawa K (1998) Two visual pigments in a single photoreceptor cell: identification and histological localization of three mRNAs encoding visual pigment opsins in the retinal of the butterfly *Papilio Xuthus. J Exp Biol* 301: 1255–1261
16 Chen N, Ma J-x Hazard ES, Corson DW, Crouch RK (1996) Molecular cloning of a rhodopsin gene from salamander red rods. *Invest Ophthalmol Visual Sci* 37: 1907–1913
17 Sakmar TP, Franke RR, Khorana HG (1989) Glutamic acid 113 serves as the retinylidene Schiff base counterion in bovine rhodopsin. *Proc Natl Acad Sci USA* 86: 8309–8313
18 Zhukovsky EA, Oprian DD (1989) Effect of carboxylic acid side chains on the absorption maximum of visual pigments. *Science* 246: 928–930
19 Xu L, Lockman K, Hazard ES, Crouch RK, Ma J-x (1998) Molecular cloning and sequencing of a salamander red and blue cone pigments. *Molecular Vision* 4:<http://www.emory. edu/molvis/v4/ma>
20 Unger VM, Hargrave PA, Baldwin JM, Schertler GF (1997) Arrangement of rhodopsin transmembrane α-helices. *Nature* 389: 203–206
21 Zhang H, Lerro KA, Yamamoto T, Lieu TH, Sastry L, Gowinowicz MA, Nakanishi K (1994) The location of the chromophore in rhodops*In*: a photoaffinity study. *J Amer Chem Soc* 116: 10 165–10 173
22 Nakanishi K, Zhang H, Lerro KA, Takekuma S, Yamamoto T, Lien TH, Sastry L, Baek DJ, Moquin-Pattey C, Boehm MF (1995) Photoaffinity labeling of rhodopsin and bacteriorhodopsin. *Biophys Chem* 56: 13–22
23 Baylor D (1996) How photons start vision. *Proc Natl Acad Sci USA* 93: 560–565
24 Leibrock CS, Reuter T, Lamb TD (1998) Molecular basis of dark adaptation in rod photoreceptors. *Eye* 12: 511–520
25 Surya AK, Foster W, Knox BE (1995) Transducin activation by the bovine opsin apoprotein. *J Biol Chem* 270: 5024–5031
26 Buczylko J, Saari J, Crouch RK, Palczewski K (1996) Mechanisms of opsin activation. *J Biol Chem* 271: 20 621–20 630
27 Cornwall MC, Fain GL (1994) Bleached pigment activates transduction in isolated rods of the salamander retina. *J Physiol* 480: 261–279
28 Melia TJ, Cowan CW, Angleson JK, Wensel TG (1997) A comparison of the efficiency of G protein activation by ligand-free and light-activated forms of rhodopsin. *Biophys J* 73: 3182–91
29 Cornwall MC, Matthews HR, Crouch RK, Fain GL (1995) Bleached pigment activates transduction in salamander cones. *J Gen Physiol* 106: 543–557
30 Fukada Y, Yoshizawa T (1981) Activation of phosphodiesterase in frog rod outer segments by an intermediate of rhodopsin photolysis. II. *Biochim Biophys Acta* 675: 195–200
31 Palczewski K, Jäger S, Buczylko J, Crouch RK, Bredberg DL, Hofmann KP, Asson-Batres MA, Saari JC (1994) Rod outer segment retinol dehydrogenase: Substrate specificity and role in phototransduction. *Biochemistry* 33: 13 741–13 750
32 Hofmann KP, Pulvermüller A, Buczylko J, van Hooser P, Palczewski K (1992) The role of arrestin and

retinoids in the regeneration pathway of rhodopsin. *J Biol Chem* 267: 15 701–15 706

33 Jäger S, Palczewski K, Hofmann KP (1996) Opsin/all-*trans* retinal complex activates transducin by different mechanisms than photolyzed rhodopsin. *Biochemistry* 35: 2901–2908

34 Surya A, Knox BE (1998) Enhancement of opsin activity by all-*trans* retinal *Exp Eye Res* 66: 599–603

35 Robinson PR, Cohen GB, Zhukoyzky EA, Oprian DD (1992) Constitutively active mutants of rhodopsin. *Neuron* 9: 719–725

36 Dryja TP, Berson EL, Rao VR, Oprian DD (1993) Heterozygous missense mutation in the rhodopsin gene as a cause of congenital stationary night blindness. *Nat Genet* 4: 280–283

37 Rao VR, Cohen GB, Oprian DD (1994) Rhodopsin mutation G90D and a molecular mechanism for congenital night blindness. *Nature* 367: 639–642

38 Rim J, Oprian DD (1995) Constitutive activation of ops*In*: Interaction of mutant with rhodopsin kinase and arrestin. *Biochemistry* 34: 11 938–11 945

39 Berson EL (1996) Retinitis pigmentosa: unfolding its mystery. *Proc Natl Acad Sci USA* 93: 4526–4528

40 Jäger F, Jäger S, Kräutle O, Friedman N, Sheves M, Hofmann KP, Siebert F (1994) Interactions of the b-ionone ring with the protein in the visual pigment rhodopsin control the activation mechanism. An FTIR and fluorescence study on artificial vertebrate rhodopsins. *Biochemistry* 88: 7389–7397

41 Kefalov VJ, Cornwall MC, Crouch RK (1999) Occupancy of the chromophore binding site of opsin activates visual transduction in rod photoreceptors. *J Physiology* 113: 491–503

42 Peng Y-W, Robishaw JD, Levine MA, Yau K-W (1992) Retinal rods and cones have distinct G protein β and γ subunits. *Proc Natl Acad Sci USA* 89: 10 882–10 886

43 Ong OC, Yamane HK, Phan KB, Fong HKW, Bok D, Lee RH, Fung BK-K (1995) Molecular cloning and characterization of the G protein γ subunit of cone photoreceptors. *J Biol Chem* 270: 8495–8500

44 Lerea CL, Bunt-Milam AH, Hurley JB (1989) A transducin is present in blue-, green-, and red-sensitive cone photoreceptors in the human retina. *Neuron* 3: 367–376

45 McDowell JH, Kühn H (1997) Light-induced phosphorylation of rhodopsin in cattle photoreceptor membranes: substrate activation and inactivation. *Biochemistry* 16: 4054–4060

46 Fukada Y, Kokame K, Okano T, Shichida Y, Yoshizawa T, McDowell JH, Hargrave PA, Palczewski K (1990) Phosphorylation of iodopsin, chicken red-sensitive cone visual pigment. *Biochemistry* 29: 10 102–10 106

47 Palczewski K, Buczylko J, Lebioda L, Crabb JW, Polans AS (1993) Identification of the N-terminal region in rhodopsin kinase involved in its interaction with rhodopsin. *J Biol Chem* 268: 6004–6013

48 Hurley JB, Spencer M, Niemi GA (1998) Rhodopsin phosphorylation and its role in photoreceptor function. *Vision Res* 38: 1341–1352

49 Morrison DF, Ting TD, Vallury V, Ho YK, Crouch RK, Corson DW, Mangini NJ, Pepperberg DR (1995) Reduced light dependent phosphorylation of an analog visual pigment containing 9-demethylretinal as its chromophore. *J Biol Chem* 270: 6718–6721

50 Corson DW, Cornwall MC, MacNichol EF, Tsang S, Derguini F, Crouch RK, Nakanishi K (1994) Relief of opsin desensitization and prolonged excitation of rod photoreceptors by 9-desmethylretinal. *Proc Natl Acad Sci USA* 91: 6958–6962

51 Yamamoto S, Sippel KC, Berson EL (1997) Defects in the rhodopsin kinase gene in the Oguchi form of stationary night blindness. *Nat Genet* 15: 175–178

52 Hofmann KP, Pulvermüller A, Buczylko P, van Hooser P, Palczewski K (1992) The role of arrestin and retinoids in the regeneration pathway of rhodopsin. *J Biol Chem* 267: 15 701–15 706

53 Ishiguro S, Suzuki Y, Tamai M, Mizuno K (1991) Purification of retinol dehydrogenase from bovine retinal rod outer segments. *J Biol Chem* 266: 15 520–15 524

54 Saari JC, Garwin GG, vanHooser JP, Palczewski K (1998) Reduction of all-*trans* retinal limits regeneration of visual pigment in mice. *Vision Res* 38: 1325–1333

55 Pepperberg DR, Okajima TL, Wiggert B, Ripps H, Crouch RK, Chader GJ (1993) The role of interphotoreceptor retinoid-binding protein (IRBP) in the visual cycle of rhodopsin. *Mol Neurobiol* 7: 61–84

56 Chen Y, Noy N (1994) Retinoid specificity of interphotoreceptor retinoid-binding protein. *Biochemistry* 33: 10 658–10 665

57 Lin Z-Y, Li G-R, Takizawa N, Si J-S, Gross EA, Richardson K, Nickerson J (1997) Structure-function relationships in interphotoreceptor retinoid-binding protein (IRBP). *Molecular Vision* 3: <http://www.emory.edu/molvis/v3/lin>

58 Okajima T-I, Pepperberg D, Ripps H, Wiggert B, Chader G (1990) Interphotoreceptor retinoid-binding protein promotes rhodopsin regeneration in toad photoreceptors. *Proc Natl Acad Sci USA* 87: 6907–6911

59 Jones G, Crouch R, Wiggert B, Cornwall MC, Chader GJ (1989) Retinoid requirements for recovery of sensitivity after visual-pigment bleaching in isolated photoreceptors. *Proc Natl Acad Sci USA* 86: 9606–9610

60 Ho M-T, Massey J, Pownall H, Anderson R, Hollyfield J (1989) Mechanism of vitamin A movement between rod outer segments, interphotoreceptor retinoid-binding protein and liposomes. *J Biol Chem* 264: 928–935

61 Liou GI, Fei Y, Peachey NS, Matragoon S, Wei S, Blaner WS, Wang Y, Liu C, Gottesman ME, Ripps H (1998) Early onset photoreceptor abnormalities induced by targeted disruption of the interphotoreceptor retinoid-binding protein gene. *J Neurosci* 18: 4511–4520

62 Livrea MA, Tesoriere L, Bongiorno A (1991) All-*trans* to 11-*cis* retinol isomerization in nuclear membrane fraction from bovine retinal pigment epithelium. *Exp Eye Res* 52: 451–459
63 Rando RR (1992) Molecular mechanisms in visual pigment regeneration. *Photochem Photobiol* 56: 1145–1156
64 Saari J, Bredberg L, Noy N (1994) Control of substrate flow at a branch point in the visual cycle. *Biochemistry* 33: 3106–3112
65 Winston A, Rando RR (1998) Regulation of isomerohydrolase activity in the visual cycle. *Biochemistry* 37: 2044–2050
66 Maw MA, Kennedy B, Knight A, Bridges R, Roth KE, Mani EJ, Mukkadan JK, Nancarrow D, Crabb JW, Denton MJ (1997) Mutation of the gene encoding cellular retinaldehyde-binding protein in autosomal recessive retinitis pigmentosa. *Nat Genet* 127: 198–200
67 Crabb JW, Nie Z, Chen Y, Hulmes JD, West KA, Kapron JT, Ruuska SE, Noy N, Saari JC (1998) Cellular retinaldehyde-binding protein ligand interactions. *J Biol Chem* 273: 20 712–20 720
68 Bridges CDB (1976) Vitamin A and the role of the pigment epithelium during bleaching and regeneration of rhodopsin in the frog eye. *Exp Eye Res* 24: 435–455
69 Mata NL, Tsin AT, Chambers JP (1992) Hydrolysis of 11-*cis* and all-*trans*-retinyl palmitate by retinal pigment epithelium microsomes. *J Biol Chem* 267: 9794–9799
70 Mata NL, Villazana ET, Tsin AT (1998) Colocalization of 11-*cis* retinyl esters and retinyl ester hydrolase activityin retinal pigment epithelium plasma membrane. *Invest Ophthalmol Vis Sci* 39: 1312–1319
71 Saari J, Bredberg L (1989) Lecith*In*:retinol acetyltransferase in retinal pigment epithelium. *J Biol Chem* 264: 8636–8647
72 Hamel CE, Tsilou E, Pfeffer B, Hooks J, Detrick B, Redmond TM (1993) Molecular cloning and expression of RPE65, a novel retinal pigment epithelium-specific microsomal protein that is post-translationally regulated *in vitro*. *J Biol Chem* 268: 15 751–15 757
73 Marlhens F, Bareil C, Griffoin JM, Zrenner E, Amalric P, Eliaou C, Liu SY, Harris E, Redmond TM, Arnaud B et al (1997) Mutations in RPE65 cause Leber's congenital amaurosis. *Nat Genet* 17: 139–141
74 Gu SM, Thompson DA, Srikumari CR, Lorenz B, Finckh U, Nicoletti A, Murthy KR, Rathmann M, Kumaramanckavel G, Denton MJ (1997) Mutations in RPE65 cause autosomal recessive childhood-onset severe retinal dystrophy. *Nat Genet* 17: 194–197
75 Morimura H, Fishman GA, Grover SA, Fulton AB, Berson EL, Dryja TP (1998) Mutations in the RPE65 gene in patients with autosomal recessive retinitis pigmentosa or leber congenital amaurosis. *Proc Natl Acad Sci USA* 95: 3088–3093
76 Aguirre GD, Baldwin V, Pearce-Kelling S, Narfström K, Ray K, Acland GM (1998) Congenital stationary night blindness in the dog: common mutation in the RPE65 gene indicates founder effect. *Molecular Vision* 4: <http://www.emory.edu/molvis/v4/aguirre>
77 Redmond TM, Yu S, Lee E, Hamasaki D, Chen N, Goletz P, Ma J-x Crouch RK, Pfeifer K (1998) Rpe65 is necessary for production of 11-*cis* vitamin A in the retinal visual cycle. *Nat Genet* 20: 344–350
78 Ma J-X, Xu L, Lockman K, Redmond TM, Crouch RK (1998) Cloning and localization of RPE65 mRNA is in salamander cone photoreceptor cells. *Biochim Biophys Acta* 1443: 255–261
79 Jin J, Jones GJ, Cornwall MC (1994) Movement of retinal along cone and rod photoreceptors. *Visual Neurosci* 11: 389–399
80 Goldstein EB (1970) Cone pigment regeneration in the isolated frog retina. *Vision Res* 10: 1943–1951
81 Bunt-Milan AH, Saari JC (1983) Immunocytochemical localization of two retinoid-binding proteins in vertebrate retina. *J Cell Biol* 97: 703–712
82 Das SR, Bhardwaj N, Kjeldbye H, Gouras P (1992) Muller cells of chicken retina synthesize 11-*cis* retinol. *Biochem J* 285: 907–913

Vitamin A and retinoids: an update of biological aspects and clinical applications
M.A. Livrea (ed.)
© 2000 Birkhäuser Verlag Basel/Switzerland

The role and evolutionary development of retinoic-acid signalling in the eye

U.C. Dräger[1], E. Wagner[1], A. Andreadis[2] and P. McCaffery[1]

E. Kennedy Shriver Center and Departments of Psychiatry[1] and Neurology[2], Harvard Medical School, 200 Trapelo Road, Waltham, MA 02452, USA

Introduction

From the earliest time of its discovery as an essential nutrient, vitamin A has been known to be vital for the eye, both for its development and for its adult function. In the mature organism, the earliest sign of vitamin A deficiency is night-blindness, and in the developing embryo vitamin A deficiency causes the ventral eye to be defective, leading to micro- or anophthalmia [1, 2]. Because the oxidation of retinaldehyde to retinoic acid (RA) is an irreversible reaction, it was possible to assay their biological roles separately [3]. These assays revealed that when adult rats are fed a diet completely lacking in vitamin A (retinol and β-carotene) and are given RA, they survive but turn blind due to photoreceptor degeneration. This illustrates the two distinct functions of the retinoids: 11-*cis* retinaldehyde forms the visual chromophore of rhodopsin, and RA regulates gene transcription throughout the body. In this brief review we will summarize evidence that these two functions, which are commonly studied in different fields of science, are connected in the eye. We discovered highly intricate expression patterns of different RA-generating aldehyde dehydrogenases in the developing and mature eye, and in the functioning eye we find that light causes an increase in RA. These observations indicate that the transcriptional role of RA may have its origin in vision.

Identification of RA-generating aldehyde dehydrogenases

Our studies on retinoids in the eye began with a chance observation in the context of a screen for factors involved in the axial polarization of the embryonic retina. We identified a class-1 aldehyde dehydrogenase, named AHD2 in the mouse, as a very abundant protein in the dorsal third of the embryonic and adult retina [4]. *In situ* hybridizations for this enzyme on heads of mouse embryos are illustrated in Figure 1. Aldehyde dehydrogenases are very ancient, present in bacteria, yeast and plants, and mammalian tissues contain a large number of different members of this enzyme family. Most aldehyde dehydrogenases can oxidize a wide range of aldehydes and they function mainly in the detoxification of many different compounds [5]. The anterior segment of the eye was known to contain several aldehyde dehydrogenases believed to serve in the removal of toxic photo-oxidative products [6]. The observation that an aldehyde dehydrogenase is expressed only in part of the embryonic retina and long before the eye is exposed to light, however, points to a function different from detoxification.

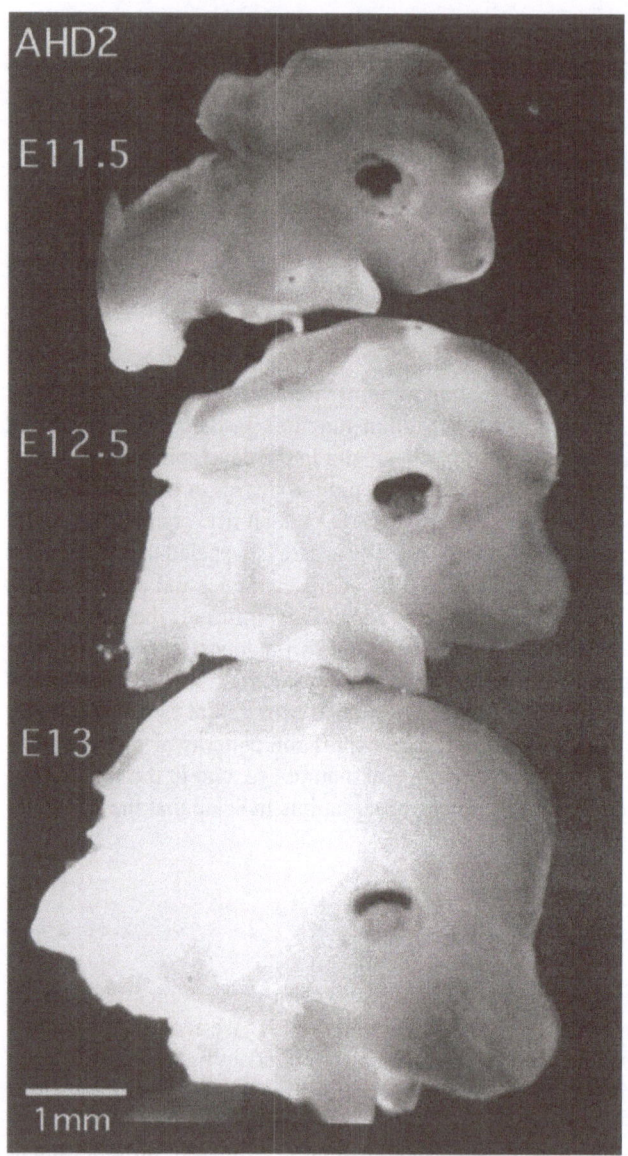

Figure 1. *In situ* hybridizations for AHD2 mRNA in heads of mouse embryos at embryonic day 11.5 (E11.5), E12.5 and E13. For these prepararions we used albino mice which lack the pigmentation of the retinal pigment epithelium. Note the strong labeling in the dorsal part of the eye, which corresponds to the dorsal third of the retina.

In a study on adult mouse liver, AHD2 was identified as the principal enzyme for RA synthesis [7], and we confirmed this function for the embryonic retina [8]. Apart from AHD2, little information had been available on which other aldehyde dehydrogenases can oxidize retinaldehyde to RA. Information was sparse because of the difficulty in dissolving the retinaldehyde substrate at a concentration sufficient for the standard aldehyde dehydrogenase assay. In order to analyze RA synthesis in the tiny embryonic retina, we developed a micro-zymography assay for RA-generating aldehyde dehydrogenases in which protein fractions, sep-arated by native gel electrophoresis, are tested for RA synthesis with a RA-reporter cell line [9, 10]. With this assay we found in addition to the dorsal AHD2, a second, novel RA-generating aldehyde dehydrogenase, which we named V1. V1 generates even higher RA levels in the ventral retina than AHD2 does in the dorsal retina. A third novel enzyme, named RALDH2, was identified in the non-neural tissue behind the retina [8, 11–13].

We cloned RALDH2 by amplifying aldehyde-dehydrogenase coding sequences with degenerate polymerase chain reaction primers to two very conserved protein sequences in this class of enzymes, and we distinguished it from other aldehyde dehydrogenases by the fact that it is induced by RA in P19 cells [14]. The three RA-generating dehydrogenases differ in several enzymatic criteria, including substrate selectivity. AHD2 oxidizes a broad range of aliphatic aldehydes; V1 is much more selective for retinaldehyde but also acts on some other aldehydes such as octanal; and RALDH2 is the most specific enzyme for retinaldehyde under physiological substrate conditions [12]. RALDH2 is the earliest enzyme expressed in the eye anlage (see below) and the embryo overall [11, 15, 16].

Aldehyde-dehydrogenase expression patterns explain retinoid teratology

We analyzed the expression patterns of the different aldehyde dehydrogenases in the developing embryo and the adult nervous system. The eye stands out in displaying the highest concentration and diversity, and the most complex patterns of these enzymes. The most outstanding feature in general is a rather sparse and highly focussed expression. For instance in the embryonic spinal cord, only the motor neurons that innervate the limbs express very high RALDH2 levels and all other neurons are devoid of RA synthesis [14, 17]; in the developing hindbrain, RALDH2 in the non-neuronal surrounding tissue synthesizes high RA levels, but practically no RA synthesis is detectable in the cerebellar cortex [18]; similarly, most RA in the adult brain is synthesized by RALDH2 in the meninges, and AHD2 is exclusively expressed in a subpopulation of dopaminergic neurons that project to the dorsal corpus striatum [19]; and in the pituitary, only the anterior lobe, but not the intermediate and posterior lobes, synthesizes RA (P. McCaffery and U.C. Dräger, unpublished observations). Since the lipid RA diffuses readily through cell membranes and solid tissue, the highly focussed aldehyde-dehydrogenase expression sites represent the origins of RA diffusion gradients [14, 17]. Moreover, in several locations we find expression of RA-degrading enzymes (P450-linked cytochrome oxidases) in a complementary pattern to the RA-synthesizing enzymes, an arrangement that steepens the local differential in RA concentrations [17, 20, 21]. By comparison, the different RA receptors are expressed much more widely, in that all tissues contain at least one and usually several receptor types [22, 23].

This arrangement of strictly localized RA synthesis and diffusion gradients, in combination with the more widespread RA effector system, explains why RA is highly teratogenic [13, 14, 17, 18]. The regional RA differences carry morphogenetic information which is interpreted by locally distinct gene expression, since transcription of particular genes is known to be activated by different and narrow ranges of RA levels [24]. We find regional RA differences as a main characteristic of systems that depend on RA and that are vulnerable to RA disturbances. In the embryonic nervous system these RA differences, ranging from undetectable to micromolar levels, are much more pronounced than in the adult, mainly because the highest levels in the embryo far exceed any adult levels. When the embryo is flooded with exogenous RA, the differences are leveled, resulting in spatially and temporally inappropriate gene expression, as we will describe next for the eye.

Eye formation

The developing eye is not only highly vulnerable to vitamin A deficiency, but also an excess of exogenous RA is known to cause severe eye malformations [1, 2, 13, 25]. We analyzed the mechanism of retinoid-related ocular teratology and its normal correlate in early zebrafish and mouse embryos, and we conclude that for the proper formation of the ventro-dorsal retina dimension, the eye anlage has to contain both a RA-rich and a RA-poor region.

Determinations of RA-generating aldehyde dehydrogenases in the normal embryonic zebrafish retina, by zymography of dissected tissue samples [9], reveal a spatial expression pattern similar to that in the mouse. A class-1 aldehyde dehydrogenase, similar to mouse AHD2, is expressed in the dorsal third of the retina, and a different aldehyde dehydrogenase, similar to the ventral V1 enzyme in the mouse, occupies the ventral two-thirds of the zebrafish retina [26]. When early zebrafish embryos at the eye anlage stage are exposed to citral, a competitive inhibitor of RA-synthesizing aldehyde dehydrogenases, they grow into larvae that lack the ventral parts of their eyes [26]. In the reverse experiment, Hyatt et al. had reported that exposure to an excess of RA during the same early eye-anlage stage causes a malformation that appears to be a duplication of the eye along the dorso-ventral axis [27].

We tested the duplicated retinas, created by exposure to 1 µM RA in the tank water, for dorsal and ventral markers [28] and found that the dorsal aldehyde dehydrogenase disappears and only the ventral dehydrogenase persists in the entire duplicated retina. Moreover, the transcription factor *msh[c]*, normally expressed in the dorsal part of the embryonic retina, is barely or not at all detectable, and *Pax[b]*, normally restricted to ventral retina and optic stalk, expands dorsally. These observations indicate that RA ventralizes the eye anlage. This notion was confirmed by a separate experiment using local RA exposure. When a RA-soaked bead is placed in the eyefield of embryonic fish, the larvae develop an ectopic optic fissure facing the bead, which resembles the normally ventral optic fissure. The effects of both global treatment and the ventralizing response to local RA treatment are limited to a brief critical period at the eye anlage stage, and the morphological changes become visible only a few days later.

In the eye anlage of the early mouse embryo, at a stage corresponding to the RA-sensitive critical period of the zebrafish eye anlage, we detect RALDH2 as the first RA-generating enzyme in zymographic assays [11]. Visualization by *in situ* hybridization shows RALDH2 mRNA to be located at one edge of the eye field [16]. This eccentric expression of RALDH2, we postulate, generates a RA gradient across the eye field during the brief period when the dorso-ventral organization is responsive to RA. In the RA-rich region next to RALDH2, transcription of ventral retina genes, including the V1 dehydrogenase, will be activated, and the RA-poorer eyefield region will be left to the default, dorsal fate, marked by the AHD2 dehydrogenase. An early and transient RA gradient is thus converted into a bipartite organization, dorsal and ventral retinal compartments marked by the two different aldehyde dehydrogenases. These enzymes propagate the orientation of the early determination event throughout development up to adulthood.

It is important to stress that the RA receptors are not early eye genes themselves, but they control expression of some of the eye genes. The sequence of eye development begins prior to RA synthesis as a single continuous eyefield, demarcated by the early eye genes *Rx*, *ET*, *Pax6* and others [29–31]. The single field is then subdivided by the *Sonic hedgehog* system, which suppresses the early eye genes in the midline region, into the bilateral eye anlagen [32]. At this stage RA synthesis begins and the retina orientation becomes responsive and is physiologically regulated by RA. For a brief time window RA-activated receptors control transcription of genes for the dorso-ventral retina axis by activating ventral and, probably, repressing dorsal factors. RA responsiveness of the eyefield appears a little earlier, however, prior to the beginning of endogenous RA synthesis in the head region: when very young embryos are exposed to RA, the rostral head including the eyes fails to form [28]. This illustrates the already mentioned point relevant to RA teratogenicity in general, which is that the spatial and temporal expression of RA responsiveness that reflects the RA receptor distribution tends to be wider than the endogenous RA availability.

Light-generated RA synthesis

A series of experiments aimed at investigating RA synthesis during postnatal ocular development resulted in a number of unexpected observations [33]. A comparison of RA synthesis in ocular tissues, normalized for tissue protein, show that the highest synthesis shifts postnatally from the neural retina to the tissue behind it, including the retinal pigment epithelium (RPE) and the choroid. Comparisons of RA synthesis and aldehyde dehydrogenase levels in preparations of dissected RPE/choroid and neural retinas at different ages reveal a discrepancy whose magnitude grows with maturation of the outer segments: in the juvenile and adult eyes, both the RPE and the retina synthesize much too much RA for the amounts of aldehyde dehydrogenases present. When the dissections are done under dim red light, the discrepancy disappears, indicating as the most likely explanation that some of the all-*trans* retinaldehyde, which is released from opsin by light, becomes accessible to the aldehyde dehydrogenases. We confirmed this explanation by dissecting adult retinas into dorsal and ventral halves, and showing that only illumination of dorsal halves results in RA synthesis, consistent with the dorsal AHD2 being the only RA-generating enzyme expressed in the adult neural retina. For the RPE/choroid, by comparison, no dorso-ventral difference is detectable, as RALDH2 expression is even throughout the posterior eye.

In order to address the possibility that the light effect might be artificially created by the tissue culture assay, in which the normal retina/RPE apposition is disrupted and the interphotoreceptor retinoid binding protein is washed away, we did experiments in which the only light exposure occurred while the mice were alive. Their eyes were then dissected under dim far-red illumination and compared to eyes of control dark-adapted mice. Measurements on eye cups show significantly more extractable RA in the light-exposed than dark-adapted eyes. In addition, the effect is stable: when the eyes are dissected in tissue culture medium, thoroughly washed for at least an hour, and cultured in fresh medium for RA detection by the reporter cells, the RPE/choroid preparations of the mice that saw light for an hour show a significant increase in RA synthesis over the RPE of dark-adapted controls. No such delayed light effect on RA synthesis is detectable in the neural retina. This phenomenon suggests that the RPE is able to bind light-generated all-*trans* retinaldehyde, a portion of which is channeled into RA production.

These observations indicate that light initiates not only an electrical but potentially a local transcriptional signal in the eye as well. Although the visual cycle is designed to preserve retinaldehyde as completely as possible, only a negligible fraction has to be oxidized, because the retina contains low millimolar levels of bound retinaldehyde, whereas most RA actions are in the low nanomolar range. Possible targets of light-generated RA are genes of several photo-transduction proteins whose expression is known to be influenced by light or to vary with the diurnal light cycle [34]. We tested expression of arrestin (= S-antigen) [35], whose transcription is known to be upregulated by light in a way that is independent of the circadian clock [36, 37]. First we did intraperitoneal RA injections and measured RA in the eye. The compound reaches the eye practically instantaneously, but ocular levels also decline rapidly — within one to a few hours, depending on the injected dose — following a single injection. We then dissected retinas of young adult mice which were dark-adapted or exposed to light for different lengths of time, and tested them for arrestin mRNA levels by quantitative Northern blotting. We confirmed that light causes an increase in arrestin mRNA, and we determined that RA injected in the dark has a similar effect. By systematically varying the dose of injected RA and the time between injection and killing of the mice, we found that a single injection of 1 μmole of RA has a maximal effect and causes a rapid upregulation in a slope similar to light-induced upregulation: mRNA levels reach a maximum after three hours and then decline. From this work we conclude that RA can mimic some of the light effects locally in the retina [35]. A recent study by others, showing that light-induced spinule formation by horizontal cells in fish is mimicked by RA application in the dark, arrives at the same conclusion [38].

A hypothesis on the visual origin of RA

The RA receptors represent by far the largest and most pervasive subgroup of the nuclear receptor transcription factors in vertebrates [39]. A large number of genes are known to be regulated by RA, including many growth factors and receptors, other transcription factors such as homeobox proteins, and a long list of ion-channel, structural and extracellular matrix proteins [40]. While RA can both up- or down-regulate transcription of some of these proteins, depending on the context, RA effects generally follow similar themes. In many cases RA terminates cell proliferation and activates differentiation programs. Early in embryogenesis, RA induces a precursor region to advance developmentally towards its destined fate. In addition, the compound can impart vectorial information onto an organ anlage, such as the antero-posterior axis onto the limb bud [41] and, as explained above, the dorso-ventral axis onto the eye anlage [28].

While the use of 11-*cis* retinaldehyde as the light-sensitive component of rhodopsins is very ancient, found even in the eyespot of the unicellular *Euglena*, an odd feature of RA-regulated transcription is that it is evolutionarily rather recent, detected mainly in the chordate/vertebrate branch, and is fully evolved only in vertebrates [42]. *Drosophila* has homologs of most of the proteins regulated by RA in vertebrates, but their transcription is generally not RA-dependent, and no RA receptors have yet been identified in *Drosophila* [39]. Beginnings of RA-mediated transcription, through as yet unidentified mechanisms, have been detected, however, in the *Drosophila* eye [43, 44]. The explanation for the differential usage of RA by invertebrates and vertebrates, we propose, lies in a crucial difference of the visual cycle: only in vertebrate photoreceptors is the photo-isomerized chromophore all-*trans* retinaldehyde released from illuminated rhodopsin [45]. As described above, some of it is oxidized to RA by the aldehyde dehydrogenases [33], which probably initially served a protective role against harmful free aldehyde. We postulate that RA-synthesizing aldehyde dehydrogenases evolved originally from the common detoxification function of the enzymes necessary in the visual cycle, but when RA gained a significant role in modulation of gene transcription, the functional emphasis shifted from substrate removal to the generation of product.

We propose the hypothesis that the free RA in the eye may have associated with existing orphan receptors, stimulating their rapid adaptation into RA receptors [13, 33, 46]. A recent observation is consistent with this hypothesis of the visual origin of RA-mediated transcription: the 11-*cis* retinol dehydrogenase [47], an enzyme function required exclusively in the eye, turns out to serve also as 9-*cis* retinol dehydrogenase, an enzyme involved in the generation of 9-*cis* RA throughout the body [48, 49].

The compound eye of insects shows very little resemblance to the vertebrate eye, which led to the general belief that the two eye types had evolved in parallel. Surprisingly, recent molecular analyses reveal that homologs to practically every eye gene in *Drosophila* also function as eye genes in vertebrates [50]. Moreover, the embryonic *Drosophila* eye disc is organized into similar dorsal and ventral compartments as we find in the vertebrate eye [51]. At this point the formation of the dorso-ventral eye axis in vertebrates by RA-generating aldehyde dehydrogenases remains the main feature for which a close homology has yet to be identified in *Drosophila*. Genetic elimination of mouse RA receptors, in particular the RXR class, results in ventral eye defects [52], as does elimination of the RXR homolog *ultraspiracle* in *Drosophila* [53]. *Ultraspiracle*, however, is not activated by retinoids, but it may function as the juvenile-hormone receptor [54].

We suggest that RA-mediated transcription, originating from vision, has evolved into a new layer of control that became super-imposed over more ancient molecular mechanisms, first in the eye by taking control over a fully functional eye morphogenesis system, and then the rest of the body. This would explain the already mentioned characteristic that RA-mediated transcription tends to activate complex programs. Fascinating features of RA-related malformations, linked both to excess and deficiency, are their complexity and their resemblance to naturally occurring malformations of unknown etiology. This applies to retinoid-related heart defects, which resemble most of the idiopathic inborn heart malformations, to the cerebellar defects, in that RA teratology phenocopies the Dandy-Walker syndrome [55], and to the eye defects: many of the severe inborn eye malformations originate from the ventral eye/optic fissure [56], similar to the malformations resulting from vitamin A deficiency [2] and inhibition of RA synthesis [26]. The explanation is, of course, not that vitamin A deficiency is common, but that retinoid-related teratology reveals building blocks or mechanisms in embryonic development which are sensitive to RA regulation.

Acknowledgements
This work was supported by National Institutes of Health grants EY01938, HD05515 and HD01179.

References

1   Hale F (1937) The relation of maternal vitamin A deficiency to microphthalmia in pigs. *Texas State J Med* 33: 228–232
2   Warkany J, Schraffenberger E (1946) Congenital malformations induced in rats by maternal vitamin A deficiency. I. Defects of the eye. *Arch Ophthalmol* 35: 150–169
3   Dowling JE, Wald G (1960) The biological function of vitamin A acid. *Proc Natl Acad Sci USA* 46: 587–608
4   McCaffery P, Tempst P, Lara G, Dräger UC (1991) Aldehyde dehydrogenase is a positional marker in the retina. *Development* 112: 693–702
5   Petersen D, Lindahl R (1997) Aldehyde dehydrogenases. *In*: FP Guengerich (ed): *Comprehensive Toxicology: Biotransformations*, Vol 3. Pergamon Press, New York, 97–118
6   Holmes RS (1988) Alcohol dehydrogenases and aldehyde dehydrogenases of anterior eye tissues from humans and other mammals. *In*: K Kuriyama, A Takada, H Ishii (eds): *Biomedical and Social Aspects of Alcohol and Alcoholism*. Elsevier Science Publishers, Amsterdam, 51–57
7   Lee M-O, Manthey CL, Sladek NE (1991) Identification of mouse liver aldehyde dehydrogenases that catalyze the oxidation of retinaldehyde to retinoic acid. *Biochem Pharmacol* 42: 1279–1285
8   McCaffery P, Lee M-O, Wagner MA, Sladek NE, Dräger UC (1992) Asymmetrical retinoic acid synthesis in the dorso-ventral axis of the retina. *Development* 115: 371–382
9   McCaffery M, Dräger UC (1997) A sensitive bioassay for enzymes that synthesize retinoic acid. *Brain Res Protocols* 1: 232–236
10 Wagner M, Han B, Jessell TM (1992) Regional differences in retinoid release from embryonic neural tissue detected by an *in vitro* reporter assay. *Development* 116: 55–66
11 McCaffery P, Posch KC, Napoli JL, Gudas L, Dräger UC (1993) Changing patterns of the retinoic acid system in the developing retina. *Dev Biol* 158: 390–399
12 McCaffery P, Dräger UC (1993) Retinoic acid synthesis in the developing retina. *Adv Exp Med Biol* 328: 181–190
13 Dräger UC, McCaffery P (1997) Retinoic acid and development of the retina. *Progr Retin Eye Res* 16: 323–346
14 Zhao D, McCaffery P, Ivins KJ, Neve RL, Hogan P, Chin WW, Dräger UC (1996) Molecular identification of a major retinoic-acid synthesizing enzyme: a retinaldehyde-specific dehydrogenase. *Eur J Biochem* 240: 15–22
15 Dräger UC, McCaffery P (1995) Retinoic-acid synthesis in the developing spinal cord. *Adv Exp Med Biol* 372: 185–192
16 Niederreither K, McCaffery P, Dräger UC, Chambon P, Dollé P (1997) Restricted expression and retinoic acid-induced downregulation of the retinaldehyde dehydrogenase type 2 (RALDH-2) gene during mouse development. *Mech Develop* 62: 67–78
17 McCaffery P, Dräger UC (1994) Hotspots of retinoic acid synthesis in the developing spinal cord. *Proc Natl Acad Sci USA* 91: 7194–7197
18 Yamamoto M, McCaffery P, Dräger UC (1996) Influence of the choroid plexus on cerebellar development: analysis of retinoic acid synthesis. *Develop Brain Res* 93: 182–190
19 McCaffery P, Dräger UC (1994) High levels of a retinoic-acid generating dehydrogenase in the meso-telencephalic dopamine system. *Proc Natl Acad Sci USA* 91: 7772–7776
20 Moss JB, Xavier-Neto J, Shapiro M, Nayeem M, McCaffery P, Dräger UC, Rosenthal N (1998) Dynamic patterns of retinoic acid synthesis and response in the developing mammalian heart. *Dev Biol* 199: 55–71
21 Yamamoto M, Dräger UC, McCaffery P (1998) A novel assay for retinoic acid catabolic enzymes shows high expression in the developing hindbrain. *Develop Brain Res* 107: 103–111
22 Dollé P, Ruberte E, Leroy P, Morriss-Kay G, Chambon P (1990) Retinoic acid receptors and cellular retinoid binding proteins. I. A systematic study of their differential pattern of transcription during mouse organogen-

esis. *Development* 110: 1133–1151
23 Dollé P, Fraulob V, Kastner P, Chambon P (1994) Developmental expression of murine retinoid X receptor (RXR) genes. *Mech Develop* 45: 91–104
24 Simeone A, Acampora D, Arcioni L, Andrews PW, Boncinelli E, Mavilio F (1990) Sequential activation of Hox2 homeobox genes by retinoic acid in human embryonal carcinoma cells. *Nature* 346: 763–766
25 Shenefelt RE (1972) Morphogenesis of malformations in hamsters caused by retinoic acid. *Teratology* 5: 403–418
26 Marsh-Armstrong N, McCaffery P, Dowling JE, Gilbert W, Dräger UC (1994) Retinoic acid is necessary for development of the ventral retina in zebrafish. *Proc Natl Acad Sci USA* 91: 7286–7290
27 Hyatt GA, Schmitt EA, Marsh-Armstrong NA, Dowling JE (1992) Retinoic acid-induced duplication of the zebrafish retina. *Proc Natl Acad Sci USA* 89: 8293–8297
28 Hyatt G, Schmitt EA, Marsh-Armstrong N, McCaffery P, Dräger UC, Dowling JE (1996) Retinoic acid establishes ventral retinal characteristics. *Development* 122: 195–204
29 Mathers PH, Grinberg A, Mahon KA, Jamrich M (1997) The Rx homeobox gene is essential for vertebrate eye development. *Nature* 387: 603–607
30 Li H, Tierney C, Wen L, Wu JY, Rao Y (1997) A single morphogenetic field gives rise to two retina primordia under the influence of the prechordal plate. *Development* 124: 603–615
31 Quiring R, Walldorf U, Kloter U, Gehring WJ (1994) Homology of the *eyeless* gene of *Drosophila* to the *Small eye* gene in mice and *Aniridia* in humans. *Science* 265: 785–789
32 Chiang C, Litingtung Y, Lee E, Young K, Corden J, Westphal H, Beachy P (1996) Cyclopia and defective axial patterning in mice lacking Sonic hedgehog gene function. *Nature* 383: 407–413
33 McCaffery P, Mey J, Dräger UC (1996) Light-mediated retinoic acid production. *Proc Natl Acad Sci USA* 93: 12 570–12 574
34 Cahill GM, Besharse JC (1995) Circadian rhythmicity in vertebrate retinas: regulation by a photoreceptor oscillator. *Progr Retin Eye Res* 14: 267–291
35 Wagner E, McCaffery P, Mey J, Farhangfar F, Applebury ML, Dräger UC (1997) Retinoic acid increases arrestin mRNA levels in the mouse retina. *FASEB J* 11: 271–275
36 McGinnis JF, Austin BJ, Stepanik PL, Lerious V (1994) Light-dependent regulation of the transcriptional activity of the mammalian gene for arrestin. *J Neurosci Res* 38: 479–482
37 Farber DB, Danciger JS, Organisciak DT (1991) Levels of mRNA encoding proteins of the cGMP cascade as a function of light environment. *Exp Eye Res* 53: 781–786
38 Weiler R, Schultz K, Pottek M, Tieding S, Janssen-Bienhold U (1998) Retinoic acid has light-adaptive effects on horizontal cells in the retina. *Proc Natl Acad Sci USA* 95: 7139–7144
39 Mangelsdorf DJ, Thummel C, Beato M, Umesono K, Blumberg B, Kastner P, Mark M, Chambon P, Evans RM (1995) The nuclear receptor superfamily: the second decade. *Cell* 83: 835–839
40 Gudas LJ, Sporn MB, Roberts AB (1994) Cellular biology and biochemistry of the retinoids. *In*: MB Sporn, AB Roberts, DS Goodman (eds): *The Retinoids: Biology, Chemistry, and Medicine*. Raven Press, New York, 443–520
41 Tickle C, Alberts B, Wolpert L, Lee J (1982) Local application of retinoic acid to the limb bud mimics the action of the polarizing region. *Nature* 296: 564–566
42 Laudet V (1997) Evolution of the nuclear receptor superfamily: early diversification from an ancestral orphan receptor. *J Mol Endocrinol* 19: 207–226
43 Picking WL, Chen D-M, Lee RD, Vogt ME, Polizzi JL, Marietta RG, Stark WS (1996) Control of *Drosophila* opsin gene expression by carotenoids and retinoic acid: Northern and Western analyses. *Exp Eye Res* 63: 493–500
44 Shim K, Picking WL, Kutty RK, Thomas CF, Wiggert BN, Stark WS (1997) Control of *Drosophila* retinoid and fatty acid binding glycoprotein expression by retinoids and retinoic acid: northern, western and immunocytochemical analyses. *Exp Eye Res* 65: 717–727
45 Saari JC (1994) Retinoids in photosensitive systems. *In*: MB Sporn, AB Roberts, DS Goodman (eds): *The Retinoids: Biology, Chemistry, and Medicine*. Raven Press, New York, 351–385
46 Dräger UC, Wagner E, McCaffery P (1998) Aldehyde dehydrogenases in the generation of retinoic acid in the developing vertebrate: a central role of the eye. *J Nutr* 128: 463S–466S
47 Simon A, Hellman U, Wernstedt C, Eriksson U (1995) The retinal pigment epithelial-specific 11-*cis* retinol dehydrogenase belongs to the family of short chain alcohol dehydrogenases. *J Biol Chem* 270: 1107–1112
48 Mertz JR, Shang E, Piantedosi R, Wei S, Wolgemuth DJ, Blaner WS (1997) Identification and characterization of a stereospecific human enzyme that catalyzes 9-*cis*-retinol oxidation. A possible role in 9-*cis*-retinoic acid formation. *J Biol Chem* 272: 11 744–11 749
49 Driessen CA, Winkens HJ, Kuhlmann ED, Janssen AP, van Vugt AH, Deutman AF, Janssen JJ (1998) The visual cycle retinol dehydrogenase: possible involvement in the 9-*cis* retinoic acid biosynthetic pathway. *FEBS Lett* 428: 135–140
50 Desplan C (1997) Eye development: governed by a dictator or a junta? *Cell* 91: 861–864
51 Brodsky MH, Steller H (1996) Positional information along the dorsal-ventral axis of the *Drosophila* eye: graded expression of the *four-jointed* gene. *Dev Biol* 173: 428–446
52 Kastner P, Grondona JM, Mark M, Gansmuller A, LeMeur M, Decimo D, Vonesch J-L, Dollé P, Chambon P

(1994) Genetic analysis of RXRα developmental function: convergence of RXR and RAR signalling pathways in heart and eye morphogenesis. *Cell* 78: 987–1003

53 Oro AE, McKeown M, Evans RM (1992) The *Drosophila* retinoid X receptor homolog *ultraspiracle* functions in both female reproduction and eye morphogenesis. *Development,* 115: 449–462

54 Jones G, Sharp PA (1997) Ultraspiracle: an invertebrate nuclear receptor for juvenile hormones. *Proc Natl Acad Sci USA* 94: 13 499–13 503

55 Lammer EJ, Armstrong DL (1992) Malformations of hindbrain structures among humans exposed to isotretinoin (13-*cis*-retinoic acid) during early embryogenesis. *In*: GM Morris-Kay (ed.): *Retinoids in normal development and teratogenesis.* Oxford University Press, Oxford, 281–295

56 Mann I (1937) *Developmental abnormalities of the eye.* Cambridge University Press, Cambridge

M.A. Livrea (ed.)

Vitamin A, retinoids and immune responses

A.C. Ross

Department of Nutrition and Department of Veterinary Science, The Pennsylvania State University, University Park, PA 16802, USA

Introduction

The development of effective immunity depends on highly coordinated and complex interactions amongst organs, circulating and tissue-fixed cells, and molecular factors that are elicited in very specific patterns that differ with the species, and even type, of pathogen or immune stimuli that the host encounters. Early investigators of vitamin A (VA) recognized this nutrient as essential for the prevention of infectious disease; indeed, Green and Mellanby referred to VA as the "anti-infective agent" [1]. Research that has been reviewed previously showed requirements for VA for lymphoid organ and cell integrity, antigen-specific antibody responses, cell-mediated immunity, and innate immunity. Experimental studies conducted in a variety of *in vivo* and cell models demonstrated that retinoid depletion results in dysregulation of T lymphocyte functions, impaired antibody responses to T cell-regulated antigens, and reduced non-specific immune responses. Other studies demonstrated that retinoids can, in some circumstances, enhance immune responses. With previous reviews [2–7] as a point of departure, this chapter provides an update focused on recently gained knowledge of the roles of retinoids in lymphocyte activation, antibody production, and the production of cytokines, cytokine receptors and intracellular signalling molecules that are involved in immune and inflammatory responses.

T cell activation and helper T cell function

T cells enter an activation program when they are stimulated through the T cell receptor (TCR)-CD3 complex by antigen bound to major histocompatibility complex (MHC) class II molecules on antigen-presenting cells; additional signals are provided when the T cell surface CD28 molecule is simultaneously ligated. These signals initiate the processes of interleukin (IL)-2 production, expression of the IL-2 receptor (IL-2R) and other activation-induced changes that lead to cell cycle progression and clonal expansion of antigen-activated T cells. Depending on circumstances (for example, see [8]), activation may result in apoptosis and T cell deletion. The ability of retinoids to modulate T cell activation has been explored *in vivo*, in cultured primary cells and in long-term cell culture models. A number of studies have used stimulation by anti-CD3, with or without anti-CD28, as a means to study the effects of retinoids on activation-induced T cell proliferation or apoptosis.

Retinoid requirements

Whether retinol has a unique function in lymphocyte activation that is not shared by retinoic acid (RA) is still an open question. However, in a number of situations, including *in vivo* antibody production (see below), RA is sufficient to activate T cell dependent responses in VA-deficient animals and/or to enhance responses in normal animals. In contrast, Garbe et al. [9] reported that mouse thymocytes require either retinol or a metabolite of retinol, 14-hydroxy-retro-retinol (14-HRR), to proliferate. Cells without retinol or 14-HRR were reported to die; interestingly, RA did not prevent cell death. In these studies, cells were stimulated by anti-CD3 in serum-free medium. However, subsequent results suggest that a requirement for retinol may not be general. Allende et al. [10] similarly investigated the proliferation of peripheral blood mononuclear cells (PBMC) from healthy human donors and patients with T cell related immunodeficiencies. Proliferation induced by anti-CD3 was increased ~2-fold by retinol $(10^{-7}-10^{-8}$ M), but only when retinol was added 48 h after anti-CD3. Time windows for retinoid action were also suggested by a study of Buck et al. [11] in which anhydroretinol acted as an antagonist of retinol only if added to human B cell lines or murine primary thymocytes prior to retinol. In contrast to the lack of effect of RA reported for murine thymocytes [9], Allende et al. [10] reported that RA increased CD3-mediated cell proliferation in human PBMC, similar to the effect of retinol. Retinol significantly increased the proliferative response to CD3 and increased proliferation in retinol-treated cells stimulated with anti-CD3 plus anti-CD28. However, retinol decreased the T cell response to Concanavalin A and had no significant effect on activation induced by a variety of other molecules including enterotoxins; phorbol myristyl acetate and IL-2 individually, together, or in combination with anti-CD3; or by anti-CD2 or phytohaemagglutinin. These data provide evidence that exogenous retinol (or RA) is able to prolong or strengthen the activation signal(s) previously initiated through the TCR/CD3 ± CD28 complexes, but retinoids are not specifically required for T cell activation by other activators. In experiments using primary rat thymocytes, anti-CD3 as stimulus and serum-free incubation conditions as described previously [9], RA as well as retinol increased anti-CD3-stimulated T cell proliferation, but cells that were more strongly activated with both anti-CD3 and anti-CD28 showed little effect of added retinol or RA (T. Shimada and A. C. Ross, unpublished observations). Although retinol significantly potentiated the expression of IL-2 and interferon (IFN)-γ mRNAs by CD3-stimulated human PBMC, there was no effect on IL-2, IFN-γ, IL-4, IL-10 or IL-6 secretion measured in culture supernatants [10].

A comparison of the results above suggests that subtle differences in experimental conditions could have pronounced effects on the apparent requirements for T cell activation. Lessard and Dupuis [12] reported that the effects of retinol and RA on T cell blastogenesis induced by Concanavalin A differed markedly depending on the preparation of the retinoid-containing medium added to cells. Blastogenesis was reduced when retinol was first diluted in RPMI medium and then added to serum-containing medium, but in contrast was increased when retinol was first diluted in fetal calf serum. The association of retinoids with serum proteins may be critical, and/or the effective concentrations or stability of retinoids may differ importantly depending on preparation. In any case, these disparate results illuminate the importance of multiple experimental controls and of using great caution when interpreting biological activities based on *in vitro* studies.

Retinoids in activation-induced apoptosis

Activation-induced apoptosis is thought to be the mechanism for negative selection of thymocytes and deletion of mature, activated peripheral blood T cells. Szondy et al. [13] examined the ability of various retinoic acid receptor (RAR) or retinoic x receptor (RXR) agonists to inhibit thymocyte apoptosis in intact mice treated with anti-CD3 monoclonal antibodies or in thymocytes activated *ex vivo* by phorbol dibutyrate and calcium ionophore. In *ex vivo* thymocytes all-*trans*-RA, 9-*cis*-RA, or all-*trans*-RA plus an RXR agonist reduced apoptosis in a dose-dependent manner, but did not prevent it entirely. In intact mice treated with anti-CD3 alone thymocyte apoptosis was induced (66.1% CD4$^+$/CD8$^+$ cells in CD3-treated mice versus 83.7% in controls). Apoptosis was inhibited in mice treated with anti-CD3 and a RAR-α agonist (80.8% CD4$^+$/CD8$^+$ cells), as was the induction of tissue transglutaminase which was correlated with anti-CD3-induced apoptosis. This investigation also revealed complexities in the interactions between RAR-α, RAR-γ and RXR pathways in determining the fate of CD3-activated thymocytes.

Vitamin A deficiency and T helper cytokines

Several lines of evidence support the concept that the pattern of cytokines expressed at the time of initial T cell activation "shapes" the T cell response, modulating the initial response to antigen and often affecting later responses mediated by memory cells. Numerous immunological disorders have now been characterized by an imbalance in the number or functional activity of helper T cells, expressed by differences in the proportions of CD4$^+$ or CD8$^+$ T cells characterized as Th0, Th1, Th2, or Th3 cells according to the patterns of cytokines they produce. An imbalance favoring Th1 cells has been postulated to occur in VA-deficient mice [14]. VA deficiency was associated with higher levels of mRNA transcripts in unprimed animals for IL-12 and IFN-γ, cytokines of the Th1 type, but not for Th2 cytokines IL-4 or IL-10 [14]. Other studies have shown the ability of IFN-γ to inhibit antigen-specific B cell proliferation and IgM production [15]. Unfractionated splenocytes of VA-deficient mice produced about twice as much IFN-γ in culture compared to cells from RA-supplemented mice, and addition of RA in culture reduced IFN-γ secretion. Infection of VA-deficient mice with the helminth *Trichinella spiralis*, which normally elicits a Th2-type response, also resulted in higher IFN-γ production per secreting cell and less IL-5 and IL-10 [16].

As discussed below, a recent analysis of the IFN-γ promoter revealed an RA response element, and promoter-reporter constructs responded positively to RA [17]. It is unknown, however, whether the IFN-γ gene is a primary target of RA *in vivo*. At present, the physiological regulation of IFN-γ production during VA deficiency and by RA is not understood.

IL-2 receptor (IL-2R) expression and IL-2 gene regulation

The high affinity IL-2R is comprised of an inducible α subunit, p55, to which IL-2 binds with high affinity; a β subunit, p75, capable of binding IL-2 with intermediate affinity which is expressed constitutively at low levels on T cells and at higher levels on natural killer (NK) cells; and a γ subunit that is common among several cytokine receptors. RA up-regulated the expression of the IL-2R-α in human B cell lines and induced IL-2R-α promoter activity in transfected B cells [18], and slightly increased IL-2R-α expression by human adenoidal T cells [19]. Promoter analysis suggested cooperation between an RA responsive region and a negative regulatory element that resulted in complex regulation of IL-2R-α gene promoter transcription. Conversely to the IL-2R, the promoter of the IL-2 gene was down-regulated by RAR ligands [20]. If these gene analysis results are indicative of the physiological regulation of IL-2R and IL-2 expression in intact cells, it is conceivable that retinoids could increase the specificity of IL-2's biological actions by inducing the high-affinity IL-2R (CD25) selectively on antigen-activated cells while keeping IL-2 production low so that only cell expressing the high-affinity form of the IL-2R will be stimulated by IL-2.

Antibody production

Studies of B cell differentiation and regulation of antibody classes

Research reviewed previously [3, 4] established that antibody responses to T cell dependent (TD) antigens are typically low in VA-deficient animals and that supplementation with retinol or all-*trans*-RA reverses this impairment. Recent studies have explored the effects of retinoids, particularly RA *in vitro*, on B cell differentiation and the selective production of IgM, IgA, IgG, and IgE.

Ballow and coworkers [21, 22] studied the differentiation of B cells from adult PBMC and cord blood mononuclear cells (CBMC) in response to stimulation with Cowan I strain *Staphylococcus aureus* (SAC), a TD factor that induces immunoglobulin synthesis. RA increased the synthesis of IgM by CBMC and the synthesis of IgG by adult PBMC. RA-incubated T cells from either CBMC or PBMC provided comparable help for the differentiation of B cells. These results [21] and others from Epstein-Barr virus-transformed B cells [23] suggested an increase in the production of IL-6 in the mechanism of RA-enhanced immunoglobulin synthesis.

In contrast to the increases in IgM and IgG above, RA decreased IgE synthesis by human PBMC and B cells from healthy donors that were stimulated *in vitro* with anti-CD40 and IL-4 [24]. Very low concentrations of RA ($<10^{-10}$ M) inhibited production of IgE and reduced epsilon germline gene transcription.

Previous research has suggested that VA deficiency is associated with a reduction in the production of IgA; such a decrease may be related to the often-increased severity of infections observed in VA-deficient children and the poor recovery of mucosal epithelia following infection in animal models (see chapter in this book by R. Semba). The IgA response to influenza A in VA-deficient mice was reduced in lung tissue, yet influenza A-specific IgG titers in tracheal and lung lavage fluids did not differ between VA-deficient and control mice [25].

Vitamin A-deficient mice produced less influenza A-specific IgA and had fewer influenza-A-specific IgA secreting cells in their salivary glands and cervical lymph nodes than did control mice [26]. As noted earlier for total IgG [7, 27], total IgA was elevated in VA-deficient animals although influenza A-specific IgA production was significantly reduced. VA-deficient mice also had reduced expression of the polymeric IgA receptor/transporter system, pIgR, which is required for the transport of secretory IgA from the basal to the apical side of epithelia. However, such effects may be tissue or organ specific because results from an intestinal cell system showed up-regulation of pIgR by RA [28]. Sarkar et al. [28] examined pIgR expression in a human epithelial cell line HT-29 as a model for the intestinal epithelium. Expression of pIgR was up-regulated by IFN-γ and IL-4 in cells grown in normal medium, but was significantly reduced in intestinal epithelial cells grown in medium depleted of VA. However, normal expression of pIgR mRNA was induced in a dose and time-dependent manner after culture with RA, and cell-associated pIgR immunofluorescence was stronger in RA-treated intestinal epithelial cells. Thus IFN-γ and IL-4 acted cooperatively and interacted positively with RA to increase pIgR expression. If, like HT-29 cells, VA-deficient intestinal epithelia are insensitive to cytokine up-regulation of pIgR, a reduction in immune IgA transport to apical surfaces could result in reduced antibody responses to bacteria and viruses at these surface barriers. Further studies are needed to resolve the apparently different regulation by VA status *in vivo* or RA treatment with respect to pIgR expression and antigen-specific IgA transport that seems to exist between intact mice and cell culture systems, or possibly between epithelia from different tissues. The differential effects of VA supplementation that have been observed in young children on diarrheal disease versus pneumonia (see chapter by R. Semba for review) also suggest that there may be significant tissue-specific differences in the response to VA deficiency and VA or retinoid supplementation.

Several cytokines have been shown to be factors for B cell differentiation, maturation and immunoglobulin class switching. These effects may be further modulated by retinoids. Tokuyama and coworkers [29–31] examined retinoid- and hormone-induced modulation of immunoglobulin synthesis by murine B cells induced to differentiate by lipopolysaccharide (LPS). It should be noted that murine B cells may be unusual in their ability to secrete immunoglobulin in response to LPS alone. RA had little effect on the production of IgM, IgG2a, IgG2b and IgG3, but IgA increased and IgG1 decreased reciprocally as RA dose increased [29]. Either RA or a 170-fold higher concentration of retinol increased production of IgA and decreased IgG1. IgA production was linked to increased expression of transforming growth factor (TGF)-β [31] and was partially inhibited by the addition of antibodies to TGF-β1 and TGF-β2 [29]. RA and 17-β-estradiol increased IgA production synergistically, while male sex hormones were without effect. However, glucocorticoids at concentrations >1 nM nearly completely suppressed IgA production induced by LPS plus RA, affecting IgG1 oppositely [30]. Data from an *in vivo* study support these regulatory effects of RA on IgG1 and IgA. In VA-adequate rats with collagen II-induced arthritis, Am-80, an RAR agonist, significantly reduced arthritis symptoms as well as the serum levels of anti-collagen antibodies of the IgG1 and (rat) IgG2a classes [32]. Interestingly, IgM was unchanged and IgA increased *in vivo* [32], similar to results in cell culture [29].

IL-4 is known to be a differentiation factor for IgG1 and IgE production. IgG1 and IgE production induced *in vitro* by IL-4 and LPS was inhibited by RA [33]. Although TGF-β is known to antagonize IL-4-dependent IgG1 and IgE production, and anti-TGF-β1 and TGF-β2 were shown to neutralize IL-4-dependent Ig production, antibodies to TGF-β did not neutralize the inhibitory activity that was induced by RA. In considering their observations that RA enhanced IgA [29–31] and inhibited IgE production [33], the authors speculated [33] that RA in colostrum might, by limiting IgE production in the infant intestine, serve to control production of proinflammatory allergic responses while facilitating mucosal defense mechanisms that involve IgA.

In vivo *antibody responses to TD antigens*

Vitamin A deficiency has been shown to reduce the response to many TD antigens, including common vaccines such as tetanus toxoid ([7, 34] for reviews). Since these reports, Molrine et al. [35] studied the anti-tetanus response of VA-deficient immunodeficient *scid* mice reconstituted with human peripheral blood lymphocytes from tetanus toxoid immune donors. On reimmunization, mice produced a secondary human anti-tetanus response. Vitamin A-deficient mice produced a much lower secondary anti-tetanus response and mice repleted with VA before immunization produced a normal secondary response. Thus the results of this study agree well with previous results from wild-type rodents.

As well as being suppressed by VA deficiency, immune responses may be suppressed by excessive dietary VA. Sklan et al. [36] assayed a TD antibody response and antigen-specific T cell proliferation in young chicks fed diets supplemented with 0 to 13 200 µg VA/kg diet. Antibody production and T cell proliferation increased progressively with dietary VA up to 6600 µg/kg and then declined slightly. Thus, an inverted U-shaped curve may best describe the relationship of antibody production to dietary VA intake.

Numerous VA supplementation trials have shown convincingly that improving the VA status of children at risk of VA deficiency is effective in reducing mortality and morbidity from infectious disease (see chapter by R. Semba). It would be practical to deliver VA supplements to young children at the time of immunization contact, which may occur during the first few weeks of life. In experimental studies [37], young rats were immunized with tetanus toxoid and treated simultaneously with oral VA in a dose proportional to that used to prevent VA deficiency in children. The animals produced a low but detectable primary anti-tetanus response. The development of antigen-specific memory was demonstrated by a much stronger anti-tetanus response to secondary immunization of 40-day-old rats as compared to age-matched rats immunized for the first time. Retinol-treated and control rats produced equivalent anti-tetanus IgG responses indicating that this level of orally-administered VA neither had an adjuvant-like effect nor was inhibitory to antigen-specific antibody production [37].

Enhancement of antibody responses in VA-depleted animals

Previous studies indicated that even though the antibody response to TD antigens is typically reduced in VA deficiency, retinoid-deficient animals are still able to produce a strong antibody response if other stimuli are provided. High antigen-specific antibody responses were induced in VA-deficient rats by administering LPS along with antigen (tetanus toxoid [38] or pneumo-coccal polysaccharide [39]). Treatment with LPS induced a rapid release of tumor necrosis fac-tor (TNF)-α into plasma, which was nearly equal in VA-deficient and VA-sufficient rats [38]. In other studies, mice lacking TNF-receptor 1 (TNF-R1) produced little IgM and essentially no IgG following immunization [40, 41], implying a positive role for TNF signalling in antibody responses. In VA-deficient rats, the administration of recombinant TNF-α at the time of pri-mary immunization with tetanus toxoid significantly enhanced the primary anti-tetanus IgG response [38]. Moreover, although TNF-α was administered only once, the secondary anti-tetanus IgG response was also markedly elevated. These results illustrate the powerful effect that cytokine stimulation can have during the initial response to antigen, as well as long-last-ing stimulation that became apparent after subsequent antigenic re-challenge. Like LPS, poly-I:C, an inducer of IFNs [42], strongly increased both the primary and the secondary anti-tetanus IgG responses of VA-deficient rats even though poly-I:C was administered only once with the first immunization [43]. Poly-I:C and orally administered RA acted individually and synergized together in VA-deficient rats to increase both primary and secondary anti-tetanus IgG production.

Effects of retinoic acid on virus replication

Besides the role of retinoids in host immune defenses to viral infections (see [6] for review), retinoids may also affect virus replication. Angulo et al. [44] characterized three RA response elements in the promoter region of human cytomegalovirus (hCMV) and demonstrated the necessity for RAR and RXR in the viral promoter's positive response to RA. In contrast, the replication of herpes simplex virus-1 (HSV-1) in cultured Vero cells was inhibited by isomers of RA, but not retinol; inhibition occurred without evident induction of IFN-α or IFN-β gene expression [45]. RA also protected HL-60 and WISH cells from infection with vesicular stom-atitis virus (VSV) [46]; however, in this case RA significantly increased the ability of IFN-α to decrease virus replication. These apparently contrasting effects of RA on hCMV as compared to HSV-1 or VSV replication further illustrate the potential for retinoids to act either positive-ly or negatively in host resistance to viral infection and anti-viral immunity. Retinoid–IFN inter-actions are further discussed later in this chapter.

Retinoids in innate immunity and inflammatory responses

Natural killer (NK) cells

Studies reviewed previously [6, 34] showed that VA deficiency is associated with a reduction in NK cell cytotoxic activity and the number of NK cells. The ability of RA to modulate NK cell number was studied in a VA-deficient rat model. Peripheral blood and spleen cells bearing the NKR-P1A protein, a marker of rat NK cells [47], were reduced in number in VA-deficient

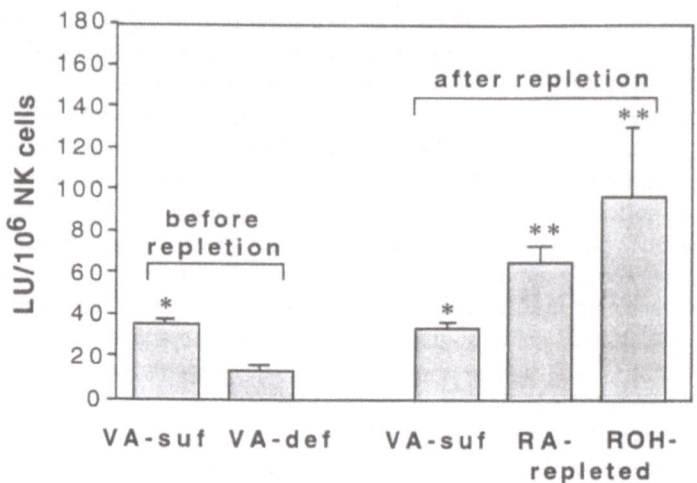

Figure 1. Effects of an 8-day dietary repletion with all-*trans*-retinoic acid or retinol (ROH) on natural killer (NK) cell cytotoxicity in rats previously deficient in vitamin A (VA). Values are means ± SE of NK cell cytolytic activity (lytic units) per 10⁶ NK cells (anti-NKR-P1A⁺ cells by flow cytometry). At the beginning of the repletion period, groups of five VA-deficient and five VA-sufficient rats were killed to confirm the low splenic NK cell activity of VA-deficient rats. Other VA-deficient animals were switched at this time to the RA-containing diet or to the retinol-sufficient diet. Eight days later, splenic NK cell cytotoxic activity and NK cell number were examined in four VA-deficient and five RA- and five ROH-repleted rats. * Significantly different (P < 0.05) from VA-deficient rats. ** Significantly different (P < 0.05) from VA-deficient and VA-sufficient rats (Mann-Whitney U test). From reference [48] with permission.

rats to 30–40% of control values (Fig. 1) [48]. After repletion with RA or retinol *in vivo*, cell number increased and lytic efficiency (lytic activity per NK cell) was also increased, exceeding that of VA-sufficient animals. Since RA-repleted animals still lacked retinol, this *in vivo* result implies that RA is sufficient to support NK cell maturation and function. Oral treatment with 4-hydroxyphenylretinamide (4-HPR) also restored NK cell number and lytic activity in VA-deficient rats [49]. Moreover, even normal rats responded to RA and 4-HPR with a significant increase in NK cell cytotoxic activity [49]. Because NK cells are important producers of immunomodulatory cytokines (e.g. IFN-γ and TNF-α) and are capable of directly lysing tumor target cells, the ability of RA and 4-HPR to increase NK cell number and enhance lytic activity may contribute to the antitumorigenicity that has been observed for these retinoids in numerous chemoprevention experiments. The regulation of NK cell number seems to be especially sensitive to VA status: the number of NK cells was low in both young and old rats fed a diet low in VA that produced chronic marginal VA status (reduced plasma retinol but no clinical signs or symptoms of VA deficiency), and was elevated in rats fed a VA-supplemented diet [50].

Lymphokine-activated killer (LAK) cells can be derived from NK cells or T lymphocytes by incubating them with IL-2. LAK activity was increased in the presence of all-*trans*-RA along with expression of TNF, IFN-γ, and enhanced expression of the IL-2R β subunit [51].

The greater cytotoxic activity of LAK cells compared to NK or unstimulated cytotoxic T lymphocytes may contribute to the antitumor activity of RA and various retinoids.

Regulation of proinflammatory cytokines

RA may regulate the expression of proinflammatory cytokines or their receptors in myeloid and monocytic cell lines, increase phagocytic activity and potentiate the activity of IL-1 [2, 52]. Retinoic acid enhanced the expression of IL-1β protein and mRNA in activated THP-1 myeloid leukemia cells and human monocytes [53]. A study of the regulation of IL-1β, IL-1-α and TNF-α by RA [54] demonstrated biphasic induction kinetics of the mRNAs for IL-1β and TNF-α. The increase in IL-1β mRNA depended on gene transcription and was sensitive to RA dose. This study provided novel evidence that RA can regulate the processing of IL-1β pre-mRNA into mature transcripts. Besides the demonstration that RA can regulate TNF-α gene expression [54], it has been shown that TNF-α can repress the activity of the RXR-β gene promoter [55]. Moreover, in cultured cells, RA negatively regulated TNF-α binding due to down-regulation of TNF- receptors 1 and 2 [56]. The potential for very complex interactions between RA and TNF signalling was further demonstrated by their synergistic activation of IL-8 gene transcription in melanoma cells [57]. Although the IL-8 gene promoter does not contain a classical RA response element, RA affected IL-8 gene transcription indirectly through changes in the composition and binding of a nuclear factor (NF)-κ B complex.

Retinoids can also modulate the production of TNF-α and nitric oxide (NO) by macrophages. TNF release was strongly induced in murine peritoneal macrophages by LPS and IFN-γ but was abrogated dose-dependently by RA, and less potently by 4-hydroxy-RA, 13-*cis*-RA, and retinol [52]. All-*trans*-RA also reduced, but did not abolish, NO production in LPS + IFN-γ-stimulated macrophages [52]. However, when murine RAW264.7 macrophage cells were pretreated with all-*trans*-RA, more NO was produced in response to stimulation with low concentrations of IFN-γ or LPS [58].

Retinoid-interferon interactions

Both retinoids and IFNs have antiproliferative activity in normal and malignant cells as seen in clinical studies. The clinical use of retinoid-IFN combinations is reviewed in this book by R. Lotan. Molecular studies have partially elucidated how retinoids can facilitate IFN's biological activities through up-regulation of transcription factors involved in the signalling pathways of type I and type II IFNs. Matikainen et al. [59–61] demonstrated the ability of RA to rapidly induce the expression of mRNA for IFN regulatory protein (IRF)-1 in NB4, U-937 and THP-1 cell lines and primary monocyte-macrophages. IRF-1 functions as a positive transcription factor for the IFN-β gene and other genes inducible by IFNs. The mRNA for IRF-2, a related factor thought to oppose IRF-1, also increased although to a lesser extent and somewhat later. IRF-1 gene expression was shown to be a primary transcriptional response to RA that did not require new protein synthesis; indeed, IRF-1 mRNA was superinduced in cells treated with both RA and cycloheximide. RA and IFN-γ together enhanced IRF-1 gene expression synergistically in NB4 cells [60]. The mRNAs for two constitutively expressed transcription factors of the STAT (signal transducer and activator of transcription) family were also increased by all-*trans*-RA, although induction was later for STAT than for IRF-1. However, Pelicano et al. [46] provided evidence that RA can stimulate IRF-1 gene expression directly through an IFN-γ-activated sequence (GAS) motif in the IRF-1 promoter, since the putative RA response motif could be removed and promoter activation by RA still retained. The promoter of the murine STAT-1 contains an RA response element that was activated by binding RAR-RXR heterodimers [62], whereas the effect of RA on Stat1, Stat2 and Stat3 protein expression in three cell lines (HL-60, WISH and HeLa) varied from a strong induction (Stat1 in HL-60) to no change in HeLa cells [46]. Thus, predictions can not yet be made of the RA inducibility of IRF-1 and STAT genes and proteins during lymphopoiesis and immune responses. These factors are of considerable interest with respect to immune responses because IRF-1 and Stat-1 proteins modulate the response of several genes mediating cytokine responses [46], and mice lacking genes for IRF-1 and STAT have been shown to be defective in NK and T cell maturation [63] and signalling in response to type I and type II IFNs [64], respectively. Some of the mechanisms through which RA and IFNs potentially interact are summarized in Table 1. A working model [46] of the effects of RA on IFN-stimulated genes in various cell types could involve: the induction of IRF-1 and STAT proteins by RA, the reciprocal induction of IFN-α and IRF-1, and the ability of IRF-1 and STAT factors to regulate the expression of various IFN-regulated proteins that are thought to be the proximal effectors of IFN's biological activities. If RA up-regulates expression of the gene for IFN-γ, as suggested by promoter studies [17], a mechanism could exist for the heightened amplification of IFN-γ signalling with RA increasing both the IFN-γ ligand and key elements of the intracellular IFN signal transduction pathway.

Table 1. Potential interactions between interferon and retinoid signalling pathways in immune function

Direct effects of RA on viral replication (enhancing or inhibiting), modulating virus-induced IFN production	[44–46]
Up-regulation of IRF-1 gene transcription by RA	[46, 60]
Induction of type I IFN gene transcription by IRF-1	[65]
Up-regulation of IFN-γ gene transcription by RA	[17]
Increased activation of Stat proteins by IFNs following IFN binding to their cell-surface receptors	[46, 64]
Induction of STAT-1 and/or STAT-2 genes by RA	[46, 61]
Cooperation of RA, IFN-γ and IL-4 in inducing polymeric IgA transporter expression	[28]

IFN, interferon; IL, interleukin; IRF, interferon regulatory factor; RA, retinoic acid; STAT, signal transducing activator of transcription.

Summary

Recent cellular and molecular studies have pointed to numerous mechanisms by which VA and related retinoids can potentially regulate the immune system. Evidence exists for retinoid-modulation of antigenic activation of T lymphocytes, activation of monocytes, and B cell differentiation. Retinoids are likely to be involved in a direct manner in regulating the production of some of the cytokines, cytokine receptors, and transcription factors involved in cytokine signalling. Due to the particularities and complexity of immune responses that are elicited by natural pathogens, and differences in host susceptibility due to genetics and environment, including nutrition, detailed empirical studies in models of natural infection will be required to truly understand the physiological role(s) of VA and retinoids in the immune system.

Acknowledgements
Funds from NIH grant DK-41479 and the Howard Heinz Endowment are gratefully acknowledged.

References

1   Green HN, Mellanby E (1928) Vitamin A as an anti-infective agent. *Brit Med J* 2: 691–696
2   Yamamoto M (1991) Retinoids in the host defense system. *World Rev Nutr Diet* 64: 58–84
3   Ross AC (1992) Vitamin A status: relationship to immunity and the antibody response. *Proc Soc Exp Biol Med* 200: 303–330
4   Ross AC, Hämmerling UG (1994) Retinoids and the immune system. *In*: MB Sporn, AB Roberts, DS Goodman (eds): *The Retinoids: Biology, Chemistry and Medicine.* Raven Press, New York, 521–544
5   Ross AC (1992) Vitamin A and protective immunity. *Nutr Today* 27: 18–26
6   Ross AC, Stephensen CB (1996) Vitamin A and retinoids in antiviral responses. *FASEB J* 10: 979–985
7   Ross AC (1996) The relationship between immunocompetence and vitamin A status. *In*: A Sommer, KP West Jr (eds): *Vitamin A deficiency: health, survival, and vision.* Oxford University Press, New York, 251–273
8   Zamorano J, Wang HY, Wang R, Shi Y, Longmore GD, Keegan AD (1998) Regulation of cell growth by IL-2: Role of STAT5 in protection from apoptosis but not in cell cycle progression. *J Immunol* 160: 3502–3512
9   Garbe A, Buck J, Hämmerling U (1992) Retinoids are important cofactors in T cell activation. *J Exp Med* 176: 109–117
10 Allende LM, Corell A, Madroño R, Góngora R, Rodríguez-Gallego C, López-Goyanes A, Rosal M, Arnaiz-Villena A (1997) Retinol (vitamin A) is a cofactor in CD3-induced human T-lymphocyte activation. *Immunology* 90: 388–396
11 Buck J, Grün F, Derguini F, Chen Y, Kimura S, Noy N, Hämmerling U (1993) Anhydroretinol: a naturally occurring inhibitor of lymphocyte physiology. *J Exp Med* 178: 675–680
12 Lessard M, Dupuis M (1994) Differential modulation of chicken lymphocyte blastogenesis and cytotoxic activity of natural killer cells *in vitro* by retinol, retinoic acid and β-carotene. *Nutr Res* 14: 1201–1217
13 Szondy Z, Reichert U, Bernardon JM, Michel S, Tóth R, Karászi E, Fésüs L (1998) Inhibition of activation-induced apoptosis of thymocytes by all-*trans*- and 9-*cis*-retinoic acid is mediated via retinoic acid receptor α. *Biochem J* 331: 767–774
14 Cantorna MT, Nashold FE, Hayes CE (1995) Vitamin A deficiency results in a priming environment conducive for Th1 cell development. *Eur J Immunol* 25: 1673–1679
15 Vogel L, Pike BL (1995) Interferon-γ downregulates the proliferative response of hapten-specific B cells stimulated by antigen and cytokines. *Immun Cell Biol* 73: 52–56
16 Cantorna MT, Nashold FE, Hayes CE (1994) In vitamin A deficiency multiple mechanisms establish a regulatory T helper cell imbalance with excess Th1 and insufficient Th2 function. *J Immunol* 152: 1515–1522
17 Cippitelli M, Ye J, Viggiano V, Sica A, Ghosh P, Gulino A, Santoni A, Young HA (1996) Retinoic acid-induced transcriptional modulation of the human interferon-γ promoter. *J Biol Chem* 271: 26 783–26 793
18 Bhatti L, Sidell N (1994) Transcriptional regulation by retinoic acid of interleukin-2α receptors in human B cells. *Immunology* 81: 273–279
19 Ballow M, Xiang SN, Greenberg SJ, Brodsky L, Allen C, Rich G (1997) Retinoic acid-induced modulation of IL-2 mRNA production and IL-2 receptor expression on T cells. *Int Arch Allergy Immunol* 113: 167–169
20 Felli MP, Vacca A, Meco D, Screpanti I, Farina AR, Maroder M, Martinotti S, Petrageli E, Frati L, Gulino A (1994) Retinoic acid-induced down-regulation of the interleukin-2 promoter via *cis*-regulatory sequences containing an octamer motif. *Mol Cell Biol* 11: 4771–4778
21 Ballow M, Wang WP, Xiang SN (1996) Modulation of B-cell immunoglobulin synthesis by retinoic acid. *Clin Immunol Immunopathol* 80:S73–S81
22 Israel H, Odziemiec C, Ballow M (1991) The effects of retinoic acid on immunoglobulin synthesis by human cord blood mononuclear cells. *Clin Immunol Immunopathol* 59: 417–425
23 Ballow M, Xiang S, Wang W, Brodsky L (1996) The effects of retinoic acid on immunoglobulin synthesis: role of interleukin 6. *J Clin Immunol* 16: 171–179

24 Worm M, Krah JM, Manz RA, Henz BM (1998) Retinoic acid inhibits CD40 plus interleukin-4-mediated IgE production *in vitro*. *Blood* 92: 1713–1720

25 Stephensen CB, Blount SR, Schoeb TR, Park JY (1993) Vitamin A deficiency impairs some aspects of the host response to influenza A virus infection in BALB/c mice. *J Nutr* 123: 823–833

26 Gangopadhyay NN, Moldoveanu Z, Stephensen CB (1996) Vitamin A deficiency has different effects on immunoglobulin A production and transport during influenza A infection in BALB/c mice. *J Nutr* 126: 2960–2967

27 Kinoshita M, Ross AC (1993) Vitamin A status and immunoglobulin G subclasses in rats immunized with tetanus toxoid. *FASEB J* 7: 1277–1282

28 Sarkar J, Gangopadhyay NN, Moldoveanu Z, Mestecky J, Stephensen CB (1998) Vitamin A is required for regulation of polymeric immunoglobulin receptor (pIgR) expression by interleukin-4 and interferon-γ in a human intestinal epithelial cell line. *J Nutr* 128: 1063–1069

29 Tokuyama H, Tokuyama Y (1993) Retinoids enhance IgA production by lipopolysaccharide-stimulated murine spleen cells. *Cell Immunol* 150: 353–363

30 Tokuyama Y, Tokuyama H (1994) Retinoic acid and steroid hormones regulate IgA production by LPS-stimulated murine spleen cells. *Immunopharmacology* 28: 145–151

31 Tokuyama H, Tokuyama Y (1995) Endogenous cytokine expression profiles in retinoic acid-induced IgA production by LPS-stimulated murine splenocytes. *Cell Immunol* 166: 247–253

32 Kuwabara K, Shudo K, Hori Y (1996) Novel synthetic retinoic acid inhibits rat collagen arthritis and differentially affects serum immunoglobulin subclass levels. *FEBS Lett* 378: 153–156

33 Tokuyama H, Tokuyama Y, Nakanishi K (1995) Retinoids inhibit IL-4-dependent IgE and IgG1 production by LPS-stimulated murine splenic B cells. *Cell Immunol* 162: 153–158

34 Ross AC, Zhao Z, Arora D, Pasatiempo AMG, Kinoshita M, Gardner EM, Sri Kantha S, Taylor CE (1994) Retinoids in specific and nonspecific immunity: studies on antibody production and natural killer cells. *In*: MA Livrea, G Vidali (eds): *Retinoids: from basic science to clinical applications*. Birkhäuser, Basel, 197–213

35 Molrine DC, Polk DB, Ciamarra A, Phillips N, Ambrosino DM (1995) Impaired human responses to tetanus toxoid in vitamin A-deficient SCID mice reconstituted with human peripheral blood lymphocytes. *Infect Immunity* 63: 2867–2872

36 Sklan D, Melamed D, Friedman A (1994) The effect of varying levels of dietary vitamin A on immune response in the chick. *Poultry Sci* 73: 843–847

37 Gardner EM, Ross AC (1995) Immunologic memory is established in nursling rats immunized with tetanus toxoid, but is not affected by concurrent supplementation with vitamin A. *Amer J Clin Nutr* 62: 1007–1012

38 Arora D, Ross AC (1994) Antibody response against tetanus toxoid is enhanced by lipopolysaccharide or tumor necrosis factor-α in vitamin A-sufficient and -deficient rats. *Amer J Clin Nutr* 59: 922–928

39 Pasatiempo AMG, Kinoshita M, Foulke DT, Ross AC (1994) The antibody response of vitamin A-deficient rats to pneumococcal polysaccharide is enhanced through co-immunization with lipopolysaccharide. *J Infect Dis* 169: 441–444

40 Le Hir M, Bluethmann H, Kosco-Vilbois MH, Müller M, Di Padova F, Moore M, Ryffel B, Eugster HP (1996) Tumor necrosis factor receptor-1 signalling is required for differentiation of follicular dendritic cells, germinal center formation, and full antibody responses. *J Inflamm* 47: 76–80

41 Pasparakis M, Alexopoulou L, Episkopou V, Kollias G (1996) Immune and inflammatory responses in TNFα-deficient mice: A critical requirement for TNFα in the formation of primary B cell follicles, follicular dendritic cell networks and germinal centers, and in the maturation of the humoral immune response. *J Exp Med* 184: 1397–1411

42 Zhao Z, Murasko DM, Ross AC (1994) The role of vitamin A in natural killer cell cytotoxicity, number and activation in the rat. *Nat Immun* 13: 29–41

43 DeCicco KL, Zolfaghari R, Ross AC (1999) IL-2Rβ, IRF-1, and STAT-1 mRNA are reduced in vitamin A deficiency but recovered by treatment with retinoic acid (RA) and poly-I:C-L,C (PIC) *in vivo*. *FASEB J* 13: Abstract 674.26

44 Angulo A, Suto C, Heyman RA, Ghazal P (1996) Characterization of the sequences of the human cytomegalovirus enhancer that mediate differential regulation by natural and synthetic retinoids. *Mol Endocrinol* 10: 781–793

45 Isaacs CE, Kascsak R, Pullarkat RK, Xu W, Schneidman K (1997) Inhibition of herpes simplex virus replication by retinoic acid. *Antivir Res* 33: 117–127

46 Pelicano L, Li FS, Schindler C, Chelbi-Alix MK (1997) Retinoic acid enhances the expression of interferon-induced proteins: evidence for multiple mechanisms of action. *Oncogene* 15: 2349–2359

47 Chambers WH, Adamkiewicz T, Houchins JP (1993) Type II integral membrane proteins with characteristics of C-type animal lectins expressed by natural killer (NK) cells. *Glycobiology* 3: 9–14

48 Zhao Z, Ross AC (1995) Retinoic acid repletion restores the number of leukocytes and their subsets and stimulates natural cytotoxicity in vitamin A-deficient rats. *J Nutr* 125: 2064–2073

49 Zhao Z, Matsuura T, Popoff K, Ross AC (1994) Effects of N-(4-hydroxyphenyl) retinamide on the number and cytotoxicity of natural killer cells in vitamin A-sufficient and -deficient rats. *Nat Immun* 13: 280–288

50 Dawson HD, Li N-Q, Ross AC (1997) Differential effects of chronic vitamin A status on the age-related changes in natural killer (NK) cell number and cytotoxic function. *FASEB J* 11:A818

51 Fegan C, Bailey-Wood R, Coleman S, Phillips SA, Neale L, Hoy T, Whittaker JA (1995) All-*trans* retinoic acid enhances human LAK activity. *Eur J Haematologia* 54: 95–100
52 Mehta K, McQueen T, Tucker S, Pandita R, Aggarwal BB (1994) Inhibition by all-*trans*-retinoic acid of tumor necrosis factor and nitric oxide production by peritoneal macrophages. *J Leukocyte Biol* 55: 336–342
53 Matikainen S, Serkkola E, Hurme M (1991) Retinoic acid enhances IL-1β expression in myeloid leukemia cells and in human monocytes. *J Immunol* 147: 162–167
54 Jarrous N, Kaempfer R (1994) Induction of human interleukin-1 gene expression by retinoic acid and its regulation at processing of precursor transcripts. *J Biol Chem* 269: 23 141–23 149
55 Sugawara A, Uruno A, Nagata T, Taketo MM, Takeuchi K, Ito S (1998) Characterization of mouse retinoid X receptor (RXR)-β gene promoter: negative regulation by tumor necrosis factor (TNF)-α. *Endocrinology* 139: 3030–3033
56 Totpal K, Chaturvedi MM, LaPushin R, Aggarwal BB (1995) Retinoids downregulate both p60 and p80 forms of tumor necrosis factor receptors in human histiocytic lymphoma U-937 cells. *Blood* 85: 3547–3555
57 Harant H, de Martin R, Andrew PJ, Foglar E, Dittrich C, Lindley IJD (1996) Synergystic activation of interleukin-8 gene transcription by all-*trans*-retinoic acid and tumor necrosis factor-α involves the transcription factor NF-kB. *J Biol Chem* 271: 26 954–26 961
58 Austenaa LMI, Ross AC (1999) Retinoic acid causes an increase in the production of nitric oxide from the murine macrophage cell line RAW264.7 during stimulation with interferon-γ. *FASEB J* 13: Abstract 674.25
59 Matikainen S, Ronni T, Lehtonen A, Sareneva T, Melén K, Nordling S, Levy DE, Julkunen I (1997) Retinoic acid induces signal transducer and activator of transcription (STAT) 1, STAT2, and p48 expression in myeloid leukemia cells and enhances their responsiveness to interferons. *Cell Growth Differ* 8: 687–698
60 Matikainen S, Ronni T, Hurme M, Pine R, Julkunen I (1996) Retinoic acid activates interferon regulatory factor-1 gene expression in myeloid cells. *Blood* 88: 114–123
61 Matikainen S, Lehtonen A, Sareneva T, Julkunen I (1998) Regulation of IRF and STAT gene expression by retinoic acid. *Leuk Lymphoma* 30: 63–71
62 Weihua X, Kolla V, Kalvakolanu DV (1997) Modulation of interferon action by retinoids. *J Biol Chem* 272: 9742–9748
63 Ohteki T, Yoshida H, Matsuyama T, Duncan GS, Mak TW, Ohashi PS (1998) The transcription factor interferon regulatory factor 1 (IRF-1) is important during the maturation of natural killer 1.1[+] T cell receptor-α/β[+] (NK1[+] T) cells, natural killer cells, and intestinal intraepithelial T cells. *J Exp Med* 187: 967–972
64 Leonard WJ, O'Shea JJ (1998) Jaks and STATS: Biological implications. *Annu Rev Immunol* 16: 293–322
65 Kimura T, Nakayama K, Penninger J, Kitagawa M, Harada H, Matsuyama T, Tanaka N, Kamijo R, Vilcek J, Mak TW et al (1994) Involvement of the IRF-1 transcription factor in antiviral responses to interferons. *Science* 264: 1921–1924

Vitamin A and infectious diseases

R.D. Semba

Johns Hopkins University School of Medicine, Baltimore, Maryland, USA

Introduction

One of the most important therapeutic applications of vitamin A is its use in reducing the morbidity and mortality of infectious diseases. It is estimated that over 100 million preschool children and a large proportion of women of reproductive age suffer from clinical and subclinical vitamin A deficiency worldwide [1]. A recent series of large, randomized, controlled clinical trials conducted in the last two decades has shown that vitamin A supplementation can reduce child mortality in developing countries by 30–50% [2]. Vitamin A supplementation is now part of public health programs to improve child survival in many developing countries, and high dose vitamin A supplementation is recommended therapy for acute measles infection, both in industrialized and developing countries [3]. Vitamin A supplementation is recognized for being as important as vaccines as a public health intervention aimed at saving children's lives [4].

The recognition that vitamin A could be used as a treatment for infections dates to antiquity. Fish liver oils, a potent source of vitamin A, were used as a treatment for infections in Greek and Roman medicine. Experiments in the 1920 s suggested that vitamin A-deficient animals were more susceptible to infections, and Green and Mellanby dubbed vitamin A the "anti-infective" vitamin in 1928 [5]. At least thirty trials were conducted between 1920 and 1940 which examined the use of vitamin A as a therapy for a wide variety of infections in humans. Although these early trials were encouraging, the emergence of sulfa antibiotics in the late 1930 s and the disappearance of malnutrition in industrialized countries led to a caesura in research on vitamin A as an anti-infective therapy. The description of an association between mild vitamin A deficiency and increased child mortality renewed interest in vitamin A [6], and further controlled clinical trials were conducted in the 1980 s which provided more definitive evidence for vitamin A as a public health intervention [2].

Vitamin A and immunity

Vitamin A appears to play a role in modulating a wide spectrum of immune responses, including T and B lymphocyte function, generation of antibody responses, function of other immune effector cells such as monocytes, neutrophils, and Langerhans cells, and maintenance of mucosal immunity. The role of vitamin A in immune function has been reviewed in more detail elsewhere [7–9]. Protective immunity to infectious diseases varies widely, and the role of vitamin A in enhancing immunity to different infections remains to be elucidated. For example, in measles infection, T-helper type 2-like responses play a stronger role in protective immunity, whereas in tuberculosis, T-helper type 1-like responses are generally considered to be more important. Although clinical trials in humans have provided evidence that vitamin A enhances immunity, much less is known about how vitamin A is modulating immunity on the cellular and molecular level during different types of infections.

Vitamin A and infections

Epidemiological studies and animal models support the idea that vitamin A deficiency and infection are related in a "vicious cycle" [3, 8]. During an episode of infection, anorexia, malabsorption, increased utilization of vitamin A, and accelerated urinary loss of vitamin A can increase the risk of developing a vitamin A deficiency [3, 10]. Vitamin A deficiency can cause pathological alterations of mucosal epithelia of the respiratory, gastrointestinal, and genitourinary tracts, as well as functional defects in immunity, thus increasing the risk of developing more severe infections. Vitamin A deficiency and infection are closely synergistic, often occurring together [3]. Among children, the morbidity of vitamin A deficiency and infections is often expressed as alterations in growth, such as stunting and/or wasting [11].

Whether vitamin A deficiency occurs in populations with infections usually requires the consideration of different indicators of nutritional status, including the prevalence of clinical vitamin A deficiency (i.e. night blindness, Bitot's spots) in the population, dietary assessment, socioeconomic status, the prevalence of malnutrition, a population-based frequency distribution of plasma retinol levels, dark adaptometry, modified relative dose response, and breastmilk vitamin A concentrations [12]. The acute phase response may depress plasma retinol levels, and the amount of decrease is probably related to the plasma retinol level before the infection, the type and severity of infection, and the degree of malnutrition. Acute phase response proteins, such as C-reactive protein or α-1-glycoprotein, may indicate the presence of an acute phase response [13], but do not distinguish low retinol levels due to vitamin A deficiency from low retinol levels due to the acute phase response. Both the acute phase response and vitamin A deficiency can occur simultaneously [14]. The retinol-binding protein: transthyretin (RBP:TTR) ratio may be useful in distinguishing low retinol levels due to vitamin A deficiency from low retinol levels due to the acute phase response [15], but further work is needed to establish the validity of this method. Ultimately, policy recommendations regarding vitamin A supplementation are generally based upon evidence provided by controlled clinical trials and the population involved. For example, high dose vitamin A supplementation is recommended by the American Academy of Pediatrics for acute measles in the United States [16] and the Australian Academy of Pediatrics [17], based upon general nutritional status.

Vitamin A has been evaluated as a therapy for infectious diseases largely through controlled clinical trials (Tabs. 1 and 2), and several factors must be considered in the interpretation of these studies. It is important to distinguish between the *incidence* versus the *severity* of infections as outcomes of such studies. For example, some studies have been conducted under the premise that vitamin A supplementation will prevent infection altogether, i.e. alter the *inci-*

Table 1. Large community-based controlled clinical trials of vitamin A, 1986–1998

Location	Age (months)	n	Dose*	Result	Reference
Child mortality					
Indonesia	1–60	25 939	60 mg RE q 6 months	34% ↓ mortality	[18]
Indonesia	1–60	11 000	vit A-fortified MSG	45% ↓ mortality	[19]
India	1–72	15 419	weekly RDA	54% ↓ mortality	[20]
India	12–60	15 775	60 mg RE q 6 months	6% ↓ mortality	[21]
India	1–72	15 247	30 mg RE q 4 months	17% ↓ mortality	[22]
Nepal	6–72	28 630	60 mg RE q 4 months	30% ↓ mortality	[23]
Nepal	1–60	7 197	60 mg RE once	29% ↓ mortality	[24]
Sudan	9–72	28 753	60 mg RE q 6 months	6% ↑ mortality	[25]
Ghana	6–90	21 906	60 mg RE q 4 months	19% ↓ mortality	[26]
Child morbidity					
Brazil	6–48	1 240	60 mg RE q 4 months	↓ morbidity diarrhea	[32]
Indonesia	6–47	1 407	60 mg RE q 4 months	no impact, diarrhea	[33]

*Dose given for children > 12 months. Most trials had age-adjusted vitamin A doses for children < 12 months. MSG, monosodium glutamate; RE, retinol equivalent; RDA, recommended dietary allowance; q 4 months, every 4 months

dence, while others have examined the impact of vitamin A supplementation on the duration of disease and case-fatality rates, or *severity*. The age of the subjects, the prevalence of vitamin A deficiency in the subjects and community in general, the type of infections involved, and the level of ancillary medical support need to be considered. Some clinical trials have been conducted with sample sizes which are so limited as to have insufficient statistical power to examine major clinical outcomes. Finally, the doses and frequency of dosing of vitamin A have varied widely between clinical trials.

Child survival

A large series of community-based clinical trials has addressed the use of periodic high dose vitamin A supplementation, weekly vitamin A supplementation, or vitamin A fortification in reducing preschool child mortality in developing countries [18–26]. The first study, conducted in Aceh, Indonesia, by Sommer and colleagues [18], suggested that periodic, high dose vitamin A supplementation reduces preschool child mortality by 34%. Other studies soon followed which attempted to confirm these observations among preschool children in other populations in southeast Asia, south Asia, and Africa [19–26]. Most of these studies confirmed that vitamin A reduces child mortality [19, 20, 22–24, 26], and two studies showed no statistically significant effect [21, 25]. A meta-analysis of these child survival studies showed that vitamin A intervention reduces child mortality by about one third [2]. These large community trials have led to major program and policy recommendations of periodic vitamin A supplementation for preschool children in developing countries around the world [1].

Table 2. Other controlled clinical trials of vitamin A, 1986–1998

Location	Age (months)	n	Dose*	Result	Reference
Diarrheal disease					
China	6–36	172	60 mg RE twice	↓ incidence of diarrhea	[34]
India	12–71	174	60 mg RE once	↓ duration of diarrhea	[35]
Brazil	unspecified	25	60 mg RE once	↓ duration of diarrhea	[36]
Bangladesh	12–60	83	60 mg RE once	no effect	[37]
India	6–60	216	60 mg RE once	↓ duration of diarrhea	[38]
Bangladesh	12–84	83	60 mg RE once	↓ morbidity shigellosis	[39]
Acute respiratory infections (ARI)					
Bangladesh	2.5	165	15 mg RE thrice	↓ duration of disease	[42]
Guatemala	3–48	263	60 mg RE once	no effect	[43]
Brazil	6–59	472	60 mg RE twice	no effect	[44]
Tanzania	6–60	687	60 mg RE twice	no effect	[45]
Peru	3–120	95	60 mg RE twice	↑ morbidity	[46]
Respiratory syncytial virus infection					
Chile	1–72	180	60 mg RE once	no effect	[47]
USA	1–72	239	60 mg RE once	no effect	[48]
USA	2–58	32	30 mg RE once	no effect	[49]
Measles					
Tanzania	9–60	180	60 mg RE twice	↓ mortality	[56]
South Africa	6–24	189	60 mg RE twice	↓ mortality	[57]
South Africa	4–24	60	60 mg RE thrice	↓ morbidity	[58]
Kenya	<60	294	60 mg RE once	↓ morbidity	[59]
Zambia	6–120	200	60 mg RE once	no effect	[60]
Malaria					
Papua New Guinea	6–72	484	60 mg RE q 3 mos	↓ morbidity	[64]
Tuberculosis					
South Africa	3–152	85	60 mg RE twice	no effect	[65]
HIV infection					
South Africa	infants	118	periodic high doses	↓ morbidity	[68]
USA	adults	120	60 mg RE once	no effect HIV load	[69]
Tanzania	pregnant women	1075	daily dose	no effect birth outcomes	[70]
Tanzania	6–60	72	60 mg RE twice	↓ mortality	[71]
Maternal mortality					
Nepal	pregnant women	19 131	7.5 mg RE q week	30% ↓ mortality	[72]
Infant morbidity and mortality					
3 countries	infants/mothers	9424	periodic high doses	no effect on mortality	[52]
Indonesia	infants	2067	15 mg RE at birth	↓ mortality	[73]
India	infants/mothers	909	60 mg RE at birth	no effect on morbidity	[74]
Infections in elderly					
USA	elderly adults	109	60 mg RE once	no effect	[75]

*Dose given for children > 12 months. Most trials had age-adjusted vitamin A doses for children < 12 months.

Diarrheal disease

Worldwide, diarrheal disease is estimated to account for 19% of children's deaths and a heavy burden of morbidity [1]. Children under age five have the highest risk for diarrhea and malnutrition. In developing countries, the main pathogens involved in diarrheal diseases among children are rotavirus, *E. coli*, and *Shigella*, and a significant proportion of disease is due to *Vibrio cholerae*, *Salmonella*, and *Entamoeba histolytica*. A high prevalence of clinical vitamin A deficiency (nightblindness and xerophthalmia) has been described among children with diarrheal disease [27–30]. Children with preexisting xerophthalmia are at a higher risk of having a subsequent episode of diarrheal disease [31]. Recent large community-based clinical trials of vitamin A supplementation in Tamil Nadu, Nepal, and Ghana show that vitamin A has a major impact upon the overall mortality of diarrheal disease [3].

Whether vitamin A supplementation reduces morbidity and mortality from all types of diarrheal disease or has more pathogen-specific effects is unclear. In Brazil, a community-based clinical trial showed that vitamin A supplementation reduced both the incidence and severity of diarrheal disease [32], but another trial in central Java, Indonesia, did not show an effect of vitamin A on incidence of diarrheal disease [33]. Three smaller, community-based trials suggest that vitamin A may reduce the incidence and/or duration of diarrheal disease [34–36].

Some clinical trials have specifically addressed the use of vitamin A as adjunct therapy for children presenting with diarrheal disease [37–39]. Vitamin A supplementation had no apparent effect upon the duration of acute, watery, non-cholera, diarrhea in Bangladesh [37]. The study may have been limited in detecting any effect of vitamin A in reducing duration of diarrhea, as the average episode of diarrhea lasted two days. A hospital-based study in New Delhi was suggestive that vitamin A supplementation reduced the duration of diarrhea, and the effect of vitamin A was strongest in children with pre-existing vitamin A deficiency [38]. Vitamin A supplementation was shown to reduce the severity of acute shigellosis in children in Dhaka, Bangladesh [39].

Respiratory disease

Worldwide, lower respiratory infections account for 19% of all child deaths [1]. Vitamin A deficiency causes pathological alterations in the mucosal epithelium of the respiratory tract, including keratinization and loss of ciliated cells, mucus, and goblet cells. Epidemiological studies demonstrate that clinical vitamin A deficiency is associated with lower respiratory infections [29, 40]. A large longitudinal study in Indonesia suggested that preschool children with mild xerophthalmia are at an increased risk of having a subsequent episode of respiratory infection [31]. These observations raised two important questions. Community-based clinical trials have attempted to determine whether periodic high dose vitamin A supplementation reduces the incidence and/or severity of acute respiratory infections in children. Hospital-based clinical trials have addressed whether high dose vitamin A supplementation can reduce morbidity when given at the time of an acute attack of respiratory disease.

The community-based trials of vitamin A supplementation did not seem to reduce the incidence of acute lower respiratory infections, although there was an overall beneficial effect in reducing child mortality [41]. One community-based clinical trial addressed the issue of vitamin A supplementation through childhood immunization programs, and vitamin A was shown to reduce the duration of acute respiratory infections among infants [42]. Hospital-based studies have shown that high dose vitamin A supplementation has no therapeutic effect upon the morbidity of acute lower respiratory infections in children [43–46]. In Chile and the USA, hospital-based trials showed that vitamin A supplementation had little impact upon respiratory syncytial virus (RSV) infection among infants and young children [47–49].

Although vitamin A status is related to the severity of acute respiratory infection in children [50], it is unclear why vitamin A therapy has no apparent effect upon the morbidity of acute respiratory infections among preschool children. Young age might be one contributing factor to the lack of an effect, as large community-based studies suggest that vitamin A supplementation has little effect on morbidity and mortality of infants [51, 52]. For example, RSV infection tends to occur primarily among young infants. Other micronutrient deficiencies, such as that of vitamin D and zinc, may have more influential roles in immunity to respiratory infections [53, 54].

Measles

Measles causes an estimated two million deaths per year, and despite measles immunization, periodic and serious outbreaks are occurring due to lapses in immunization programs, vaccine failure, and problems related to optimal timing of immunization. Case-fatality rates during acute, complicated measles infection are often 10–30%, depending on age and nutritional status of the children. Low serum vitamin A levels are associated with higher mortality in acute, complicated measles infection [55]. Randomized, placebo-controlled clinical trials show that vitamin A supplementation can reduce the mortality of measles by 50% or more [56, 57] and reduce the morbidity associated with measles [58, 59]. Although clinical trials show that two consecutive oral doses of 60 mg RE reduces measles mortality, the current WHO/UNICEF recommendations are for a single 60 mg RE dose for children greater than twelve months of age [3]. One clinical trial using a single 60 mg RE dose of vitamin A showed no impact of vitamin A supplementation upon measles morbidity [60]. The exact biological mechanism(s) by which vitamin A reduces the morbidity and mortality of measles infection is uncertain, although studies with live measles virus vaccine suggest that vitamin A supplementation may possibly limit measles virus replication [61].

Malaria

Malaria is currently undergoing a resurgence worldwide. An estimated 400 million individuals are affected with malaria each year with one to two million deaths [62]. Low serum or plasma vitamin A levels have been described during *Plasmodium falciparum* infection [63]. Little attention has been paid to the nutritional status of those with malaria, as traditional strategies have focused on vector control and anti-malarial drugs. A recent randomized, placebo-controlled clinical trial conducted in Papua New Guinea suggests that vitamin A supplementation, 60 mg RE every three months, can reduce malarial morbidity in preschool children by about one third [64].

This study suggests that vitamin A supplementation might obtain the same impact which has been the goal of new malaria vaccines. But, in contrast, vitamin A supplementation only costs a few cents per year.

Tuberculosis

About 1.8 billion individuals, or about one-third of the world's population, are infected with *Mycobacterium tuberculosis*, and most of these individuals have latent infection. Although malnutrition is a major risk factor for the progression of tuberculosis, tuberculosis control programs have tended to focus upon chemoprophylaxis and chemotherapy alone, rather than upon improvement of host nutritional status. For over one hundred years, cod-liver oil, a rich source of vitamins A and D, was used as a treatment for tuberculosis. The role of nutrition and tuberculosis remains a major area of neglect, despite the promise that micronutrients have shown as therapy for other types of infections and the long record of the use of vitamins A and D for treatment of pulmonary and miliary tuberculosis in both Europe and the United States. A recent clinical trial suggests that high dose vitamin A supplementation does not alter the morbidity of tuberculosis in children [65]. Studies have not been conducted which address the use of multivitamins and minerals or vitamins A plus D as an adjunct therapy for tuberculosis.

Human immunodeficiency virus infection

During HIV infection, microenteropathy, malabsorption, steatorrhea, and anorexia are not uncommon, and low plasma or serum vitamin A levels have been described in all stages of HIV infection and among many different risk groups [66]. Plasma or serum levels or intake of vitamin A has been associated with increased disease progression, mortality, and higher mother-to-child transmission of HIV [66, 67]. Vitamin A supplementation for HIV infection has been explored in a few clinical trials [68–71]. Periodic high dose vitamin A supplementation seems to reduce morbidity among children born to HIV-infected mothers [68], but vitamin A supplementation itself does not appear to influence HIV load in the blood [69]. Multivitamin supplementation to HIV-infected women during pregnancy reduces fetal mortality and low birth weight by about 40%, but vitamin A supplementation alone had no significant impact [70]. High dose vitamin A supplementation appeared to reduce mortality among HIV-infected children who presented with acute lower respiratory infection [71].

Maternal mortality

Maternal mortality rates in some developing countries have been reported to be over one hundred times higher than that in many developed countries, and the traditional thinking has been that these differences are due to the lack of sophisticated, emergency obstetrical care in developing countries. A recent large clinical trial from Nepal suggests that vitamin A status may be an important contributing factor to maternal mortality [72]. Women of childbearing age were supplemented with vitamin A, 7.5 mg RE per week, or the equivalent as β-carotene, and during the study many women became pregnant and continued with supplementation. Vitamin A and/or β-carotene supplementation reduced maternal mortality by one third to one half [72].

Other studies

In Indonesia, a recent study suggests that vitamin A supplementation to neonates at birth may reduce infant mortality from infections by over one half [73]. A single oral supplement of 15 mg RE appeared to have a sustained effect on mortality over the first six months of life, and the possible immune basis for such an effect is unclear. Other studies suggest that vitamin A supplementation does not influence mortality in infants under six months of age [51, 52]. A clinical trial in southern India is currently in progress to replicate the findings of the Indonesian study. Another recent study from Tamil Nadu, South India, suggests that vitamin A supplementation to mother and infant within two weeks of delivery does not influence the incidence and/or duration of diarrheal and respiratory disease [74]. This study is notable for the 60 mg RE dose of vitamin A given to young infants. One study examined the effect of vitamin A supplementation on the incidence of infection among elderly nursing home residents [75]. A single dose of vitamin A, 60 mg RE, had no effect on the incidence of infections in the elderly, but the small sample size and low incidence rates may have limited the statistical power in this study.

Conclusions and recommendations

In the last two decades, over forty clinical trials have examined the use of vitamin A as an intervention to reduce the morbidity and mortality from infectious disease. Large community-based controlled clinical trials have provided persuasive scientific evidence that preschool child morbidity and mortality can be lowered by improving vitamin A status. The impact of vitamin A supplementation upon case fatality rates in measles has been impressive. The World Health Organization, UNICEF, and the International Vitamin A Consultative Group now recommend the use of vitamin A supplementation as a means of improving child survival in developing countries where vitamin A deficiency is a public health problem and for acute complicated measles. These measures are expected to save thousands of children's lives in developing countries.

The recent clinical trials suggest that vitamin A supplementation has more of an impact upon the severity of infections rather than the incidence of infections. In addition, vitamin A appears to have more effect upon the morbidity and mortality of diarrheal rather than respiratory disease. As a specific disease-targeted therapy, high dose vitamin A supplementation seems to have little effect upon acute lower respiratory infections. The reasons for these differences are unclear since vitamin A deficiency is known to affect mucosal epithelia of both the gastrointestinal and respiratory tracts [3]. Vitamin A supplementation also seems to have age-specific effects, as most trials have not shown an impact of vitamin A therapy on morbidity and mortality on infants under six months of age.

Vitamin A is currently under investigation as an "anti-infective" therapy for different infectious diseases, such as HIV infection and malaria. Given the low cost of vitamin A, this therapeutic intervention may be appropriate for health budgets in developing countries. Further studies are needed to address the use of vitamin A in multi-micronutrient supplements, as there is increasing evidence that other coexisting micronutrient deficiencies may limit the efficacy of vitamin A. Despite many recent clinical trials, little has been done to investigate how vitamin A supplementation modulates the immune response and the potential role of immunogenetics in humans. The possible use of other related retinoids in immunomodulation during infection remains to be addressed.

References

1   UNICEF (1998) *The State of the World's Children 1998.* Oxford University Press, Oxford and New York
2   Beaton GH, Martorell R, L'Abbe KA, Edmonston B, McCabe G, Ross AC, Harvey B (1993) *Effectiveness of Vitamin A Supplementation in the Control of Young Child Morbidity and Morality in Developing Countries.* ACC/SCN State-of-the-Art Nutrition Policy Discussion Paper No. 13, United Nations
3   Sommer A, West KP Jr (1996) *Vitamin A Deficiency: Health, Survival, and Vision.* Oxford University Press, New York
4   World Bank (1993) *World Bank Development Report 1993: Investing in Health.* University Press, New York
5   Green HN, Mellanby E (1928) Vitamin A as an anti-infective agent. *Brit Med J* 2: 691–696
6   Sommer A, Tarwotjo I, Hussaini G, Susanto D (1983) Increased mortality in children with mild vitamin A deficiency. *Lancet* 2: 585–588
7   Ross AC (1992) Vitamin A status: relationship to immunity and the antibody response. *Proc Soc Exp Biol Med* 200: 303–320
8   Semba RD (1994) Vitamin A, immunity, and infection. *Clin Infect Dis* 19: 489–499
9   Semba RD (1998) The role of vitamin A and related retinoids in immune function. *Nutr Rev* 56: S38–S48
10 Mitra AK, Alvarez JO, Stephensen CB (1998) Increased urinary retinol loss in children with severe infections. *Lancet* 351: 1033–1034
11 West KP Jr, Djunaedi E, Pandji A, Kusdiono, Tarwotjo I, Sommer A, and the Aceh Study Group (1988) Vitamin A supplementation and growth: a randomized community trial. *Amer J Clin Nutr* 48: 1257–1264
12 Underwood BA, Olson JA (eds) (1993) *A Brief Guide to Current Methods of Assessing Vitamin A Status.* International Vitamin A Consultative Group, Nutrition Foundation, Washington, D.C
13 Filteau SM, Morris SS, Abbott RA, Tomkins AM, Kirkwood BR, Arthur P, Ross DA, Gyapong JO, Raynes JG (1993) Influence of morbidity on serum retinol of children in a community-based study in northern Ghana. *Amer J Clin Nutr* 58: 192–197
14 Christian P, Schulze K, Stoltzfus RJ, West KP Jr (1998) Hyporetinolemia, illness symptoms, and acute phase protein response in pregnant women with and without night blindness. *Amer J Clin Nutr* 67: 1237–1243
15 Rosales FJ, Ross AC (1998) A low molar ratio of retinol binding protein to transthyretin indicates vitamin A deficiency during inflammation: studies in rats and a posteriori analysis of vitamin A-supplemented children with measles. *J Nutr* 128: 1681–1687
16 Committee on Infectious Diseases (1993) Vitamin A treatment of measles. *Pediatrics* 91: 1014–1015
17 Australian College of Paediatrics (1996) Policy statement. Vitamin A supplementation in measles. *J Paediat Child Health* 32: 209–210

18 Sommer A, Tarwotjo I, Djunaedi E, West KP Jr, Loeden AA, Tilden R, Mele L, and the Aceh Study Group (1986) Impact of vitamin A supplementation on childhood mortality. A randomised controlled community trial. *Lancet* 1: 1169–1173

19 Muhilal, Permeisih D, Idjradinata YR, Muherdiyantiningsih, Karyadi D (1988) Vitamin A-fortified monosodium glutamate and health, growth, and survival of children: a controlled field trial. *Amer J Clin Nutr* 48: 1271–1276

20 Rahmathullah L, Underwood BA, Thulasiraj RD, Milton RC, Ramaswamy K, Rahmathullah R, Babu G (1990) Reduced mortality among children in southern India receiving a small weekly dose of vitamin A. *New Engl J Med* 323: 929–935

21 Vijayaraghavan K, Radhaiah G, Prakasam BS, Sarma KVR, Reddy V (1990) Effect of massive dose vitamin A on morbidity and mortality in Indian children. *Lancet* 336: 1342–1345

22 Agarwal DK, Pandey CM, Agarwal KN (1995) Vitamin A administration and preschool child mortality. *Nutr Res* 15: 669–680

23 West KP Jr, Pokhrel RP, Katz J, LeClerq SC, Khatry SK, Shrestha SR, Pradhan EK, Tielsch JM, Pandey MR, Sommer A (1991) Efficacy of vitamin A in reducing preschool child mortality in Nepal. *Lancet* 338: 67–71

24 Daulaire NM, Starbuck ES, Houston RM, Church MS, Stukel TA, Pandey MR (1992) Childhood mortality after a high dose of vitamin A in a high risk population. *Brit J Med* 304: 207–210

25 Herrera MG, Nestel P, El Amin A, Fawzi WW, Mohamed KA, Weld L (1992) Vitamin A supplementation and child survival. *Lancet* 340: 267–271

26 Ghana VAST Study Team (1993) Vitamin A supplementation in northern Ghana: effects on clinic attendances, hospital admissions, and child mortality. *Lancet* 342: 7–12

27 Brilliant LB, Pokhrel RP, Grasset NC, Lepkowski JM, Kolstad A, Hawks W, Pararajasegaram R, Brilliant GE, Gilbert S, Shrestha SR et al (1985) Epidemiology of blindness in Nepal. *Bull WHO* 63: 375–386

28 Gujiral S, Abbi R, Gopaldas T (1993) Xerophthalmia, vitamin A supplementation and morbidity in children. *J Trop Pediat* 39: 89–92

29 DeSole G, Belay Y, Zegeye B (1987) Vitamin A deficiency in southern Ethiopia. *Amer J Clin Nutr* 45: 780–784

30 Schaumberg DA, O'Connor J, Semba RD (1996) Risk factors for xerophthalmia in the Republic of Kiribati. *Eur J Clin Nutr* 50: 761–764

31 Sommer A, Katz J, Tarwotjo I (1984) Increased risk of respiratory disease and diarrhea in children with pre-existing mild vitamin A deficiency. *Amer J Clin Nutr* 40: 1090–1095

32 Barreto ML, Santos LMP, Assis AMO, Purificação M, Araújo N, Farenzena GG, Santos PAB, Fiaccone RL (1994) Effect of vitamin A supplementation on diarrhoea and acute lower-respiratory-tract infections in young children in Brazil. *Lancet* 344: 228–231

33 Dibley MJ, Sadjimin T, Kjolhede CL, Moulton LH (1996) Vitamin A supplementation fails to reduce incidence of acute respiratory illness and diarrhea in preschool-age Indonesian children. *J Nutr* 126: 434–442

34 Lie C, Ying C, Wang EL, Brun T, Geissler C (1993) Impact of large dose vitamin A supplementation on childhood diarrhoea, respiratory disease and growth. *Eur J Clin Nutr* 47: 88–96

35 Biswas R, Biswas AB, Manna B, Bhattacharya SK, Dey R, Sarkar S (1994) Effect of vitamin A supplementation on diarrhoea and acute respiratory tract infection in children. *Eur J Epidemiol* 10: 57–61

36 Walser BL, Lima AAM, Guerrant RL (1996) Effects of high-dose oral vitamin A on diarrheal episodes among children with persistent diarrhea in a northeast Brazilian community. *Amer J Trop Med Hyg* 54: 582–585

37 Henning B, Stewart K, Zaman K, Alam AN, Brown KH, Black RE (1992) Lack of therapeutic efficacy of vitamin A for non-cholera, watery diarrhoea in Bangladeshi children. *Eur J Clin Nutr* 46: 437–443

38 Dewan V, Patwari AK, Jain M, Dewan N (1995) A randomized controlled trial of vitamin A supplementation in acute diarrhea. *Indian Pediatr* 32: 21–25

39 Hossain S, Biswas R, Kabir I, Sarker S, Dibley M, Fuchs G, Mahalanabis D (1998) Single dose vitamin A treatment in acute shigellosis in Bangladeshi children: randomised double blind controlled trial. *Brit J Med* 316: 422–426

40 Tielsch JM, West KP, Katz J, Chirambo MC, Schwab L, Johnson GJ, Tizazu T, Swartwood J, Sommer A (1986) Prevalence and severity of xerophthalmia in southern Malawi. *Amer J Epidemiol* 124: 561–586

41 The Vitamin A and Pneumonia Working Group (1995) Potential interventions for the prevention of childhood pneumonia in developing countries: a meta-analysis of data from field trials to assess the impact of vitamin A supplementation on pneumonia morbidity and mortality. *Bull WHO* 73: 609–619

42 Rahman MM, Mahalanabis D, Alvarez JO, Wahed MA, Islam MA, Habte D, Khaled MA (1996) Acute respiratory infections prevent improvement of vitamin A status in young infants supplemented with vitamin A. *J Nutr* 126; 628–633

43 Kjolhede CL, Chew FJ, Gadomski AM, Marroquin DP (1995) Clinical trial of vitamin A as adjuvant treatment for lower respiratory tract infections. *J Pediat* 126: 807–812

44 Nacul LC, Kirkwood BR, Arthur P, Morris SS, Magalhães M, Fink MCDS (1997) Randomised, double blind, placebo controlled clinical trial of efficacy of vitamin A treatment in non-measles childhood pneumonia. *Brit J Med* 315: 505–510

45 Fawzi WW, Mbise RL, Fataki MR, Herrera MG, Kawau F, Hertzmark E, Spiegelman D, Ndossi G (1998) Vitamin A supplementation and severity of pneumonia in children admitted to the hospital in Dar es Salaam,

Tanzania. *Amer J Clin Nutr* 68: 187–192
46   Stephensen CB, Franchi LM, Hernandez H, Campos M, Gilman RH, Alvarez JO (1998) Adverse effects of high-dose vitamin A supplements in children hospitalized with pneumonia. *Pediatrics* 101: 915–916
47   Dowell SF, Papic Z, Bresee JS, Larrañaga C, Mendez M, Sowell AL, Gary HE, Anderson LJ, Avendaño LF (1996) Treatment of respiratory syncytial virus infection with vitamin A: a randomized, placebo-controlled trial in Santiago, Chile. *Pediat Infect Dis J* 15: 782–786
48   Bresee JS, Fischer M, Dowell SF, Johnston BD, Biggs VM, Levine RS, Lingappa JR, Keyserling HL, Petersen KM, Bak JR et al (1996) Vitamin A therapy for children with respiratory syncytial virus infection: a multi-center trial in the United States. *Pediat Infect Dis J* 15: 777–782
49   Quinlan KP, Hayani KC (1996) Vitamin A and respiratory syncytial virus infection. Serum levels and supplementation trial. *Arch Pediat Adolesc Med* 150: 25–30
50   Dudley L, Hussey G, Huskissen J, Kessow G (1997) Vitamin A status, other risk factors and acute respiratory infection morbidity in children. *S Afr Med J* 87; 65–70
51   West KP Jr, Katz J, Shrestha SR, LeClerq SC, Khatry SK, Pradhan EK, Adhikari R, Wu LS, Pokhrel RP, Sommer A (1995) Mortality of infants <6 mo of age supplemented with vitamin A: a randomized, double-masked trial in Nepal. *Amer J Clin Nutr* 62: 143–148
52   WHO/CHD Immunisation-Linked Vitamin A Supplementation Study Group (1998) Randomised trial to assess benefits and safety of vitamin A supplementation linked to immunisation in early infancy. *Lancet* 352: 1257–1263
53   Muhe L, Lulseged S, Mason KE, Simoes EAF (1997) Case-control study of the role of nutritional rickets in the risk of developing pneumonia in Ethiopian children. *Lancet* 349: 1801–1804
54   Ruel MT, Rivera JA, Santizo MC, Lonnerdal B, Brown KH (1997) Impact of zinc supplementation on morbidity from diarrhoea and respiratory infections among Guatemalan children. *Pediatrics* 99: 808–813
55   Markowitz L, Nzilambi N, Driskell WJ, Sension MG, Rovira EZ, Nieburg P, Ryder RW (1989) Vitamin A levels and mortality among hospitalized measles patients, Kinshasa, Zaire. *J Trop Pediat* 35: 109–112
56   Barclay AJG, Foster A, Sommer A (1987) Vitamin A supplements and mortality related to measles: a ran-domised clinical trial. *Brit J Med* 323: 160–164
57   Hussey GD, Klein M (1990) A randomized, controlled trial of vitamin A in children with severe measles. *New Engl J Med* 323: 160–164
58   Coutsoudis A, Broughton M, Coovadia HM (1991) Vitamin A supplementation reduces measles morbidity in young African children: a randomized, placebo-controlled, double-blind trial. *Amer J Clin Nutr* 54: 890–895
59   Ogaro FO, Orinda VA, Onyango FE, Black RE (1993) Effect of vitamin A on diarrhoeal and respiratory com-plications of measles. *Trop Geograph Med* 45: 283–286
60   Rosales FJ, Kjolhede C, Goodman S (1996) Efficacy of a single dose of 200 000 IU of oil-soluble vitamin A in measles-associated morbidity. *Amer J Epidemiol* 143: 413–422
61   Semba RD, Munasir Z, Beeler J, Akib A, Muhilal, Audet S, Sommer A (1995) Reduced seroconversion to measles in infants given vitamin A with measles vaccination. *Lancet* 345: 1330–1332
62   Krogstad DJ (1996) Malaria as a reemerging disease. *Epidemiol Rev* 18: 77–89
63   Friis H, Mwaniki D, Omondi B, Muniu E, Magnussen P, Geissler W, Thiong'o F, Michaelsen KF (1997) Serum retinol concentrations and *Schistosoma mansoni* intestinal helminths, and malarial parasitemia: a cross-sec-tional study in Kenyan preschool and primary school children. *Amer J Clin Nutr* 66: 665–71
64   Shankar AH, Genton B, Semba RD, Tielsch J, West KP Jr (1997) Vitamin A supplementation as a nutrient-based intervention to reduce malaria-related morbidity. *Abstracts of the XVIII International Vitamin A Consultative Group Meeting, Cairo, Egypt, 22–26 September, 1997.* The Human Nutrition Institute, Washington, 31
65   Hanekom WA, Potgieter S, Hughes EJ, Malan H, Kessow G, Hussey GD (1997) Vitamin A status and thera-py in childhood pulmonary tuberculosis. *J Pediat* 131: 925–927
66   Semba RD, Tang AM (1999) Micronutrients and the pathogenesis of human immunodeficiency virus infec-tion. *Brit J Nutr* 81: 181–189
67   Semba RD, Miotti PG, Chiphangwi JD, Saah AJ, Canner JK, Dallabetta GA, Hoover DR (1994) Maternal vitamin A deficiency and mother-to-child transmission of HIV-1. *Lancet* 343: 1593–1597
68   Coutsoudis A, Bobat RA, Coovadia HM, Kuhn L, Tsai WY, Stein ZA (1995) The effects of vitamin A sup-plementation on the morbidity of children born to HIV-infected women. *Amer J Public Health* 85: 1076–1081
69   Semba RD, Lyles CM, Margolick JB, Caiaffa WT, Farzadegan H, Cohn S, Vlahov D (1998) Vitamin A sup-plementation and human immunodeficiency virus load in injection drug users. *J Infect Dis* 177: 611–616
70   Fawzi WW, Msamanga GI, Spiegelman D, Urassa EJN, McGrath N, Mwakagile D, Antelman G, Mbise R, Herrera G, Kapiga S et al (1998) Randomised trial of effects of vitamin supplements on pregnancy outcomes and T cell counts in HIV-1-infected women in Tanzania. *Lancet* 351: 1477–1482
71   Fawzi W, Mbise FL, Kawau F, Herrera MG, Hertzmark E, Fataki M, Spiegelman G (1998)   Vitamin   A supplements and mortality among HIV positive and negative children in Tanzania [Abstract 170/42331]. *Abstracts of the 12th World AIDS Conference,* June 28–July 3, 1998, Geneva, 840
72   West KP Jr, Khatry SK, Katz J, LeClerq SC, Pradhan EK, Shrestha SR, Connor PB, Dali S, Adhikari R, Pokhrel RP et al (1997) Impact of weekly supplementation of women with vitamin A or β-carotene on fetal, infant and maternal mortality in Nepal. *Abstracts of the XVIII International Vitamin A Consultative Group*

Meeting, Cairo, Egypt, 22–26 September, 1997. Human Nutrition Institute, Washington, 28
73 Humphrey JH, Agoestina T, Wu L, Usman A, Nurachim M, Subardja D, Hidayat S, Tielsch J, West KP Jr, Sommer A (1996) Impact of neonatal vitamin A supplementation on infant morbidity and mortality. *J Pediat* 128: 489–496
74 Venkatarao T, Ramakrishnan R, Nair NG, Radhakrishnan S, Sundaramoorthy L, Koya PK, Kumar SK (1996) Effect of vitamin A supplementation to mother and infant on morbidity in infancy. *Indian Pediatr* 33: 279–286
75 Murphy S, West KP Jr, Greenough WB III, Cherot E, Katz J, Clement L (1992) Impact of vitamin A supplementation on the incidence of infection in elderly nursing-home residents: a randomized controlled trial. *Age Ageing* 21: 435–439

Retinoid metabolism in the embryo

J.M. Creech Kraft

87 Sudden Valley, Bellingham, WA 98226, USA

Consequences of retinoid metabolism for human embryos

In 1982, isotretinoin (Accutane) was marketed as an agent for treatment of acne. As early as 1983, Franz Rosa, MD, an official of the Federal Drug Administration (FDA), reported five cases of malformations associated with use of Accutane in pregnancy [1]. The malformations were major and included hydrocephalus, cardiovascular and ear defects [2, 3]. By the most recent estimates available, malformations in a total of 94 human infants have been associated with prenatal Accutane exposure [4]. In contrast with the number of reported cases, the US FDA estimates that 900–1300 retinoid-exposed babies were born with severe birth defects in the first five years after Accutane was marketed. Because the drug is still commercially available, the Teratology Society has strongly recommended that women of childbearing age using the drug should practice contraception and that use of this drug should be thoroughly monitored [5].

Accutane is a naturally occurring metabolite of retinol (Vitamin A_1) (see Fig. 1), and is the trade name for 13-*cis*-retinoic acid (RA). After administration of 13-*cis*-RA to humans at therapeutic doses, four metabolites were identified in the serum [6]. These were 4-oxo-all-*trans*-RA, 4-oxo-13-*cis*-RA, all-*trans*-retinoyl-β-glucuronide (RAG) and all-*trans*-RA. High levels of all-*trans*-RA were detected in an exposed embryo of a woman who was being therapeutically treated with Accutane [7]. The levels of both all-*trans*-RA and 13-*cis*-RA in the abortus were ten-fold higher than average endogenous RA levels [8]. Possibly then, the teratogenic effects of 13-*cis*-RA in humans are due to metabolic conversion to all-*trans*-RA.

Studies in rodents indicate that all-*trans*-RA, 4-oxo-all-*trans*-RA and 9-*cis*-RA are direct-acting teratogens

Although 13-*cis*-RA is a potent human teratogen, this substance is only marginally teratogenic in the mouse at doses 100 times higher than those used in human therapy; the all-*trans* isomer, on the other hand, is a potent teratogen in the mouse at low doses [9]. Transplacental pharmacokinetics showed that all-*trans*-RA accumulated much more extensively than its *cis* isomer in mouse embryos [10, 11]. In those studies, the areas under the concentration-time curves in embryonic tissues were 30-fold higher for the *trans* isomer than for the *cis* isomer. The teratogenicity of 13-*cis*-RA observed after multiple high oral dosing of pregnant mice was hypothetically due to the conversion of 13-*cis*-RA to the all-*trans* isomer. The latter was subsequently

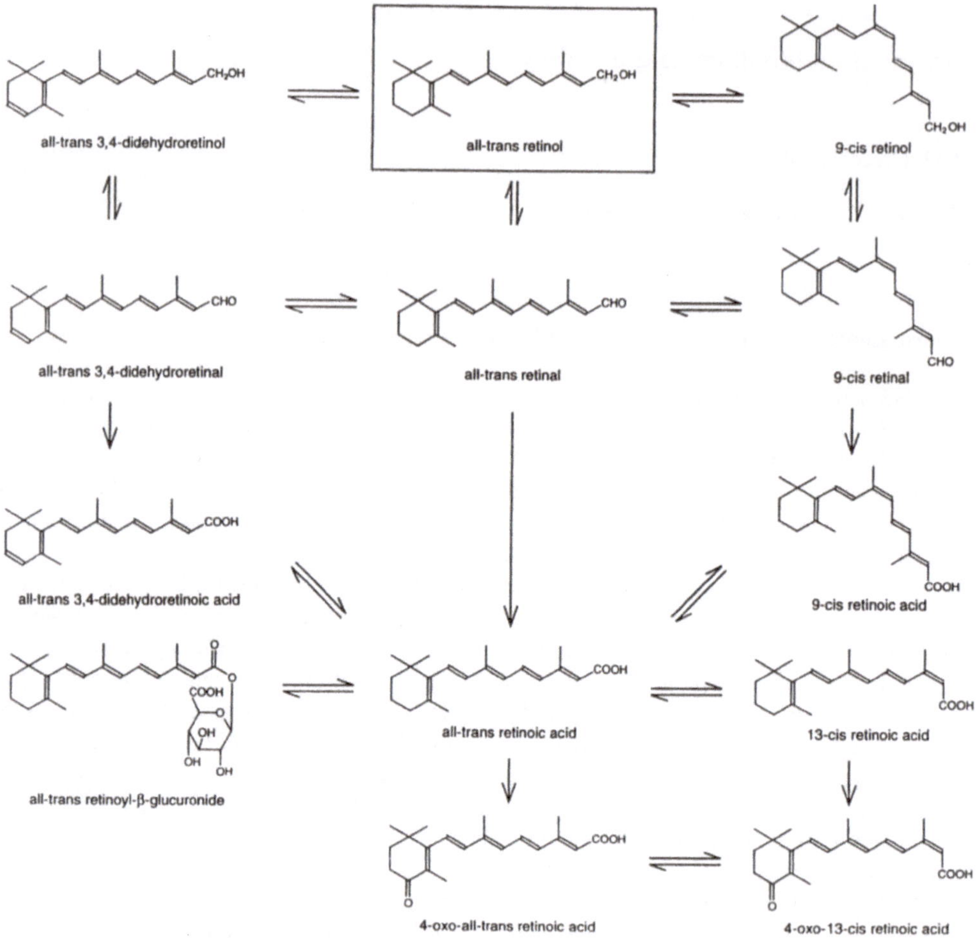

Figure 1. Metabolic pathway for vitamin A (all-*trans*-retinol) with cross-over to the vitamin A2 (all-*trans*-dd retinol) metabolic pathway.

detected at high concentrations in the embryo for ten critical hours during gestation [12]. 4-oxo-all-*trans*-RA is a potent teratogen in most species [13], whereas in mice 4-oxo-13-*cis*-RA is only marginally teratogenic [11, 14, 15]. Embryonic exposure to 4-oxo-all-*trans*-RA was far higher than to its corresponding *cis* isomer [11]. Gunning et al. reported that the sugar conjugate of RA, all-*trans*-RAG, was not teratogenic when administered orally to rats during gestation [16]. Comprehensive studies using intraamniotic microinjections in rat whole embryos revealed that all-*trans*-RA and 4-oxo-all-*trans*-RA were far more potent than their corresponding *cis* isomers and precursors, retinol and all-*trans*-RAG [17–21]. HPLC analyses indicated that biotransformation to all-*trans*-RA or to 4-oxo-all-*trans*-RA by the conceptus appeared responsible for the dysmorphogenic effects produced by all of the studied retinoids. In addition, it was observed in rat embryos that 9-*cis*-RA is an equi-potent dysmorphogen compared to all-*trans*-RA and causes a higher incidence of missing optic vesicles and cardiac defects [22]. From these studies it can be concluded that all-*trans*-RA, 4-oxo-all-*trans*-RA and 9-*cis*-RA are direct-acting teratogens in rodents.

Mechanism of action of retinoids

Retinoids regulate vertebrate development through the action of two types of receptors, the retinoic acid receptors (RARs) α, β, γ [23–25] and the retinoid-X-receptors (RXRs) [26]. All-*trans*-RA, 4-oxo-all-*trans*-RA and all-*trans*-didehydro-RA (dd-RA) bind and activate RARs α, β and γ [27] whereas their corresponding 13-*cis* isomers do not [28]. 9-*cis*-retinoic acid binds RXR with high affinity [29, 30]. Most *in vitro* experiments suggest that RXRs are not liganded but serve as auxiliary factors forming heterodimers with liganded partner receptors such as RAR. More recently, Lu has shown that liganded RXR can mediate retinoid signal transduction in the chick embryo [31].

Endogenous retinoids

Studies aimed at identifying the specific metabolites involved in retinoid metabolism during normal embryogenesis have been few. High pressure liquid chromatography (HPLC) is the standard and most objective technique used to identify endogenous retinoids in embryos of various species and has been used by the following research teams. Thaller and Eichele identified endogenous all-*trans*-RA in the chick limb bud and several years later found that the predominant RA in the chick limb bud was dd-RA [32, 33]. Scott et al. and Horton and Maden detected endogenous all-*trans* RA in mouse limb, but no didehydroretinoids [34, 35]. In 1989 Durston and colleagues reported the presence of all-*trans*- RA, 13-*cis*-RA and all-*trans*-retinal in *Xenopus* embryos [36] which was later confirmed in 1994 [37]. Likewise in *Xenopus* embryos, 4-oxo-RA was detected by two independent laboratories [38, 39] and 9-*cis*-RA by Creech Kraft et al. [37]. In addition, all-*trans*-retinal, 3,4-didehydroretinal and the sugar conjugates, 13-*cis*- and all-*trans*-RAG, have been found in abundance in *Xenopus* embryos [37, 40]. Surprisingly, Blumberg et al. found the only bioactive retinoid in *Xenopus* embryos to be 4-oxo- retinaldehyde [41], whereas, with a more sensitive HPLC system and simple ethanol precipitation of proteins, Creech Kraft detected in this same species the acid metabolites, all-*trans*-RA or 4-oxo-all-*trans* RA at concentrations ten times lower than the retinaldehydes [39]. In Zebrafish, all-*trans*-RA and dd-RA have been identified, but not 9-*cis*-RA [42].

Interestingly a new class of intracellular messenger molecules, the retro-retinoids were recently identified [43–45]. Both the growth supportive 14-hydroxyl-retro-retinol (14-HHR) pathway and the growth suppressive anhydroretinol (AR) pathway have been described. The AR-producing enzyme ARase has been purified to homogeneity. It is not yet known whether these naturally occurring agonist/antagonist pairs of retinoids act via specific nuclear, cytoplasmic or cell membrane receptors or whether they are present in embryos.

With the use of an indirect method for quantifying retinoids, namely a luciferase reporter gene construct containing two RA response elements (RAREs) from the RARβ gene, Chen et al. confirmed significant RA activity in *Xenopus* embryos [46]. Later it was shown that the reporter construct used by Chen et al. can be activated by all-*trans*-RA, 9-*cis*-RA and synthetic ligands selective for RARs in *Xenopus* embryos [47].

An important observation is that endogenous levels of retinoids in human embryos are quantitatively and qualitatively similar to those in *Xenopus* embryos, whereby the following retinoids were detected: 13-*cis*-RA, 9-*cis*-RA, all-*trans*-RA, all-*trans*-retinol, all-*trans*-dd-retinol and all-*trans*-retinal [8, 37, 39]. Until this report, dd-retinol had been detected only in chick embryos [33], but not in rodent embryos. These retinoids were detected in *Xenopus* embryos at levels that can be accurately quantified with a sensitive HPLC detection system. In contrast, levels of the same endogenous retinoids in rodents are low [19, 20, 21], resulting in less reliable quantification.

Generation of active receptor ligands

During the past ten years, several important studies have been performed on the tissue specific synthesis of RA in embryos during critical stages of development [32, 37, 46, 48, 49]. Recently, Maden demonstrated in the chick that the developing brain did not generate RA, but the spinal part of the neural tube generated it at high levels [50]. This team observed that the posterior half of the limb bud generated higher levels than the anterior half. Local application of RA to the anterior margin of the chick limb bud results in pattern duplication [32]. It was previously widely accepted that RA acts directly by conferring positional information to the limb bud cells, but more recently RA has been shown to act indirectly by creating a polarizing region in the tissue distal to the RA source. Helms et al. demonstrated that RA treatment induces sonic hedgehog, a putative zone of polarizing activity (ZPA) morphogen and *Hoxd-11*, a gene induced by the polarizing signal [51]. Retinoid receptor-specific antagonists blocked limb morphogensis and down-regulated the polarizing signal, sonic hedgehog [51, 52]. Lu et al. have shown that *Hoxb-8* can be induced by exogenous RA and down-regulated by retinoid receptor antagonists, further evidence that retinoids are required for the establishment of ZPA [52, 53].

Nevertheless, it is highly likely that local (target tissue) regulation of levels of tissue retinoids and retinoid metabolites may well be a major determinant of both the morphogenic and the dysmorphogenic effects of these substances. Retinol-induced teratogenicity can be averted by simultaneous exposure to citral, which apparently inhibits the metabolic conversion of retinol to RAs (54). It seems logical to presume that retinoid biotransformation could provide a significant component of local regulation and that tissue-expressed enzymes and binding proteins would serve, in turn, as regulators of rates of biotransformation.

Embryos depend on the maternal uptake of retinol which must be metabolized to active retinoic acid isomers in order to fulfill all of its roles in vertebrate development. During vertebrate embryogenesis RA is detected in distinct spatiotemporal patterns (32, 37) suggesting that it is produced from retinol in a regulated fashion. Enzymes involved in retinol and retinal metabolism are likely candidates for regulators of RA levels. The aim of numerous studies has been to identify the mechanism for tissue-specific generation of RA in developing embryos. Recently, using immunohistochemical localization, Bavik demonstrated the presence of retinol binding protein receptor (RPDr), cellular retinol binding protein (CRBP), retinol dehydrogenase (RoOH) and retinal dehydrogenase (RalOH) in rat presomitic embryos [55]. These proteins are responsible for the cellular uptake of retinol and its cellular transport and conversion to RA. All four proteins were detected in tissues that depend upon vitamin A for normal development including the yolk sac, heart, gut, notochord, somites, sensory placodes and the limb. The uptake of retinol into the yolk sac depends upon a retinol binding receptor [56, 57, 58].

During retinoid signalling, retinol is first converted to retinal followed by conversion to the active nuclear receptor ligands, 9-cis- and all-trans-RA. The alcohol dehydrogenase (ADH) enzyme family may function in the metabolism of retinol, the alcohol form of vitamin A. Likewise, there exists an aldehyde dehydrogenase (ALDH) family containing members preferring retinal as substrate. Certain forms of ADH and ALDH may cooperate to upregulate RA synthesis during development. Deuster has shown that the expression of ALDH-IV and ALDH-I in the cranial mesenchyme of mouse embryos could provide a source of RA for cranial neural crest cell survival [59].

The recent identification of 9-cis-retinol dehydrogenase in the mouse embryo reveals a pathway for 9-cis-RAs synthesis in this species [60]. This membrane-bound enzyme is able to oxidize 9-cis-retinol into 9-cis-retinaldehyde which can be subsequently oxidized to 9-cis-RA. The expression of this enzyme is temporally and spatially controlled during embryogenesis in parts of the nervous system, sensory organs, somites and myotomes, and several tissues of endodermal origin. Mertz et al. have also identified a stereospecific human enzyme that catalyzes 9-cis-retinol oxidation and is likewise a member of the short chain alcohol dehydrogenase protein family [61]. The mRNA for the protein is most abundant in human mammary tissues.

In the mouse, RA was not detected in egg cylinder embryos but in late primitive streak stage embryos whereas there was a ubiquitous expression of RARα and RARγ both prior to and during primitive streak formation [62]. Detection of class IV ADH mRNA coincided with the onset of RA synthesis, being absent in egg cylinder embryos but present in the posterior mesoderm of late primitive streak embryos [63]. During neurulation, class RA and class IV ADH mRNA were colocalized in the craniofacial region, trunk, and forelimb bud [63].

In situ hybridization revealed high levels of both class I and IV ADH messenger RNAs in adrenal glands of 16.5 day mouse embryos as wells as significant amounts of RA, suggesting this organ as a potential endocrine source of this hormone during mouse development [64].

Most of the oxidation of retinalydehyde to RA is mediated by several isoforms of aldehyde dehydrogenases, which are expressed in spatially and temporally regulated patterns in the developing vertebrate [65]. McCaffery and Dräger found an axial asymmetry in RA-generating aldehyde dehydrogenases along the dorsal-ventral axis of the embryonic retina [66, 67]. The retinas of all embryonic vertebrates express a class-1 ALDH in the dorsal part and a more powerful aldehyde dehydrogenase ventrally. In all early embryos tested, the ALDH arrangements result in an RA gradient along the ventro-dorsal axis of the retina [68]. When early zebrafish embryos are briefly exposed to citral, a competitive inhibitor of RA-generating dehydrogenase, they develop into larvae that lack the ventral part of the eye [69]. The opposite effect is observed when early zebrafish are exposed to RA: they grow into larvae whose eyes appear duplicated along the dorso-ventral axis [70].

A novel member of the cytochrome P450 superfamily, CYP26, has been cloned in mouse embryos and shown to be expressed in mouse and human liver, as well as in regions of the brain and the placenta [71]. Administration of all-*trans*-RA increases the levels of the transcript in the adult liver, but not in the brain. It has been suggested that CYP26 functions in some aspect of retinoid signalling or metabolism in the embryo, perhaps directly downstream of RA. Since the 4-oxo metabolites have been detected endogenously in *Xenopus* after the completion of neu-rulation, this may be an indication that cytochrome P450- mediated retinoid metabolism is involved in controlling the levels, and thus the biological effects, of RA [38, 39].

Effects of exogenous retinoids on embryonic development

Sive et al. have shown that exogenously applied all-*trans*-RA to *Xenopus* embryos prevents the proper formation of eyes, brain and heart, three structures that are also involved in human and rodent retinoid dysmorphogenesis [72]. Durston et al. reported endogenous all-*trans*-RA in *Xenopus* embryos and that it causes alterations along the anteroposterior axis when applied dur-ing early embryonic development [36]. In *Xenopus*, 9-*cis*-RA is six-fold more potent than all-*trans*-RA in causing alterations along the anteroposterior axis and more effective than all-*trans*-RA at inducing the homeobox gene *Xlim-1* in embryonic explant cultures [37].

Investigations providing data pertaining to the metabolism and disposition of exogenous 9-*cis*-RA and all-*trans*-RA during neurulation in *Xenopus* embryos have been completed [73]. Each isomer elicited malformations of the heart, eye and brain, but approximately two-fold higher concentrations of all-*trans*-RA than 9-*cis*-RA were required to produce qualitatively and quantitatively similar dysmorphogenic effects. The dysmorphogenic effects of all-*trans*-RA could not be attributed to the isomerization of all-*trans*-RA to 9-*cis*-RA and subsequent eleva-tion of embryonic levels of 9-*cis*-RA. An important finding of these studies was that exoge-nous, combined low concentrations of 9-*cis*-RA and all-*trans*-RA, in the range of their endoge-nous and normally detected concentrations, appeared to be synergistic when comparing the dysmorphogenic effects elicited by single exposures to 9-*cis*-RA or all-*trans*-RA to effects pro-duced by combined exposures. This may not be surprising because it is clear that the various heterodimer interactions between RARs and RXRs, their respective ligands, coregulators and response elements are highly complex [74] and are phenomena undoubtedly necessary for refinement of patterns of gene activation during organogenesis [75]. A synergy was also observed when the RXRα mutation was introduced into RARα or RARγ mutant backgrounds [76]. The phenotypes of these mice showed a clear convergence between RXR- and RAR-dependent signalling pathways, supporting a role for RXR/RAR heterodimers *in vivo*. Importantly, RXRα null mutants in mice die *in utero* and display myocardial and ocular mal-formations. In our studies, approximately 35% of the embryos treated with all-*trans*-RA (1000 ng/ml), 9-*cis*-RA (500 ng/ml) or a combination of 9-*cis*-RA (50 ng/ml) plus all-*trans*-

RA (50 ng/ml) were cycloptic, and over 50% of the embryos from the above treatments had cardiac defects [73].

Metabolism studies brought important evidence of enzymatic activity in embryos. For example, after *Xenopus* embryos were exposed to all-*trans*-RA during neurulation, elevated levels of 4-oxo-all-*trans*-RA, 4-oxo-13-*cis*-RA, all-*trans*-RAG and 13-*cis*-RA were detected in the embryos whereas embryonic levels of 9-*cis*-RA were actually lower than endogenous levels during early neurulation [73]. After embryos were exposed to 9-*cis*-RA during neurulation, elevated levels of 4-oxo metabolites, glucuronides and 9,13 di-*cis*-RA were observed in the embryos. These findings are good evidence for the existence of cytochrome P450 enzymes and glucuronosyltransferases in embryonic tissues, but not for the widespread presence of an all-*trans*/9-*cis* isomerase [73].

It should be emphasized that investigations of embryonic enzymatic catalysis of the four retinoid biotransforming reactions (all-*trans*/9-*cis* isomerization, retinol and retinal dehydrogenation, glucuronidation and monooxygenation) have been numerous [6, 7, 10–13, 15, 18–22, 77–85].

Subsequent to these findings, teratology studies were carried out during *Xenopus* neurulation which showed that at equivalent concentrations, 4-oxo-13-*cis*-RA, 9,13-di-*cis*-RA and 13-*cis*-RA elicited fewer severe multiple malformations than all-*trans*- or 9-*cis*- isomers. All-*trans*-RAG was only marginally teratogenic at the highest concentration tested. Since only the 9-*cis*- and *trans* isomers bind and activate the RXRs and RARs, it is probable that metabolism to less active metabolites, 13-*cis* and 9,13-di-*cis* isomers, is one way the embryo could control and prevent all-*trans*-RA and 9-*cis*-RA from activating the expression of RA-dependent genes prior to their normal time of activation which would result in developmental aberrations. HPLC analysis of embryos exposed to teratogenic amounts of 13-*cis*-RA, 9,13-di-*cis*-RA and 4-oxo-13-*cis*-RA showed that each isomer had been converted to significant levels of all-*trans* or 9-*cis*- isomers.

The correct level of the active metabolites of vitamin A which control the nuclear receptor signalling pathway is required for appropriate embryonic development. Too much or too little of the receptor ligands is equally harmful for the embryo. It has been well documented that when there is an excess of vitamin A or its metabolites during embryonic development, defects occur in the CNS. An excess of RA causes specific defects of the anterior hindbrain whereby anterior rhombomeres can be lost or respecified to more posterior areas or it can cause posteriorization of the whole CNS, whereby forebrain structures are lost [36, 50, 86–88]. More recently, Maden et al. describe defects that arise in the central nervous system of quail embryos when they develop in the absence of vitamin A [50]. There are three defects in these embryos: (1) the posterior hindbrain is completely missing; (2) the neural tube fails to extend neurites out into the periphery; (3) the neural crest cells die.

The developing embryo tightly regulates the temporal and spatial synthesis of vitamin A metabolites to insure that the proper amounts of active receptor ligands are available for sequential gene activation during embryogenesis. Embryonic dysmorhogenesis may be caused by an excess or a deficiency of these active retinoid metabolites.

References

1   Rosa FW (1983) Teratogenicity of isotretinoin. *Lancet* 2: 513
2   Braun JT, Franciosi RA, Mastri AR, Drake RM, O'Niel BL (1984) Isotretinoin dysmorphic syndrome. *Lancet* 1: 506–507
3   Lammer EJ, Chen DT, Hoar RM, Agnish ND, Benke PJ, Braun JT, Curry CJ, Fernhoff PM, Grix AW, Lott IT et al (1985) Retinoic acid embryopathy. *New Engl J Med* 313: 837–841
4   Schardein JL (1993) Human studies: Retinoic Acid Embryopathy. *In:* JL Shardein (ed.): *Chemically Induced Birth Defects.* Marcel Dekker, New York, 558–567
5   Teratology Society (1991) Recommendations for Isotretinoin use in woman of childbearing potential. *Teratology* 44: 1–6
6   Creech Kraft J, Slikker Jr, W, Bailey JR, Roberts LG, Fischer B, Wittfoht W, Nau H (1991) Plasma pharmacokinetics of 13-*cis*- and all-*trans*-retinoic acid in the Cynomolgus monkey and the identification of the conjugate metabolites 13-*cis* and all-*trans*-retinoyl-β-glucuronides: a comparison to one human case study with Isotretinoin. *Drug Metab Dispos* 19: 317–324
7   Creech Kraft J, Nau H, Lammer E, Olney A (1989) Embryonic retinoid concentrations after maternal intake of Isotretinoin. *New Engl J Med* 321: 262
8   Creech Kraft J, Shepard T, Juchau MR (1993) Tissue levels of retinoids in human embryo/fetuses unexposed to Accutane^R. *Reprod Toxicol* 7: 11–15
9   Kochhar DM, Penner JD, Tellone C (1984) Comparative teratogenic activities of two retinoids: Effects on palate and limb development. *Teratogen Carcinogen Mutagen* 4: 377–387
10 Creech Kraft J, Kochhar DM, Scott WJJr, Nau H (1987) Low teratogenicity of 13-*cis*-retinoic acid (isotretinoin) in the mouse corresponds to low embryo concentrations during organogenesis: comparison to the all-*trans* isomer. *Toxicol Appl Pharmacol* 87: 474–482
11 Creech Kraft J, Lofberg B, Chahoud I, Bochert G, Nau H (1989) Teratogenicity and placental transfer of all-*trans,* 13-*cis,* 4-oxo-all-*trans,* and 4-oxo-13-*cis*-retinoic acid after a low oral dose during organogenesis in mice. *Toxicol Appl Pharmacol* 100: 162–176
12 Creech Kraft J, Eckhoff Chr Kochhar DM, Bochert G, Chahoud I, Nau H (1991) Isotretinoin (13-*cis*-retinoic acid), metabolism, *cis-trans* isomerization, glucuronidation and transfer to the mouse embryo: Consequences for teratogenicity. *Teratogen Carcinogen Mutagen* 11: 21–30
13 Willhite CC, Wier PJ, Berry D (1989) Dose response and structure activity considerations in retinoic-induced dysmorphogenesis. *Crit Rev Toxicol* 20: 113–135
14 Kochhar DM, Penner J (1987) Developmental effects of Isotretinoin and 4-oxo-Isotretinoin: The role of metabolism in teratogenicity. *Teratology* 36: 67–75
15 Satre MA, Penner JD, Kochhar DM (1989) Pharmacokinetic assessment of teratologically effective concentrations of an endogenous retinoic acid metabolite. *Teratology* 39: 341–348
16 Gunning DB, Barua AB, Olson JA (1993) Comparative teratogenicity and metabolism of all-*trans*-retinoic acid, all-*trans*-retinoyl-β-glucose and all-*trans*-retinoyl-β-glucuronide in pregnant Sprague Dawly rats. *Teratology* 47: 29–36
17 Lee QP, Juchau MR, Creech Kraft J (1991) Microinjection of cultured rat embryos: studies with retinol, 13-*cis*- and all-*trans*-retinoic acid. *Teratology* 44: 313–323
18 Creech Kraft J, Bechter R, Lee QP, Juchau MR (1992) Microinjections of cultured rat conceptuses: Studies with 4-oxo-all-*trans*-retinoic acid, 4-oxo-13-*cis*-retinoic acid and all-*trans*-retinoyl-β-glucuronide. *Teratology* 45: 259–270
19 Creech Kraft J, Juchau MR (1992) Correlations between conceptal concentrations of all-*trans* retinoic acid and dysmorphogenesis after microinjections of all-*trans*-retinoic acid, 13-*cis*-retinoic acid, all-*trans*-retinoyl-β-glucuronide or retinol in cultured whole rat embryos. *Drug Metab Dispos* 20: 218–225
20 Creech Kraft J, Bui T, Juchau MR (1992) Elevated levels of all-*trans*-retinoic acid in cultured rat embryos 1.5 h after microinjections with 13-*cis*-retinoic acid or retinol and correlations with dysmorphogenesis. *Biochem Pharmacol* 44:R 21–24
21 Creech Kraft J, Juchau MR (1992) Conceptal biotransformation of 4-oxo-all-*trans*-retinoic acid, 4-oxo-13-*cis*-retinoic acid and all-*trans*-retinoyl-β-glucuronide in rat whole embryo culture. *Biochem Pharmacol* 43: 2289–2292
22 Creech Kraft J, Juchau MR (1993) 9-*cis*-Retinoic acid: a direct-acting dysmorphogen. *Biochem Pharmacol* 46: 709–716
23 Pekovich M, Brand NJ, Krust A, Chambon P (1987) A human retinoic acid receptor which belongs to the family of nuclear receptors. *Nature* 330: 444–450
24 Brand NJ, Petkovich M, Krust A, Chambon P, de The Machio A, Tiollais P, Dejean A (1988) Identification of a second human retinoic acid receptor. *Nature* 332: 850–853
25 Krust A, Kastner P, Petkovich M, Zelent A, Chambon P (1989) A third human retinoic acid receptor, hRAR-α. *Proc Natl Acad Sci USA* 86: 5210–5214
26 Giguere V, Ong ES, Sequi P, Evans R (1987) Identification of a receptor for morphogen retinoic acid. *Nature* 330: 624–629
27 Allenby G, Bocquel M-T, Saunders M, Kazmer S, Speck J, Rosenberger M, Lovey A, Kastner P, Grippo J,

Chambon P et al (1993) Retinoic acid receptors and retinoic X receptors: Interactions with endogenous retinoic acids. *Proc Natl Acad Sci USA* 90: 30–34

28 Apfel C, Crettaz M, LeMotte P (1992) Differential binding and activation of synthetic retinoids to retinoic acid receptors. *In:* G Morriss-Kay (ed.): *Retinoids in Normal Development and Teratogenesis.* Oxford University Press, Oxford, 65–74

29 Heyman RA, Mangelsdorf DJ, Dyck JA, Stein RB, Eichele G, Evans RM, Thaller C (1992) 9-*cis* retinoic acid is a high affinity ligand for the retinoid X receptor. *Cell* 68: 397–406

30 Levin AA, Sturzenbecker LJ, Kazmer S, Bosakowski T, Huselton C, Allenby G, Speck J, Kratzeisen C, Rosenberger M, Lovey A et al (1992) 9-*Cis* retinoic acid stereoisomer binds and activates the nuclear receptor RXRα. *Nature* 355: 359–361

31 Lu HC, Eichele G, Thaller C (1997) Ligand-bound RXR can mediate retinoid signal transduction during embryogenesis. *Development* 124: 195–203

32 Thaller C, Eichele G (1987) Identification and spatial distribution of retinoids in the developing chick limb bud. *Nature* 327: 625–628

33 Thaller C, Eichele G (1990) Isolation of 3,4-didehydroretinoic acid, a novel morphogenetic in the chick limb bud. *Nature* 345: 815–819

34 Scott WJ, Walter R, Tzimas G, Sass JO, Nau H, Collins M (1994) Endogenous status of retinoids and their cytosolic binding proteins in limb buds of chick versus mouse embryos. *Dev Biol* 165: 397–409

35 Horton C, Maden M (1995) The endogenous distribution of retinoids during normal development and teratogenesis in mouse embryo. *Develop Dyn* 202: 312–323

36 Durston AJ, Timmermans WJ, Hage HFJ, deVries NJ, Heideveld M, Nieuwkoop PD (1989) Retinoic acid causes an anteroposterior transformation in the developing central nervous system. *Nature* 340: 140–144

37 Creech Kraft J, Schuh T, Juchau MR, Kimelman D (1994) The retinoid X receptor, 9-*cis* retinoic acid, is a potential regulator of early *Xenopus* development. *Proc Natl Acad Sci USA* 91: 3067–3071

38 Pijnappel WWM, Hendriks HFJ, Folkers GE, van den Brink CE, Dekker EJ, Edellenbosch C, van der Saag PT, Durston AJ (1993) The retinoid ligand 4-oxo-retinoic acid is a highly active modulator of positional specification. *Nature* 366: 340–344

39 Creech Kraft J, Schuh T, Juchau J, Kimelman D (1994) Temporal distribution, localization and metabolism of retinol, didehydroretinol and retinal during development. *Biochem J 301*: 111–119

40 Azuma M, Seki T, Fujishita S (1990) Changes of egg retinoids during the development of *Xenopus laevis Vision Res* 30: 1395–1400

41 Blumberg B, Bolado Jr, J, Derguini F, Craig AG, Moreno T, Chakravarti D, Heyman R, Buck J, Evans R (1996) Novel retinoic acid receptor ligands in *Xenopus* embryos *Proc Natl Acad Sci USA* 93: 4873–4878

42 Costaridis P, Horton C, Zeitlinger J, Holder N, Maden M (1996) Endogenous retinoids in Zebrafish embryo and adult. *Develop Dyn* 205: 41–51

43 Buck J, Gruen F, Derguini F, Chen Y, Kimura S, Noy N, Haemmerling U (1993) Anhydroretinol: A natually occurring inhibitor of lymphocyte physiology. *J Exp Med* 178: 1675–1680

44 Derguini F, Nakanishi K, Haemmerling U (1994) Synthesis and intracellular signalling activity of (14R), (14S) and 14RS)-14-Hydroxy-4,14-Retro-Retinol (14-HRR). *Biochemistry* 33: 623–628

45 Derguini F, Nakanishi K, Buck J, Haemmerling U, Gruen F (1994) Spectroscopic studies of anhydroretinol, an endogenous mammalian and insect retro-retinoid. *Angew Chem Int Ed Eng* 33: 1837–1839

46 Chen Y-P, Huang L, Russo A, Solursch M (1992) Retinoic acid is enriched in Hensen's node and is developmentally regulated in the early chicken embryo. *Proc Natl Acad Sci USA* 89: 7194–7197

47 Minucci S, Saint-Jeannet J-P, Toyanna R, Scita G, DeLuca LM, Taira M, Levin AA, Ozato K, Dawid IB (1996) Retinoid X receptor ligands produce malformations in *Xenopus* embryos. *Proc Natl Acad Sci USA* 93: 1803–1807

48 Hogan BLM, Thaller C, Eichele G (1992) Evidence that Hensen's node is a site of retinoic acid synthesis. *Nature* 359: 237–241

49 McCaffery P, Draeger U (1994) Hot spots of retinoic acid synthesis in the developing spinal cord. *Proc Natl Acad Sci USA* 91: 7194–7197

50 Maden M, Gale E, Zile M (1998) The role of vitamin A in the development of the central nervous system. *Am Soc Nutr Sci* 471S–475S

51 Helms J, Thaller C, Eichele G (1994) Relationship between retinoic acid and sonic hedgehog, two polarizing signals in the chick wing bud. *Development* 120: 3267–3274

52 Helms J, Kim CH, Eichele G, Thaller C (1996) Retinoic acid signalling is required during chick limb development. *Development* 122: 1385–1394

53 Lu H-C, Revelli J-P, Goering L, Thaller C, Eichele G (1997) Retinod signalling is required for the establishment of a ZPA and for the expression of *Hoxb-8*, a mediator of ZPA formation. *Development* 124: 1643–1651

54 Schuh TJ, Hall BL, Creech Kraft JM, Privalsky ML, Kimelman D (1993) V-erbA and citral reduce the teratogenic effects of all-*trans* retinoic acid and retinol, respectively, in *Xenopus* embryogenesis. *Development* 119: 785–798

55 Bavik C, Ward S, Ong D (1997) Identification of a mechanism to localize generation of retinoic acid in rat embryos. *Development* 69: 155–167

56 Zheng WL, Ong DE (1998) Spatial and temporal patterns of expression of cellular retinol-binding protein and

cellular retinoic acid-binding proteins in rat uterus during early pregnancy. *Biol Reprod* 58: 963–970

57 Johansson S, Gustafson AL, Donovan M, Romert A, Eriksson U, Dencker L (1997) Retinoid binding proteins in mouse yolk sac and chorio-allantoic placentas. *Anat Embryol (Berl)* 195: 483–490

58 Ward SJ, Chambon P, Ong D, Bavik C (1997) A retinol-binding protein receptor-mediated mechanism for uptake of vitamin A to post implantation rat embryos. *Biol Reprod* 57: 751–755

59 Deuster G (1998) Alcohol Dehydrogenase as a Critical Mediator of Retinoic Acid Syhthesis from Vitamin A in the Mouse Embryo. *Am Soc Nutr Sci* 98: 459S–462S

60 Romert A, Tuvendal P, Simon A, Dencker L, Eriksson U (1998) The identification of a 9-*cis* retinol dehydrogenase in the mouse embryo reveals a pathway for 9-*cis* retinoic acid. *Proc Natl Acad Sci USA* 94: 4404–4409

61 Mertz JR, Shang E, Piantedosi R, Wei S, Wolgemuth DJ, Blaner WS (1997) Identification and characterization of a stereospecific human enzyne that catalyzes 9-*cis*-retinol oxidation. A possible role in 9-*cis* retinoic acid formation. *J Biol Chem* 272: 11 744–11 749

62 Ang HW, Duester G (1997) Initiation of Retinoid Signalling in Primitive Streak Mouse Embryos: Spatiotemporal Expression Patterns of Receptors and Metabolic Enzymes for Ligand Synthesis *Develop Dyn* 208: 536–543

63 Ang HL, Deltour L, Hayamizu TF, Zgombic-Knight M, Deuster G (1996) Retinoc acid synthesis in the mouse during gastrulation and craniofacial development linked to calss IV alcohol dehydrogenase gene expression. *Biol Chem* 271: 9526–9534

64 Haselbeck R, Ang HL, Deltour L, Deuster G (1997) Retinoic acid and alcohol/retinol dehydrogenase in the mouse adrenal gland: A potential source of retinoic acid during development. *Endocrinology* 138: 3035–3041

65 McCaffery P, Draeger UC (1993) Retinoic acid synthesis in the developing retina. *Adv Exp Med Biol* 328: 181–190

66 McCaffery P, Lee M-O, Wagner MA, Sladek NE, Draeger UC (1992) Asymmetrical retinoic acid synthesis in the dorso-ventralmaxis of the retina. *Development* 115: 371–382

67 McCaffery P, Posh KC, Napoli JL, Gudas L, Draeger UC (1993) Changing patterns of retinoic acid synthesis in the retina. *Dev Biol* 158: 390–399

68 McCaffery P, Tempst P, Lara G, Draeger UC (1991) Aldehyde dehydrogenase is a positional marker in the retina. *Development* 112: 693–701

69 Marsh-Armstrong N, McCaffery P, Dowling JE, Gilbert W, Draeger UC (1994) Retinoic acid is necessary for development of the ventral retina in zebrafish. *Proc Natl Acad Sci USA* 91: 7286–7290

70 Hyatt GA, Schmitt EA, Marsh-Armstrong NA, Dowling JE (1992) Retinoic acid-induced duplication of the zebrafish retina. *Proc Natl Acad Sci USA* 89: 8293–8297

71 Ray WJ, Bain G, Yao M, Gottlieb DI (1997) CYP26, a novel mammalian cytochrome P450, is induced by retinoic acid and defines a new family. *J Biol Chem* 272: 18 702–18 708

72 Sive H, Draper B, Harland RM, Weintraub H (1990) Identification of a retinoic acid sensitive period during axis formation in *Xenopus laevis*. *Gene Develop* 4: 932–942

73 Creech Kraft J, Juchau MR (1995) *Xenopus laevis*: A Model System for the study of embryonic retinoid metabolism. III. Isomerization and metabolism of all-*trans*-retinoic acid and 9-*cis*-retinoic acid and their dysmorphogenic effects in embryos during neurulation. *Drug Metab Dispos* 23: 1058–1072

74 Leid M, Kastner P, Lyons R, Nakshatri N, Saunders M, Zacharewski T, Chen J-Y, Staub A, Garnier J-M Mader S et al (1992) Purification, cloning and RXR identity of the HeLa cell factor with which RAR or TR heterodimerizes to bind target sequences efficiently. *Cell* 68: 377–395

75 Yu VC, Delsert C, Anderson B, Holloway JM, Devary O, Naeaer AM, Kim SY, Boulin J-M, Glass CK, Rosenfeld MG (1991) RXRβ: A coregulator that enhances binding of retinoic acid, thyroid hormone, and vitamin D receptors to their cognate response elements. *Cell* 67: 1251–1266

76 Kastner P, Grondona J, Mark M, Gansmuller A, LeMeur M, Decimo D, Vonesch J-L, Dolle P, Chambon P (1994) Genetic analysis of RXRα developmental function: convergence of RXR and RAR signalling pathways in heart and eye morphogenesis. *Cell* 78: 987–1003

77 Creech Kraft J, Kimelman D, Juchau MR (1995) *Xenopus laevis*: A Model System for the Study of Retinoid Metabolism 1. Embryonic metabolism of 9-*cis*- and all-*trans*-retinals and retinols to their corresponding acid forms *Drug Metab Dispos* 23: 72–82

78 Collins M, Tzimas G, Hummler H, Buergin H, Nau H (1994) Comparative teratology and transplacental pharmacokinetics of all-*trans*-retinoic acid, 13-*cis*-retinoic acid and retinyl palmitate following daily administration in rats. *Toxicol Appl Pharmacol* 127: 132–144

79 Collins M, Tzimas G, Burgin H, Hummler H, Nau H (1995) Single versus multiple dose administration of all-*trans*-retinoic acid during organogenesis: differential metabolism and transplacental kinetics in rat and rabbit. *Toxicol Appl Pharmacol* 130: 9–18

80 Creech Kraft J, Kimelman D, Juchau MR (1995) *Xenopus laevis*: A Model System for the Study of Retinoid Metabolism 2. Embryonic metabolism of all-*trans*-3,4-didehydroretinol to all-*trans*-3,4-didehydroretinoic acid. *Drug Metab Dispos* 23: 83–89

81 Eckhoff C, Chari S, Kromka M, Staudner H, Juhasz Rudiger H, Agnish N (1994) Teratogenicity and transplacental pharmacokinetics of 13-*cis*-retinoic acid in rabbits. *Toxicol Appl Pharmacol* 125: 34–41

82 Frolik CA, Roller P, Roberts AB, Sporn MB (1980) *In vitro* and *in vivo* metabolism of all-*trans* and 13-*cis*

retinoic acid in hamsters. Identification of 13-*cis*-4-oxo-retinoic acid. *J Biol Chem* 255: 8057–8062

83 Kalin JR, Wells MJ, Hill D (1984) Effects of phenobarbital, 3-methylcholanthrene, and retinoid pretreatment on disposition of orally administered retinoids in mice. *Drug Metab Dipos* 12: 63–67

84 Sandberg JA, Eckhoff C, Nau H, Slikker W (1994) Pharmacokinetics of 13-*cis*-, all-*trans*-, 13-*cis*-4-oxo-, and all-*trans*-4-oxo-retinoic acid after intravenous administration in the Cynomolgus monkey. *Drug Metab Dispos* 22: 154–160

85 Urbach J, Rando RR (1994) Isomerization of all-*trans*-retinoic acid to 9-*cis*-retinoic acid. *Biochem J* 299: 459–465

86 Morriss-Kay GM (1993) Retinoic acid and craniofacial development: molecules and morphogenesis. *Bioessays* 15: 9–15

87 Avantaggiato V, Acampora D, Tuorto F, Simeone A (1996) Retinoic acid induces stage specific repatterning of the rostral central nervous system. *Dev Biol* 175: 347–357

88 Cunningham ML, MacAuley A, Mirkes PE (1994) From gastrulation to neurulation: transition in retinoic acid sensitivity identifies distinct stages of neural patterning in the rat. *Develop Dyn* 200: 227–241

Vitamin A and retinoids: an update of biological aspects and clinical applications
M.A. Livrea (ed.)
© 2000 Birkhäuser Verlag Basel/Switzerland

Physiologically-based pharmacokinetic scaling in retinoid developmental toxicity

C.C. Willhite[1] and H.J. Clewell[2]

[1]State of California, 700 Heinz Street, Suite 200, Berkeley, CA 94710, USA
[2]ICF/Kaiser International, 602 East Georgia Avenue, Ruston, LA 71270, USA

Retinoids in embryonic development

Requirement of the retinoids (as the retinoic acid precursors retinol and its esters) in normal embryonic development was first demonstrated in farm and laboratory animals more than 60 years ago. While there are scattered reports of anophthalmia and other terata among offspring of malnourished women [1], the requirement for adequate dietary retinol and retinyl esters in prenatal and neonatal development has been documented most clearly from intervention trials [2–4]. Bolus administration of 200 000 IU vitamin A once per month during the first four months of life in deficient populations reduced neonatal mortality 30–60% [5]. Among low birth weight infants, low hepatic and circulating retinol levels have been demonstrated [6–8]; it is evident that maternal retinoid deficiency (plasma retinoid ~10–20 µg/100 ml) increases risk of embryonic and fetal demise and precipitates retarded intrauterine growth. One factor contributing to increased mortality among neonates born to retinoid-deficient mothers is reduced development of the fetal lung [9, 10]. The present U.S. National Research Council recommended daily allowance (RDA) of dietary retinoid for adults is 5000 IU and it is increased for pregnant/lactating women to 8000 IU (2400 retinol equivalents). The RDA values are based upon maintenance of normal dark adaptation, taking into account bioavailability and individual differences in vitamin utilization. The RDA assumes 50% of total daily intake as retinol and its esters and 50% from β-carotene.

Retinol and its esters are teratogenic in all species examined to date, due to their dose-dependent biotransformation to the retinoic acids [1, 11]. Whereas Rothman et al. [12] concluded that cranial neural crest defects are increased among offspring of mothers consuming 8000–10 000 IU retinol/day, the methods used to reach those conclusions have come into question [13, 14]. Epidemiological studies of individual malformations in relation to vitamin A supplements have either shown no adverse effect [15] or found a possible increase in certain defects [16]. Retinoid embryopathy exhibits a U-shaped dose-response relationship, with both deficiency and excess presenting risk of developmental toxicity. Even though vitamin A teratogenesis in humans has been suspect for many years [17], the conditions of excessive exposure or dose required to elicit terata remain the subject of considered debate [13, 14].

The toxicology, teratogenic profile and mechanism of retinoid action have been reviewed [1, 11, 18–20]. Among the drawbacks to currently approved retinoids, 13-*cis*-retinoic acid (isotretinoin; Accutane), all-*trans*-retinoic acid (tretinoin) and etretinate (Tegason), remain their marked teratogenic properties [21]. Whereas management of the female patient undergoing retinoid therapy can be accomplished by careful adherence to negative serum pregnancy tests, mandatory contraception and informed consent, the discovery of efficacious oral retinoids with sufficiently large therapeutic ratios so as to negate concern for the embryo has yet to be accomplished. With synthesis of >10 000 congeners that failed to separate completely retinoid efficacy from toxicity and the realization that retinoid nuclear receptors (RAR) responsible for their action in dermatology and chemoprevention are identical to those in the embryo, it is unlikely that an orally active therapeutic retinoid can be devised which is totally devoid of the characteristic spectrum of toxicity. Even among new retinoids like the conformationally restricted adamantyl naphthoic acid derivatives (adapalene; Differin) having improved local (topical) therapeutic ratios [22], the RAR selectivity is associated with teratogenic potential. Nonetheless, retinoids like adaplanene with topical efficacy are clearly less dangerous for the embryo than oral administration of teratogenic retinoids like etretinate due to their slow percutaneous absorption and limited systemic bioavailability. It is the dose (route, timing and duration of exposure) in the fertile patient which determines the potential for retinoid-induced developmental toxicity.

Physiologically-based pharmacokinetics (PBPK)

Regulatory extrapolation of dose from animal developmental toxicity data to humans relies in the main on the use of safety factors. Until recently, only classical ('box model') pharmacokinetic analyses have been available to assist in species extrapolation for preclinical safety assessment. In contrast to the abstract or classical pharmacokinetic compartments used in common preclinical [23] and clinical [24] investigations, PBPK models describe the uptake and disposition in the various organs of the parent compound, its metabolites and critical biological processes [25]. Classical compartment analysis allows correlation between peak exposure (C_{max}) or total exposure (area under the concentration:time curve; AUC) and teratogenic outcome. If C_{max} is correlated with embryotoxicity this implies that drug absorption and distribution dictate a critical (threshold) concentration that must be achieved to elicit toxicity; if AUC is correlated with response, this implies that toxicity is related inversely to drug clearance [26]. From those results, and based on administered dose, it is customary that uncertainty or safety factors be used to accomplish interspecies scaling of dose. Using knowledge of the xenobiotic biochemical mechanism of action, PBPK disposition results can be linked to develop physiologically-based pharmacodynamic (PBPD) assessments that allow quantitative description of complex dose behavior and response in the target tissue. PBPK exposure:response assessments do not require intuitive uncertainty or safety factors to scale between species. Quantitative PBPK/PD descriptors allow refinement of understanding between administered dose and target tissue dose and between target tissue dose and the observed response. PBPK/PD models are biologically and mechanistically based and can be used to extrapolate rigorously between the pharmacokinetic behavior of a substance from high to low dose, from one route of administration to another, between species and within different subpopulations of the same species (e.g.

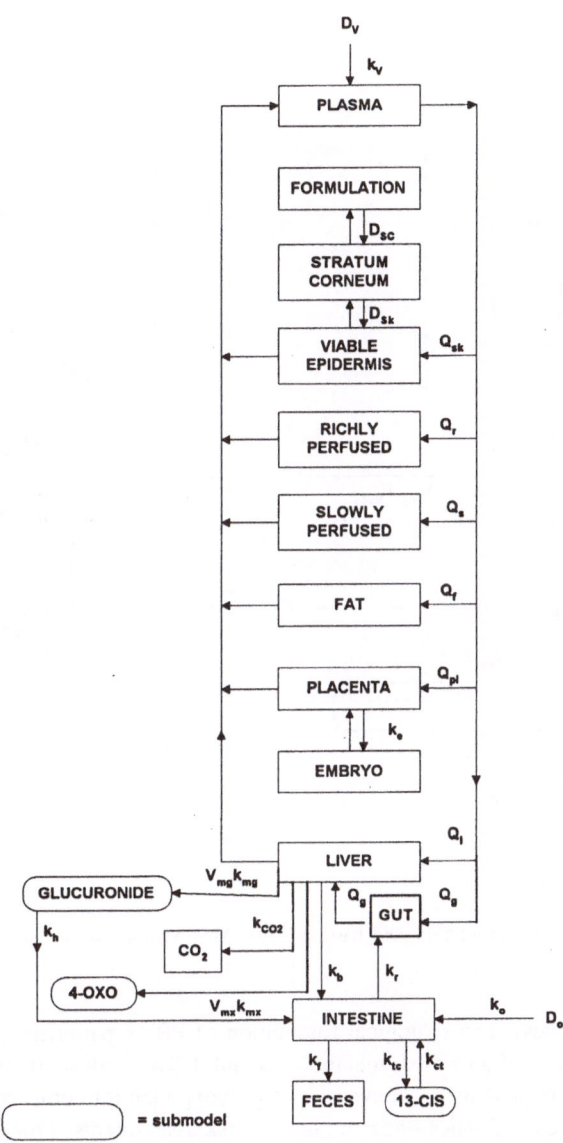

Figure 1. PBPK model for all-*trans*-retinoic acid and its metabolites. Abbreviations are: D_v intravenous dose (mg); Q or P or C, flow rate or partition coefficient or concentration (mg/l); subscripts c (cardiac), f (fat), g (gut), l (liver), pl (placenta), r (richly perfused tissues: muscle, bone), s (slowly perfused tissues: mammary gland, uterus), sk (skin); D_{sc} diffusivity in stratum corneum (cm^2/hr); k rate constant subscripts: b (biliary clearance), CO_2 (side chain oxidation to carbon dioxide/hr), ct (*cis/trans* isomerization), tc (*trans/cis* isomerization); e (diffusion limited transfer between placenta and embryo, l/hr); f (fecal excretion/hr); h (hydrolysis of glucuronide/hr); o (oral absorption); r (intestinal absorption); v (intravenous injection); K or V (affinity constant or maximum velocity where mg = apparent Michaelis-Menten glucuronidation constant and mx = apparent Michaelis-Menten oxidation constant). Rounded box indicates the submodels as diagrammed in Figures 2–5. (Reproduced with permission of Mosby-Year Book, Inc. from the American Academy of Dermatology. Clewell, [29].)

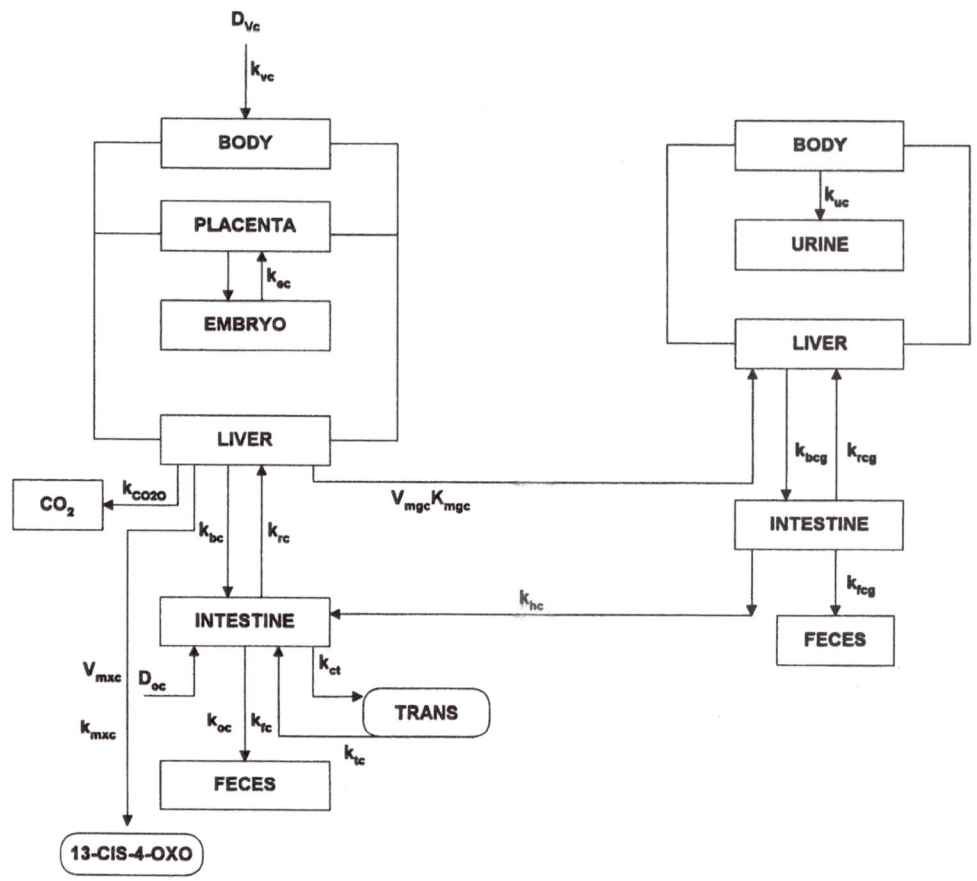

Figure 2. PBPK submodel for 13-*cis*-retinoic acid (isotretinoin). Abbreviations are as listed in the legend to Figure 1.

during various disease states). The biological foundation of PBPK parameters results in reliable interspecies predictions of absorbed, metabolized and delivered dose to target tissue (e.g. embryo) than can be generated using conventional polyexponential compartment analysis. PBPK models improve the confidence one can place in risk assessments—for example, application of traditional safety factors to retinol developmental toxicity data result in "safe" exposures that are less than the RDA.

Most PBPK models have been developed for and applied to adults and they are not valid in pregnancy. A limited number of PBPK methods have been published which take into account the continuous physiological and metabolic changes with gestation (maternal organ weight, tissue volume and perfusion and growth of the embryo/fetus) from conception to parturition. PBPK scaling in pregnancy for tetracycline, warfarin, morphine, trichloroethylene, lead, selenium,

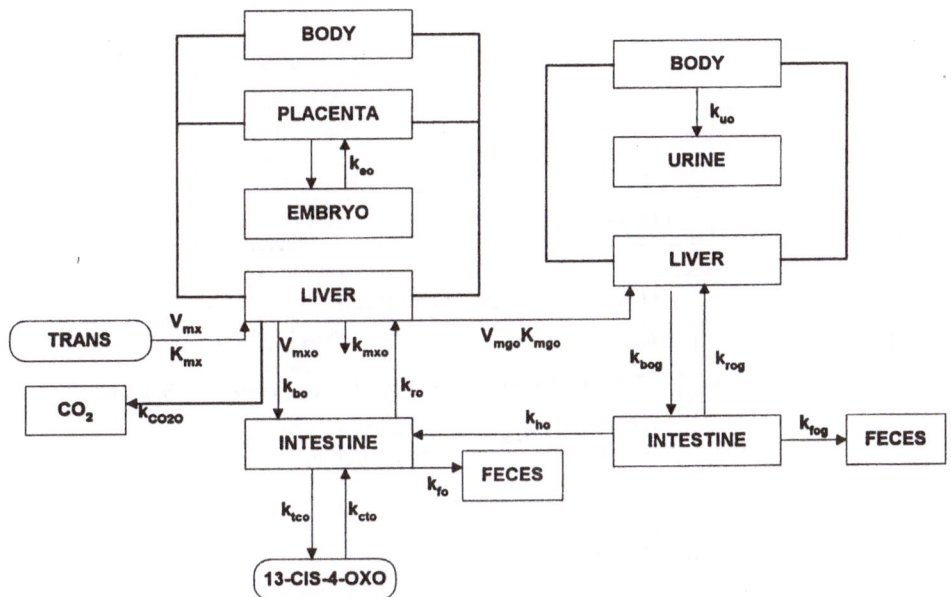

Figure 3. PBPK submodel for all-*trans*-4-oxoretinoic acid. The rounded box indicates the submodels depicted in Figure 4 (13-*cis*-4-oxoretinoic acid). Abbreviations are as listed in the legend to Figure 1.

methanol, 2-methoxyacetic acid and 2-methoxyethanol and dimethyloxazolidine-2,4-dione has been accomplished [25, 27].

Retinoid PBPK dosimetry

It is the total embryonic delivered dose of those retinoids which bind and transactivate directly the nuclear RARs which accounts for teratogenic potency [28]. A PBPK model (Figs 1–5) has been developed for intravenous, oral and topical all-*trans*-retinoic acid (Fig. 1) and 13-*cis*-retinoic acid (Fig. 2) and their metabolites to provide a biologically relevant measure of absorbed, metabolized and delivered dose across species [29]. The model describes *in vivo* *cis*/*trans* isomerization, ring oxidation (Figs 3, 4), side chain oxidation and glucuronidation (Fig. 5). Enterohepatic recirculation and fecal excretion are accounted in the submodels, and glucuronides are eliminated only in urine (Fig. 5). Retinoid distribution is flow limited except for percutaneous absorption and transplacental transfer which are diffusion-limited. Gestational parameters for rodents and primates are as published previously; partition coefficients are from *in vivo* tissue distribution studies of retinoic acid in humans, non-human primates and rodents. The dose-dependent PBPK parameters for intravenous all-*trans*-retinoic acid in the monkey were scaled allometrically to humans and the results confirmed by comparison to published clinical pharmacokinetic parameters of all-*trans*-retinoic acid after oral dosing of leukemia

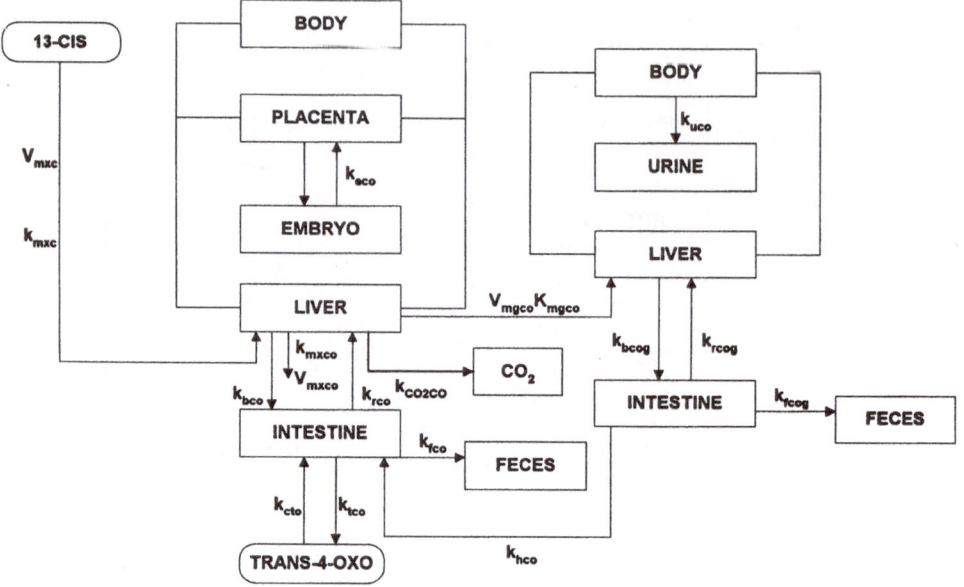

Figure 4. PBPK submodel for 13-*cis*-4-oxoretinoic acid. The rounded box indicates the submodels depicted in Figure 2 (13-*cis*-retinoic acid) and Figure 3 (all-*trans*-4-oxoretinoic acid). Abbreviations are as listed in the legend to Figure 1.

patients at 1.1 mg/kg. Since the glucuronides do not cross the placenta [28], those metabolites are not accounted in embryonic delivered dose (Fig. 5).

Comparing the results of PBPK calculated internal dose after oral 13-*cis*-retinoic acid found the effective concentrations across species to be of the same order of magnitude. The results of

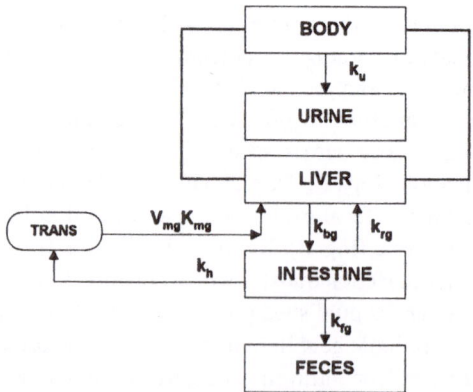

Figure 5. PBPK model for all-*trans*-retinoyl glucuronide. Rounded box indicates the submodel depicted in Figure 3 (all-*trans*-4-oxoretinoic acid). Abbreviations are as listed in the legend to Figure 1.

PBPK calculated internal dose after topical application of 0.5% all-*trans*-retinoic acid to the face, arms and chest found the circulating concentrations of retinoic acids remained four orders of magnitude less than those required for increased teratogenic risk [29]. The low absorbed and delivered dose concentrations calculated using the PBPK model are consistent with the empirical data in hamster [30] and rabbit [31] after topical compared to oral administration of this material [28, 32].

Using the *in vivo* partition coefficients of the parent material and accounting for the unique teratogenic metabolites of synthetic retinoids, the PBPK retinoid model described here can be used for interspecies scaling of other pharmacologically active congeners to improve preclinical estimation of their therapeutic ratios. With modifications to include extent and rate of dose-dependent retinol and retinyl ester biotransformation to retinoic acids, the PBPK model presented here can be used to calculate embryonic effective concentration (delivered dose) across mammalian species. For dietary retinoids as opposed to synthetic retinoids, the degree to which nutritional vitamin A status influences maternal handling of bolus retinol or its esters is not known. Understanding the range of biologically effective concentrations of dietary retinoids in prenatal development is important for quantitative derivation of the human vitamin A RDA during pregnancy.

References

1 Willhite CC, Wier PJ, Berry DL (1989) Dose-response and structure-activity considerations in retinoid-induced dysmorphogenesis. *CRC Crit Rev Toxicol* 20: 113–135
2 Sommer A, Djunaedi E, Loeden AA, Tarwotjo I, West KP, Tilden R, Mele L, Aceh Study Group (1986) Impact of vitamin A supplementation on childhood mortality. A randomized controlled community trial. *Lancet* 1: 1169–1173
3 Mulhilal Permeisih D, Idjradinata YR, Muherdiyantiningsih Karyadi D (1988) Vitamin A-fortified monosodium glutamate and health, growth and survival of children: A controlled field trial. *Amer J Clin Nutr* 48: 1271–1276
4 Rahmathullah L, Underwood BA, Thulasiraj RD, Milton RC, Ramaswamy K, Rahmathullah R, Babu G (1990) Reduced mortality among children in sourthern India receiving a small weekly dose of vitamin A. *New Engl J Med* 323: 929–935
5 West KP, Pokhrel RP, Katz J, LeClreq SC, Khatry SK, Shrestha SR, Pradhan EK, Tielsch JM, Pandey MR, Sommer A (1991) Efficacy of vitamin A in reducing preschool child mortality in Nepal. *Lancet* 338: 67–71
6 Gebre-Medhin M, Vahlquist A (1984) Vitamin A nutrition in the human fetus. A comparison of Sweden and Ethopia. *Acta Paediat Scand* 73: 333–340
7 Shenai JP, Chytil F, Stahlman MT (1985) Liver vitamin A reserves of very low birth weight neonates. *Pediat Res* 19: 892–893
8 Shah RS, Rajalakshmi R, Bhatt RV, Hazra MN, Patel BC, Swamy NB, Patel TV (1987) Liver stores of vitamin A in human fetuses in relation to gestational age, fetal size and maternal nutritional status. *Brit J Nutr* 58: 181–189
9 Sharma HS, Misra UK (1986) Postnatal distribution of vitamin A in liver, lung, heart and brain of the rat in relation to maternal vitamin A status. *Biol Neonate* 50: 345–350
10 Shenai JP, Chytil F (1990) Vitamin A storage in lungs during perinatal development in the rat. *Biol Neonate* 57: 126–132
11 Creech-Kraft JM, Willhite CC (1996) Retinoids in abnormal and normal embryonic development. *In*: GH Lambert, S Kacew (eds): *Environmental toxicology and pharmacology in human development*. Taylor and Francis, Washington, D.C., 15–49
12 Rothman KJ, Moore LL, Singer MR, Nguyen UDT, Mannino S, Milunsky A (1995) Teratogenicity of high vitamin A intake. *New Engl J Med* 333: 1369–1373
13 Werler MM, Lammer EJ, Mitchell AA (1996) Teratogenicity of high vitamin A intake. *New Engl J Med* 334: 1195
14 Brent RL, Hendrickx AG, Holmes LB, Miller RK (1996) To the editor. *New Engl J Med* 334: 1196
15 Shaw GM, Velie EM, Schaffer D, Lammer EJ (1997) Periconceptional intake of vitamin A among women and risk of neural tube defect-affected pregnancies. *Teratology* 55: 132–133

16 Werler MM, Lammer EJ, Rosenberg L, Mitchell AA (1990) Maternal vitamin A supplementation in relation to selected birth defects. *Teratology* 42: 497–503
17 Friedman JM, Polifka JE (1994) *Teratogenic effects of drugs. A resource for clinicians (TERIS)*. Johns Hopkins University Press, Baltimore, 654–656
18 Adams J, Lammer EJ (1991) Relationship between dysmorphology and neuropsychological function in children exposed to isotretinoin in utero. *In*: T Fuji, G J Boer (eds): *Functional neuroteratology of short-term exposure to drugs*. Tokyo University Press, Tokyo, 159–170
19 Armstrong RB, Ashenfelter KO, Eckhoff C, Levin AA, Shapiro SS (1994) General and reproductive toxicology of retinoids. *In*: MB Sporn, AB Roberts, DS Goodman (eds): *The retinoids. Biology, chemistry and medicine*. Raven Press, New York, 545–572
20 Morriss-Kay G (ed.) (1992) *Retinoids in normal development and teratogenesis*. Oxford University Press, Oxford
21 Orfanos CE, Ehlert R, Gollnick H (1987) The retinoids. A review of their clinical pharmacology and therapeutic use. *Drugs* 34: 459–503
22 Shroot B, Michel S (1997) Pharmacology and chemistry of adapalene. *J Amer Acad Dermatol* 36 (6 Pt 2): S96–S103
23 Kochhar DM, Kraft J, Nau H (1987) Teratogenicity and disposition of various retinoids *in vivo* and *in vitro*. *In*: H Nau, WJ Scott (eds): *Pharmacokinetics in teratogenesis. Vol. II. Experimental aspects in vivo and in vitro*. CRC Press, Boca Raton, 173–186
24 Lucek RW, Colburn WA (1985) Clinical pharmacokinetics of the retinoids. *Clin Pharmacokinet* 10: 38–62
25 Luecke RH, Wosilait WD, Pearce BA, Young JF (1994) A physiologically based pharmacokinetic computer model for human pregnancy. *Teratology* 49: 90–103
26 Nau H (1987) Species differences in pharmacokinetics, drug metabolism and teratogenesis. *In*: H Nau, WJ Scott (eds): *Pharmacokinetics in teratogenesis. Vol I Interspecies comparison and maternal-fetal drug transfer*. CRC Press, Boca Raton, 81–106
27 Welsch F, Blumenthal GM, Conolly RB (1995) Physiologically based pharmacokinetic models applicable to organogenesis: extrapolation between species and potential use in prenatal toxicity risk assessments. *Toxicol Lett* 82/83: 539–547
28 Eckhoff C, Willhite CC (1997) Embryonic delivered dose of isotretinoin (13-*cis*-retinoic acid) and its metabolites in hamsters. *Toxicol Appl Pharmacol* 146: 79–87
29 Clewell HJ, Andersen ME, Wills RJ, Latriano L (1997) A physiologically based pharmacokinetic model for retinoic acid and its metabolites. *J Amer Acad Dermatol* 36 (3 Pt 2): S77–S85
30 Willhite CC, Sharma RP, Allen PV, Berry DL (1990) Percutaneous retinoid absorption and embryotoxicity. *J Invest Dermatol* 95 (5): 523–529
31 Christian MS, Mitala JJ, Powers WJ, McKenzie BE, Latriano L (1997) A developmental toxicity study of tretinoin emollient cream (Renova) applied topically to New Zealand white rabbits. *J Amer Acad Dermatol* 36 (6 Pt 2): S67–S76
32 Eckhoff C, Chari S, Kromka M, Staudner H, Juhasz L, Rudiger H, Agnish N (1994) Teratogenicity and transplacental pharmacokinetics of 13-*cis*-retinoic acid in rabbits. *Toxicol Appl Pharmacol* 125: 34–41

Retinoic acid receptors in normal and neoplastic haematopoietic cells

F. Guidez and A. Zelent

Leukaemia Research Fund Centre at the Institute of Cancer Research, Chester Beatty Laboratories, 237 Fulham Road, London SW3 6JB, UK

Introduction

Haemopoiesis is the process by which the multiple cell lineages that constitute blood arise from a small pool of multipotent haematopoietic stem cells. Although the formation of stable transcription complexes that initiate and consolidate exclusive programmes of gene expression must play a key role in this process, exactly how a multipotential cell chooses a particular lineage remains poorly understood. Transcription factors, such as retinoid receptors and Krüppel related Zinc (Zn)-finger proteins, play important roles in a wide range of cellular processes [1–7]. Characterisation of a large number of mutations/chromosomal translocations which are associated with haematopoietic neoplasms have, in the majority of cases, identified transcription factors, including members of the above mentioned gene families, which play important roles in haemopoiesis and which serve as direct targets of oncogenic processes [8, 9]. Nevertheless, despite the molecular cloning of a plethora of translocation-generated fusion genes, which encode chimeric transcription factors, the mechanisms of leukaemogenesis remain obscure. Recent work from our and other laboratories on acute promyelocytic leukaemia (APL) has provided a major insight into the molecular pathogenesis of a haematopoietic neoplasm and the basis for its response to a number of currently-used or potential therapeutic agents (see below).

Regulation of transcription

In the past decade considerable progress has been made in the understanding of how transcriptional activators interact with the basal transcription machinery to achieve desirable levels of gene expression [10–12]. Experimental results support a general recruitment model [13] for gene activation, where strength of interaction between the activator and a component of the basal transcription machinery correlates with the degree of transcriptional enhancement [14]. Some DNA-binding factors, which themselves are unable to interact with the basal transcriptional machinery, may work through dimerization with non-DNA binding proteins (co-activators) to facilitate this interaction and recruitment [15, 16]. Activation can also be mediated through actions of activator complexes on chromatin structure which, through mechanisms such as histone acetylation, render it more accessible to other regulatory or basal factors [6, 17, 18]. Indeed, a number of co-activators possess histone acetyltransferase activities [19, 20]. In addition, SWI/SNF (switch/sucrose non-fermenting) or related complexes [21, 22], which use ATP hydrolysis to disrupt histone-DNA contacts, have been found associated with some activators [23] as well as RNA pol II holoenzyme [24].

The understanding of the mechanisms of transcriptional repression has lagged behind that of activation. Earlier studies have suggested that transcriptional repressors may interact directly with activators, sterically blocking their interactions with the basal transcriptional machinery, and/or act by displacing activators from regulatory elements through binding to overlapping DNA recognition sites. However, these do not account for the abilities of repressors to completely abolish basal transcription, even in the presence of multiple activators, suggesting a more direct action, either on the basal transcriptional machinery or chromatin structure. In this respect, a number of transcriptional repressors have recently been shown to recruit histone deacetylases which are capable of altering chromatin structure by deacetylation of the amino termini of histones [25–27].

Retinoid receptors

Retinoids encompass a continuously growing family of synthetic and natural compounds that are both structurally and functionally related to vitamin A (retinol) and its physiologically active metabolites, such as all-*trans*-retinoic acid (ATRA). Early work on the effects of vitamin A deficiency in a number of mammalian species implicated it in the control of normal differentiation and proliferation of epithelial cells [28]. Retinoid receptors, as other members of the nuclear receptor superfamily, can activate or repress transcription of target genes in a manner that is controlled by binding of small hydrophobic molecules [29–32]. To date, three different *retinoic acid receptor* (*RAR*) and *retinoid X receptor* (*RXR*) genes have been characterised, each encoding multiple N-terminal protein isoforms [29, 33–36]. RXRs serve as co-regulators for RARs, as well as many other nuclear receptors including those for thyroid hormones (TR) and vitamin D_3 (VDR), thus integrating different signalling pathways [32, 33]. RARs and RXRs contain a highly conserved cysteine-rich DNA binding domain (DBD) that forms two Zn-fingers allowing both protein-DNA and protein-protein interactions [31–33]. A second conserved region in the carboxy terminus possesses the ligand binding domain (LBD, see Fig. 1 for a schematic representation), dimerization interface, and the ligand-dependent activation function (AF-2) [31, 33]. The N-terminal A/B region possesses a ligand-independent activation function (AF-1) which can act in a cell type and promoter-specific manner [33, 37–39]. Recently a number of co-activators of the AF-2 domain have been characterised. Proteins of the SRC (Steroid Receptor Co-activator), or TIF (Transcription Intermediary Factor) family and CBP have been shown to interact in a ligand-dependent manner with a number of nuclear receptors, including RARs, and to stimulate their AF-2 activities [40–45]. SRC-1 and CREB binding protein (CBP) proteins have been shown to possess intrinsic histone acetyltransferase activities [19, 46]. In the absence of ATRA, RARs remain associated with the nuclear receptor co-repressors, N-CoR (negative co-regulator) [47] or SMRT (silencing mediator for retinoid and thyroid-hormone receptors) [48], and repress basal transcription. Both N-CoR [49, 50] and SMRT [51] have been shown to associate with the mammalian homologues (mSin3A and mSin3B) of the yeast global transcriptional repressor SIN3 [52–54] and histone deacetylase [55], and are thought to

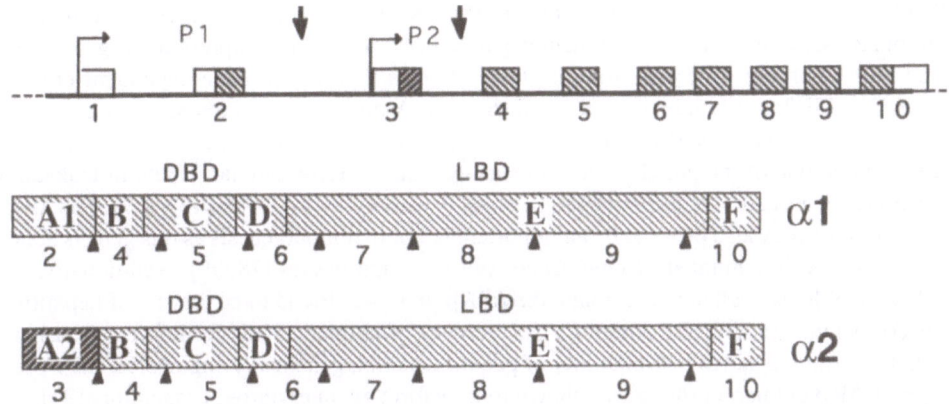

Figure 1. Schematic representation of the RARα gene and the RARα1 and α2 isoforms. Exons are represented by shaded boxes and are numbered consecutively. Regions which are not shaded represent 5' and 3' untranslated sequences. Two promoters, P1 and P2, are indicated by broken arrows. Positions of translocation breakpoints within the RARα gene are indicated with horizontal arrows. Isoforms are indicated by hatched rectangles subdivided into their conserved (B-F) and divergent (A regions) functional domains. Different patterns are used to represent RAR isoform-specific sequences. Regions encoded by different exons are indicated by arrowheads underneath each diagram. DNA (DBD) and ligand (LBD) binding domains (regions C and E, respectively) are indicated above the scheme for each isoform.

repress transcription through histone deacetylation, rendering the nearby chromatin inaccessible to transcriptional activators and/or basal transcription factors. Histone deacetylases have also been implicated in transcriptional repression by Mad/Max [50, 56, 57] or Max/Mxi [49] heterodimers, the retinoblastoma protein [58–60] and the methylcytosine binding protein 2 (MeCP2) which specifically bind methylated DNA [61].

Retinoid receptors in haemopoiesis

The action of vitamin A in controlling cellular proliferation and differentiation is not restricted to epithelial cells [62, 63]. Avitaminosis A was also associated with a reduction in the number of haematopoietic cells in the bone marrow [28] and anemia [64]. An important role for retinoids in haemopoiesis was corroborated by studies demonstrating that ATRA can stimulate the proliferation and maturation of erythroid and myeloid precursor cells *in vitro* [65], and exert beneficial effects on the function of the human immune system [66]. Inhibition and stimulation of granulocytic differentiation by dominant negative RARα mutants [67, 68] (or antagonists of RARα [69, 70] and RARα specific agonists [70, 71], respectively), suggest that granulopoiesis requires activation of RARα transcriptional activity by physiological concentrations of ATRA. Consistently with the above observations, we and others have shown that expression of the RARα gene, particularly the RARα2 isoform, increases during granulopoiesis [70].

Given their role in the control of cellular growth and proliferation, it is not surprising that some nuclear receptor loci, including *RARα* (see below), have turned out to be protooncogenes

[72, 73]. For example, the oncogenic form of the TR, v-erbA oncogene, present in the avian erythroblastosis virus, acts in a dominant negative fashion to inhibit expression of genes associated with erythroid differentiation and hence blocks thyroid hormone-dependent erythropoiesis [74, 75]. Interestingly, the ability of the v-erbA oncoprotein to transform haematopoietic cells has been correlated with its ability to suppress the retinoid response [76]. It has been suggested that one of the possible mechanisms by which v-erbA can participate in leukaemogenesis is the abrogation of the inactivation of AP-1 activity by RARs and TRs [77]. The *RARβ* gene disruption may also prove to be an important factor in tumourgenesis as this gene is abnormally expressed in a number of lung cancer cell lines and tissues [78, 79]. Additionally, in a single case of hepatocellular carcinoma, the *RARβ* gene was found to be the site of hepatitis B virus (HBV) integration [80]. Although the presence of chimeric HBV-RARβ fusion proteins was not determined due to unavailability of patient material, genetically engineered HBV pre-S antigen-RARβ chimeric protein was shown to transform avian erythroid progenitors [81].

Acute promyelocytic leukaemia

APL is most often associated with the t(15;17)(q22;q21) reciprocal chromosomal translocation which causes the fusion of the *RARα* locus with a gene of unknown function called *PML* (for promyelocytic leukaemia) and expression of PML-RARα chimeric proteins in all leukaemic cells [82–86]. Expression of the PML-RARα protein in transgenic mice results in development of APL which, as the human disease, can be induced into remission by ATRA treatment [87–90]. These results and studies with APL cells *in vitro*, or cells exogenously expressing PML-RARα, suggest that this oncoprotein is also a primary target of ATRA action [91, 92].

The wild type PML protein localises onto nuclear bodies (NBs, also called PML oncogenic domains or PODs) [93–95]. Expression of the PML-RARα chimeric protein causes delocalisation of PML, and other components of NBs, to microspeckled nuclear structures and differentiation of APL cells with ATRA restores their normal localisation pattern [86, 93–95]. At present it is not clear what relationship, if any, this phenomenon has to the pathogenesis and treatment of APL.

To date, three other APL-associated translocations of the *RARα* gene have been characterised at the molecular level. The t(5;17)(q35;q21) [96], t(11;17)(q23;q21) [97, 98] and t(11;17)(q13;q21) [99] fuse *RARα* to *Nucleophosmin* (*NPM*), *Promyelocytic Leukaemia Zinc Finger* (*PLZF*), and *Nuclear Mitotic Apparatus* (*NuMA*) genes, respectively. These translocations result in the expression of chimeric RARα fusion proteins, which all retain the DNA and ligand binding domains of the receptor and gain a dimerization domain from the fusion partner (see Fig. 2 for a schematic representation). So far, the t(5;17)(q35;q21) and t(11;17)(q13;q21) have only been reported in index cases and, as with t(15;17)-associated APL, appeared to respond to treatment with ATRA [96, 99]. In contrast, APL with t(11;17)(q23;q21) has been reported on a recurrent basis, albeit at very low frequency, and has consistently been found unresponsive to ATRA therapy [100, 101].

The PLZF gene encodes a protein with an N-terminal POZ- (pox virus and zinc-finger-) domain and nine C-terminal Krüppel-like Zn-fingers [97]. In addition to PLZF, one other POK

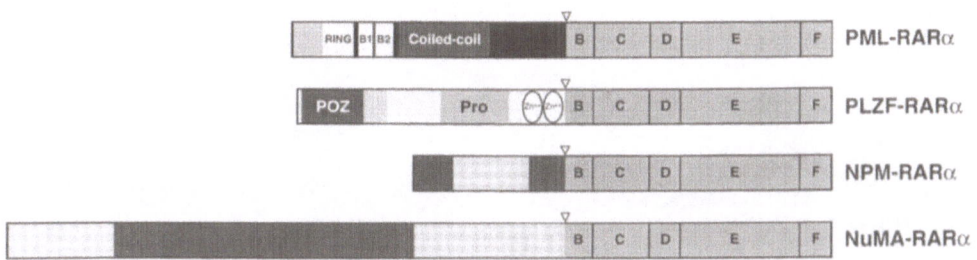

Figure 2. Schematic representation of the four APL associated RARα fusion proteins. RARα functional domains (B-F) are as indicated. Open triangles denote boundaries between sequences encoded in separate exons lying on opposite sides of a given translocation breakpoint. Different patterns are used to represent various putative functional regions of the PML, PLZF, NPM and NuMA proteins. Circled Zn^{2+} symbols and PML regions labelled as RING, B1 and B2 represent two of the nine Krüppel-like Zn-finger motifs and cystein-histidine rich domains in the PLZF and PML proteins, respectively. Dark grey regions (labelled Coiled-coil and POZ in PML and PLZF, respectively) represents protein-protein interaction motifs present in N-termini of all the RARα chimeras. It is worth noting that aside from these functionally-related structural motifs, genes which are fused with the RARα appear to have very little in common. Nevertheless, these heterologous dimerization interfaces allow the RARα chimeric proteins to form stable homodimers with altered DNA binding specificities of the RARα DBD [115, 117]. It is not clear at the present time whether the non-RARα sequences of various chimeric proteins play an active or passive role(s) in transformation. Retinoids and retinoid receptors affect not only differentiation of haematopoietic cells but also cell growth and apoptosis [121, 122]. One could therefore argue that the entire APL phenotype, block of differentiation and inhibition of programmed cell death, is a result of a disruption in retinoid signalling by the RARα chimeric proteins.

(for POZ and Krüppel) protein, called LAZ-3/BCL-6 (for Lymphoma Associated Zn-finger-3/B-Cell Lymphoma-6), has been implicated in oncogenesis [102–105]. Interestingly, both PLZF and LAZ-3/BCL-6 have recently been shown to function as DNA sequence specific transcriptional repressors which recruit the same histone deacetylase co-repressor complexes as RARs and other nuclear receptors [106–111].

The PLZF-RARα fusion protein consists of the N-terminus of the PLZF protein, including two of its nine Zn-fingers, linked to the DNA (region C) and ligand binding (region E) domains of the RARα protein (see Fig. 2 for a schematic representation). Interestingly, localisation of the PML protein and integrity of NBs remain undisturbed in APL cells expressing PLZF-RARα [112], as well as NPM-RARα [113] and NuMA-RARα [99] proteins. Nevertheless, both PML-RARα and PLZF-RARα localise to the same microspeckled structures [112, 114] and the wild type PLZF protein co-localises completely with the PML-RARα oncoprotein in APL cells [112].

Both PML-RARα and PLZF-RARα can bind to a retinoic acid response element (RARE) as homodimers or, in combination with RXR, as multimeric complexes [115–117], and in comparison to the wild type RARα possess impaired transcriptional activities in response to ATRA [82, 83, 86, 116–118]. Expression of the PLZF-RARα protein in transgenic animals leads to the development of a myeloid leukaemia which, in contrast to PML-RARα transgenic mice, is resistant to ATRA [87]. Both PML-RARα and PLZF-RARα fusion proteins act as transcriptional repressors and are able to interact with RARα transcriptional co-repressors such as SMRT and N-CoR. However, unlike PML-RARα, PLZF-RARα can form, via its PLZF moiety, co-repressor complexes which are insensitive to ATRA [87, 107–110]. These data, for the

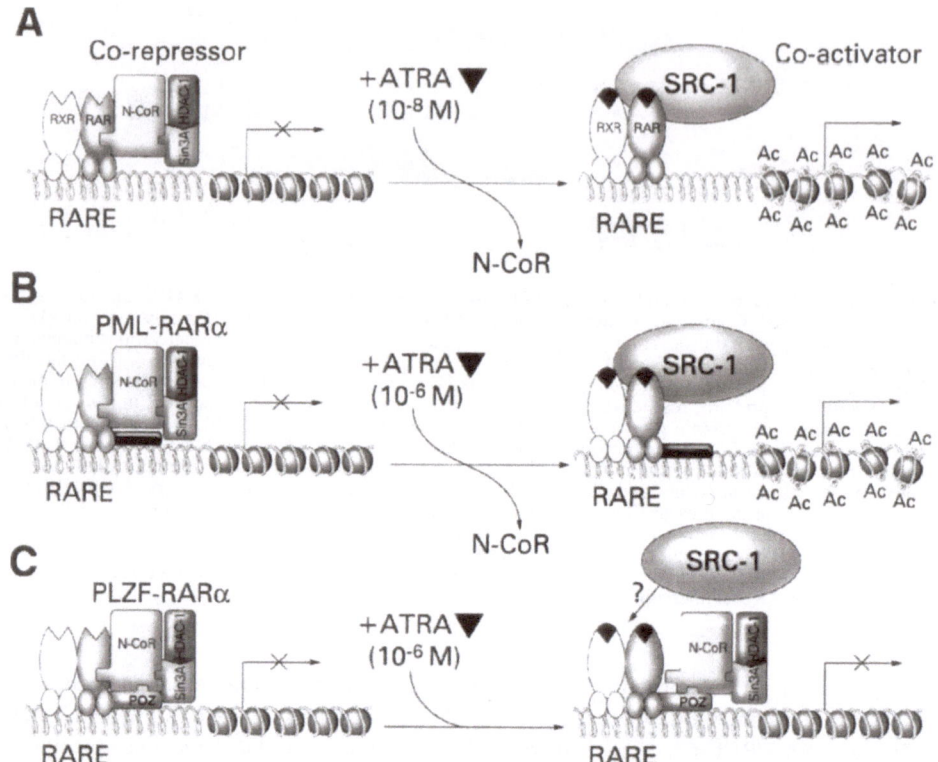

Figure 3. Model for the role of nuclear receptor co-repressor N-CoR/mSin3A/histone deacetylase 1 (HDAC1) complex and RARα fusion proteins in the pathogenesis and treatment of APL. (A) In the absence of ATRA, RARα, PML-RARα and PLZF-RARα associate with N-CoR/mSin3A/HDAC1 co-repressor complex. The associated co-repressor/RXR:RARα complex acts on chromatin structure by deacetylation of histone tails inducing its reorganisation into a repressed state which is inaccessible to basal transcription factors. Binding of ATRA (triangle) induces conformational change in the RARα causing dissociation of the co-repressor complex and association of a co-activator, such as SRC-1, with intrinsic histone acetyltransferase activity [46]. Acetylation (Ac) of histone tails disrupts tightly packed and repressed chromatin structure allowing access of the basal transcription factors and transcriptional activation. Physiological concentrations of ATRA (10^{-8}) are sufficient to induce this process. (B) In the case of the PML-RARα protein, pharmacological doses of ATRA $(10^{-6}$ M) are required to achieve efficient dissociation of the N-CoR co-repressor complex from the chimeric protein and transcriptional activation. (C) Due to additional, ligand insensitive, interactions between the PLZF moiety of the PLZF-RARα fusion protein and N-CoR (and possibly also mSin3A and HDAC1), the co-repressor/PLZF-RARα complex remains associated even in the presence of physiological concentrations of ATRA and, in the absence of chromatin remodelling by histone acetylation, transcription remains inhibited.

first time, directly implicate the nuclear receptor co-repressor/histone deacetylase complex in carcinogenesis and provide a plausible mechanism (see Fig. 2) for the molecular pathogenesis of APL and its response to treatment with ATRA. As this model predicts, histone deacetylase inhibitors, such as trichostatin A (TSA) or sodium butyrate (NaB), in combination with ATRA, can overcome the transcriptional repressive activities of PML-RARα and PLZF-RARα as well as the unresponsiveness of PLZF-RARα leukaemic cells to ATRA [87, 108–110], and therefore, could serve as useful adjuncts to retinoids in APL therapy [119]; particularly in a subset of ATRA-resistant APLs which are characterised by mutations in the RARα moiety of the PML-RARα fusion protein [110] (accounting for approximately 25% of resistant cases [120]), or which are associated with expression of PLZF-RARα [87, 108, 110].

Concluding remarks

So far, APL has been an excellent model system to study both the mechanisms of leukaemogenesis and approaches to cancer treatment. Although it remains to be seen how general the findings derived from the studies of this malignancy will become, it is worth noting that recent work suggests that a subset of leukaemias involving the AML1 (acute myeloid leukaemia 1) gene translocations (F. Guidez, A. Zelent, unpublished observations) may be associated with a similar histone deacetylase-dependent mechanism as APL; and therefore, may be good candidates for therapy with histone deacetylase (HDAC) inhibitors. Finally, diffuse large cell lymphoma, which is associated with deregulated expression of the *LAZ-3/BCL-6* gene, may also be a target for therapy with HDAC inhibitors as LAZ-3/BCL-6 has been shown to function through a HDAC-dependent mechanism [106].

Acknowledgements
We thank L. Wiedemann, T. Enver, J. Licht, S. Waxman, Z. Chen and M. Greaves for helpful comments and discussions. We are grateful to the Leukaemia Research Fund of Great Britain, National Institutes of Health (Grant No. CA59936-01) and the TMR Programme Marie Curie Research Training Grant form the European Commission for support.

References

1   Schneider-Maunoury S, Topilko P, Seitanidou T, Levi G, Cohen-Tannoudji M, Pournin S, Babinet C, Charnay P (1993) Disruption of *Krox-20* results in alteration of rhombomeres 3 and 5 in the developing hindbrain. *Cell* 75: 1199–1214
2   El-Baradi T, Pieler T (1991) Zinc finger proteins: what we know and what we would like to know. *Mech Develop* 35: 155–169
3   Lee SL, Wang Y, Milbrandt J (1996) Unimpaired macrophage differentiation and activation in mice lacking the zinc finger transplantation factor NGFI-A (EGR1). *Mol Cell Biol* 16: 4566–4572
4   Perrotti D, Melotti P, Skorski T, Casella I, Peschle C, Calabretta B (1995) Overexpression of the zinc finger protein MZF1 inhibits hematopoietic development from embryonic stem cells: correlation with negative regulation of CD34 and c-myb promoter activity. *Mol Cell Biol* 15: 6075–6087
5   Shivdasani RA, Orkin SH (1996) The transcriptional control of hematopoiesis. *Blood* 87: 4025–4039
6   Wade PA, Pruss D, Wolffe AP (1997) Histone acetylation: chromatin in action. *Trends Biochem Sci* 22: 128–132
7   Kastner P, Mark M, Chambon P (1995) Nonsteroid nuclear receptors—What Are genetic studies telling us about their role in real life. *Cell* 83: 859–869
8   Look A (1997) Oncogenic transcription factors in the human acute leukemias. *Science* 278: 1059–1064
9   Rabbitts TH (1994) Chromsomal translocations in human cancer. *Nature* 372: 143–149
10 Orphanides G, Lagrange T, Reinberg D (1996) The general transcription factors of RNA polymerase II. *Gene Develop* 10: 2657–2683
11 Roeder RG (1996) The role of general initiation factors in transcription by RNA polymerase II. *Trends Biochem Sci* 21: 327–335
12 Verrijzer CP, Tjian R (1996) TAFs mediate transcriptional activation and promoter selectivity. *Trends Biochem Sci* 21: 338–342

13 Ptashne M, Gann A (1997) Transcriptional activation by recruitment. *Nature* 386: 569–577
14 Dove SL, Joung JK, Hochschild A (1997) Activation of prokaryotic transcription through arbitrary protein-protein contacts. *Nature* 386: 627–630
15 Janknecht R, Hunter T (1996) Versatile molecular glue. Transcriptional control. *Curr Biol* 6: 951–954
16 Janknecht R, Hunter T (1996) A growing coactivator network. *Nature* 383: 22–23
17 Vettesedadey M, Grant PA, Hebbes TR, Cranerobinson C, Allis CD, Workman JL (1996) Acetylation of histone H4 plays a primary role in enhancing transcription factor-binding to nucleosomal DNA *in vitro*. *EMBO J* 15: 2508–2518
18 Grunstein M (1997) Histone acetylation in chromatin structure and transcription. *Nature* 389: 349–352
19 Montminy M (1997) Something new to hang your HAT on. *Nature* 387: 654–655
20 Torchia J, Rose DW, Inostroza J, Kamei Y, Westin S, Glass CK, Rosenfeld MG (1997) The transcriptional co-activator p/CIP binds CBP and mediates nuclear-receptor function. *Nature* 387: 677–684
21 Pazin MJ, Kadonaga JT (1997) SWI2/SNF2 and related proteins: ATP-driven motors that disrupt protein-DNA interactions? *Cell* 88: 737–740
22 Peterson CL (1996) Multiple SWItches to turn on chromatin? *Curr Opin Genet Develop* 6: 171–175
23 Yoshinaga SK, Peterson CL, Herskowitz I, Yamamoto KR (1992) Roles of SWI1, SWI2, and SWI3 proteins for transcriptional enhancement by steroid receptors. *Science* 258: 1598–1604
24 Wilson CJ, Chao DM, Imbalzano AN, Schnitzler GR, Kingston RE, Young RA (1996) RNA polymerase II holoenzyme contains SWI/SNF regulators involved in chromatin remodeling. *Cell* 84: 235–244
25 Wolffe AP (1997) Sinful repression. *Nature* 387: 16–17
26 DePinho R (1998) Transcriptional repression. The cancer-chromatin connection. *Nature* 391: 535–536
27 Bestor TH (1998) Gene silencing. Methylation meets acetylation. *Nature* 393: 311–312
28 Wolbach SB, Howe PR (1925) Tissue changes following deprivation of fat-soluble A vitamin. *J Exp Med* 42: 753–777
29 Chambon P (1996) A decade of molecular biology of retinoic acid receptors. *FASEB J* 10: 940–954
30 Wen DX, McDonnell DP (1995) Advances in our understanding of ligand-activated nuclear receptors. *Curr Opin Biotechnol* 6: 582–589
31 Mangelsdorf DJ, Evans RM (1995) The RXR heterodimers and orphan receptors. *Cell* 83: 841–850
32 Mangelsdorf DJ, Thummel C, Beato M, Herrlich P, Schutz G, Umesono K, Blumberg B, Kastner P, Mark M, Chambon P et al (1995) The nuclear receptor superfamily—the 2nd decade. *Cell* 83: 835–839
33 Leid M, Kastner P, Chambon P (1992) Multiplicity generates diversity in the retinoic acid signalling pathways. *Trends Biochem Sci* 17: 427–433
34 Leid M, Kastner P, Lyons R, Nakshatri H, Saunders M, Zacharewski T, Chen J-Y, Staub A, Garnier J-M, Mader S et al (1992) Purification, cloning, and RXR identity of the HeLa cell factor with which RAR or TR heterodimerizes to bind target sequences efficiently. *Cell* 68: 377–395
35 Zelent A (1995) Molecular mechanisms of retinoid action. *In*: L Degos, DR Parkinson (eds): *Retinoids in Oncology.* Springer-Verlag, Heidelberg, 3–25
36 Mangelsdorf DJ, Borgmeyer U, Heyman RA, Zhou JY, Ong ES, Oro AE, Kakizuka A, Evans RM (1992) Characterization of three RXR genes that mediate the action of 9-*cis* retinoic acid. *Gene Develop* 6: 329–344
37 Nagpal S, Saunders M, Kastner P, Durand B, Nakshatri H, Chambon P (1992) Promoter context- and response element-dependent specificity of the transcriptional activation and modulating functions of retinoic acid receptors. *Cell* 70: 1007–1019
38 Tora L, Gronemeyer H, Turcotte B, Gaub MP, Chambon P (1988) The N-terminal region of the chicken progesterone receptor specifies target gene activation. *Nature* 333: 185–188
39 Tasset D, Tora L, Fromental C, Scheer E, Chambon P (1990) Distinct classes of transcriptional activating domains function by different mechanisms. *Cell* 62: 1177–1187
40 Baur EV, Zechel C, Heery D, Heine MJS, Garnier JM, Vivat V, Ledouarin B, Gronemeyer H, Chambon P, Losson R (1996) Differential ligand-dependent interactions between the AF-2 activating domain of nuclear receptors and the putative transcriptional intermediary factors mSUG1 and TIF1. *EMBO J* 15: 110–124
41 Kamei Y, Xu L, Heinzel T, Torchia J, Kurokawa R, Gloss B, Lin SC, Heyman RA, Rose DW, Glass CK et al (1996) A CBP integrator complex mediates transcriptional activation and AP-1 inhibition by nuclear receptors. *Cell* 85: 403–414
42 Le Douarin B, Zechel C, Garnier JM, Lutz Y, Tora L, Pierrat B, Heery D, Gronemeyer H, Chambon P, Losson R (1995) The N-terminal part of TIF1, a putative mediator of the ligand- dependent activation function (AF-2) of nuclear receptors, is fused to B-Raf in the oncogenic protein T18. *EMBO J* 14: 2020–2033
43 Halachmi S, Marden E, Martin G, MacKay H, Abbondanza C, Brown M (1994) Estrogen receptor-associated proteins: Possible mediators of hormone induced transcription. *Science* 264: 1455–1458
44 Voegel JJ, Heine MJS, Zechel C, Chambon P, Gronemeyer H (1996) TIF2, a 160 kDa transcriptional mediator for the ligand-dependent activation function AF-2 of nuclear receptors. *EMBO J* 15: 3667–3675
45 Onate SA, Tsai SY, Tsai MJ, Omalley BW (1995) Sequence and characterization of a coactivator for the steroid-hormone receptor superfamily. *Science* 270: 1354–1357
46 Spencer TE, Jenster G, Burcin MM, Allis CD, Zhou J, Mizzen CA, McKenna NJ, Onate SA, Tsai SY, Tsai M-J et al (1997) Steroid receptor coactivator-1 is a histone acetyltransferase. *Nature* 389: 194–198
47 Horlein AJ, Naar AM, Heinzel T, Torchia J, Gloss B, Kurokawa R, Ryan A, Kamei Y, Soderstrom M, Glass

CK et al (1995) Ligand-independent repression by the thyroid hormone receptor mediated by a nuclear receptor co-repressor. *Nature* 377: 397–404

48 Chen JD, Evans RM (1995) A transcriptional co-repressor that interacts with nuclear hormone receptors. *Nature* 377: 454–457
49 Alland L, Muhle R, Hou H Jr, Potes J, Chin L, Schreiber-Agus N, DePinho RA (1997) Role for N-CoR and histone deacetylase in Sin3-mediated transcriptional repression. *Nature* 387: 49–55
50 Heinzel T, Lavinsky RM, Mullen TM, Soderstrom M, Laherty CD, Torchia J, Yang WM, Brard G, Ngo SD, Davie JR et al (1997) A complex containing N-CoR, mSin3 and histone deacetylase mediates transcriptional repression. *Nature* 387: 43–48
51 Nagy L, Kao HY, Chakravarti D, Lin RJ, Hassig CA, Ayer DE, Schreiber SL, Evans RM (1997) Nuclear receptor repression mediated by a complex containing SMRT, mSin3A, and histone deacetylase. *Cell* 89: 373–380
52 Wang H, Stillman DJ (1990) *In vitro* regulation of a SIN3-dependent DNA-binding activity by stimulatory and inhibitory factors. *Proc Natl Acad Sci USA* 87: 9761–9765
53 Wang H, Clark I, Nicholson PR, Herskowitz I, Stillman DJ (1990) The *Saccharomyces cerevisiae* SIN3 gene, a negative regulator of HO, contains four paired amphipathic helix motifs. *Mol Cell Biol* 10: 5927–5936
54 Wang H, Stillman DJ (1993) Transcriptional repression in *Saccharomyces cerevisiae* by a SIN3-LexA fusion protein. *Mol Cell Biol* 13: 1805–1814
55 Taunton J, Hassig CA, Schreiber SL (1996) A mammalian histone deacetylase related to the yeast transcriptional regulator Rpd3p. *Science* 272: 408–411
56 Laherty CD, Yang WM, Sun JM, Davie JR, Seto E, Eisenman RN (1997) Histone deacetylases associated with the mSin3 corepressor mediate Mad transcriptional repression. *Cell* 89: 349–356
57 Hassig CA, Fleischer TC, Billin AN, Schreiber SL, Ayer DE (1997) Histone deacetylase activity is required for full transcriptional repression by mSin3A. *Cell* 89: 341–347
58 Luo RX, Postigo AA, Dean DC (1998) Rb interacts with histone deacetylase to repress transcription. *Cell* 92: 463–473
59 Brehm A, Miska EA, McCance DJ, Reid JL, Bannister AJ, Kouzarides T (1998) Retinoblastoma protein recruits histone deacetylase to repress transcription. *Nature* 391: 597–601
60 Magnaghi-Jaulin L, Groisman R, Naguibneva I, Robin P, Lorain S, Le-Villain J, Troalen F, Trouche D, Harel-Bellan A (1998) Retinoblastoma protein represses transcription by recruiting a histone deacetylase. *Nature* 391: 601–605
61 Nan X, Ng HH, Johnson CA, Laherty CD, Turner BM, Eisenman RN, Bird A (1998) Transcriptional repression by the methyl-CpG-binding protein MeCP2 involves a histone deacetylase complex. *Nature* 393: 386–389
62 Lotan R (1980) Effects of vitamin A and its analogs (retinoids) on normal and neoplastic cells. *Biochim Biophys Acta* 605: 33–91
63 Roberts AB, Sporn MB (1984) Cellular biology and biochemistry of the retinoids. *In*: MB Sporn, AB Roberts, DS Goodman (eds): *The Retinoids*. Academic Press, Orlando, 209–284
64 Hodges RE, Sauberlich HE, Canham JE, Wallace DL, Rucker RB, Mejia LA, Mohanram M (1978) Hematopoietic studies in vitamin A deficiency. *Amer J Clin Nutr* 31: 876–885
65 Amatruda T, Koeffler H (1986) Retinoids and cells of the hematopoietic system. *In*: M Sherman (ed.): *Retinoids and Cell Differentiation*. CRC Press, Boca Raton, 79–103
66 Ross AC (1992) Vitamin A status: relationship to immunity and the antibody response. *Proc Soc Exp Biol Med* 200: 303–320
67 Tsai S, Collins SJ (1993) A dominant negative retinoic acid receptor blocks neutrophil differentiation at the promyelocyte stage. *Proc Natl Acad Sci USA* 90: 7153–7157
68 Tsai S, Bartelmez S, Heyman R, Damm K, Evans R, Collins SJ (1992) A mutated retinoic acid receptor-α exhibiting dominant-negative activity alters the lineage development of a multipotent hematopoietic cell line. *Gene Develop* 6: 2258–2269
69 Mehta K, McQueen T, Manshouri T, Andreeff M, Collins S, Albitar M (1997) Involvement of retinoic acid receptor-α-mediated signalling pathway in induction of CD38 cell-surface antigen. *Blood* 89: 3607–3614
70 Zelent A, Zhu J, Lanotte M, Gallagher R, Waxman S, Heyworth CM, Enver T (1997) Differential expression of retinoid receptors during multilineage differentiation of haematopoietic progenitor cells-role of the RARα2 isoform in normal granulopoiesis and leukaemia. *Blood* 90: 186
71 Chen JY, Clifford J, Zusi C, Starrett J, Tortolani D, Ostrowski J, Reczek PR, Chambon P, Gronemeyer H (1996) Two distinct actions of retinoid-receptor ligands. *Nature* 382: 819–822
72 Bishop JM (1986) Oncogenes as hormone receptors. *Nature* 321: 112–113
73 Green S, Chambon P (1986) A superfamily of potentially oncogenic hormone receptors. *Nature* 324: 615–617
74 Zenke M, Munoz A, Sap J, Vennstrom B, Beug H (1990) v-*erbA* oncogene activation entails the loss of hormone-dependent regulator activity of c-*erbA*. *Cell* 61: 1035–1049
75 Damm K, Thompson CC, Evans RM (1989) Protein encoded by v-*erbA* functions as a thyroid-hormone receptor antagonist. *Nature* 339: 593–597
76 Sharif M, Privalsky ML (1991) v-*erbA* oncogene function in neoplasia correlates with its ability to repress retinoic acid receptor action. *Cell* 66: 885–893
77 Desbois C, Aubert D, Legrand C, Pain B, Samarut J (1991) A novel mechanism of action for v-ErbA: abrogation of the inactivation of transcription factor AP-1 by retinoic acid and thyroid hormone receptors. *Cell* 67:

731–740
78 Houle B, Rochette-Egly C, Bradley WEC (1993) Tumor-suppressive effect of the retinoic acid receptor β in human epidermoid lung cancer cells. *Proc Natl Acad Sci USA* 90: 985–989
79 Gebert JF, Moghal N, Frangioni JV, Sugarbaker DJ, Neel BG (1991) High frequency of retinoic acid receptor β abnormalities in human lung cancer. *Oncogene* 6: 1859–1868
80 de The H, Marchio A, Tiollais P, Dejean A (1987) A novel steroid thyroid hormone receptor-related gene inappropriately expressed in human hepato-cellular carcinoma. *Nature* 330: 667–670
81 Garcia M, de The H, Tiollais P, Samarut J, Dejean A (1993) A hepatitis B virus pre-S-retinoic acid receptor β chimera transforms erythrocytic progenitor cells *in vitro*. *Proc Natl Acad Sci USA* 90: 89–93
82 Kakizuka A, Miller WH, Umesono K, Warrell RP, Frankel SR, Murty VVVS, Dimitrovsky E, Evans RM (1991) Chromosomal translocation t(15;17) in human acute promyelocytic leukemia fuses RARα with a novel putative transcription factor, PML. *Cell* 66: 663–674
83 de The H, Lavau C, Marchio A, Chomienne C, Degos L, Dejean A (1991) The PLM-RARα fusion mRNA generated by the t(15;17) translocation in acute promyelocytic leukemia encodes a functionally altered RAR. *Cell* 66: 675–684
84 Goddard AD, Borrow J, Freemont P, Solomon E (1991) Characterization of a zinc finger gene disrupted by the t(15;17) in acute promyelocytic leukemia. *Science* 254: 1371–1374
85 Pandolfi PP, Grignani F, Alcalay M, Mencarelli A, Biondi A, Lo Coco F, Grignani F, Pelicci PG (1991) Structure and origin of the acute promyelocytic leukemia myl/RARα cDNA and characterization of its retinoid-binding and transactivation properties. *Oncogene* 6: 1285–1292
86 Kastner P, Perez A, Lutz Y, Rochette-Egly C, Gaub M-P, Durand B, Lanotte M, Berger R, Chambon P (1992) Structure, localization and transcriptional properties of two classes of retinoic acid receptor α fusion proteins in acute promyelocytic leukemia (APL): structural similarities with a new family of oncoproteins. *EMBO J* 11: 629–642
87 He L-Z, Guidez F, Tribioli C, Peruzzi D, Ruthardt M, Zelent A, Pandolfi PP (1998) Distinct interactions of PML-RARα with transcriptional co-repressors determine differential responses to retinoic acid in APL. *Nat Genet* 18: 126–135
88 He LZ, Tribioli C, Rivi R, Peruzzi D, Pelicci PG, Soares V, Cattoretti G, Pandolfi PP (1997) Acute leukemia with promyelocytic features in PML/RARα transgenic mice. *Proc Natl Acad Sci USA* 94: 5302–5307
89 Brown D, Kogan S, Lagasse E, Weissman I, Alcalay M, Pelicci PG, Atwater S, Bishop JM (1997) A PML-RARα transgene initiates murine acute promyelocytic leukemia. *Proc Natl Acad Sci USA* 94: 2551–2556
90 Early E, Moore MA, Kakizuka A, Nason Burchenal K, Martin P, Evans RM, Dmitrovsky E (1996) Transgenic expression of PML-RARα impairs myelopoiesis. *Proc Natl Acad Sci USA* 93: 7900–7904
91 Grignani F, Testa U, Rogaia D, Ferrucci PF, Samoggia P, Pinto A, Aldinucci D, Gelmetti V, Fagioli M, Alcalay M et al (1996) Effects on differentiation by the promyelocytic leukemia PML/RARα protein depend on the fusion of the PML protein dimerization and RARα DNA-binding domains. *EMBO J* 15: 4949–4958
92 Raelson JV, Nervi C, Rosenauer A, Benedetti L, Monczak Y, Pearson M, Pelicci PG, Miller WH (1996) The PML/RARα oncoprotein is a direct molecular target of retinoic acid in acute promyelocytic leukemia cells. *Blood* 88: 2826–2832
93 Koken MHM, Puvion-Dutilleul F, Guillemin MC, Viron A, Linares-Cruz G, Stuurman N, de Jong L, Szostecki C, Clavo F, Chomienne C et al (1994) The t(15;17) translocation alters a nuclear body in a retinoic acid-reversible fashion. *EMBO J* 13: 1073–1083
94 Weis K, Rambaud S, Lavau C, Jansen J, Carvalho T, Carmo-Fonseca M, Lamond A, Dejean A (1994) Retinoic acid regulates aberrant nuclear localization of PML-RARα in acute promyelocytic leukemia cells. *Cell* 76: 345–356
95 Dyck JA, Maul GG, Miller WH Jr, Chen JD, Kakizuka A, Evans RM (1994) A novel macromolecular structure is a target of the promyelocyte-retinoic acid receptor oncoprotein. *Cell* 76: 333–343
96 Redner RL, Rush EA, Faas S, Rudert WA, Corey SJ (1996) The t(5;17) variant of acute promyelocytic leukemia expresses a nucleophosmin-retinoic acid receptor fusion. *Blood* 87: 882–886
97 Chen Z, Brand NJ, Chen A, Chen S-J, Tong J-H, Wang Z-Y, Waxman S, Zelent A (1993) Fusion between a novel *Kruppel*-like zinc finger gene and the retinoic acid receptor-α locus due to a variant t(11;17) translocation associated with acute promyelocytic leukaemia. *EMBO J* 12: 1161–1167
98 Chen S-J, Zelent A, Tong J-H, Yu H-Q, Wang Z-Y, Derre J, Berger R, Waxman S, Chen Z (1993) Rearrangements of the retinoic acid receptor α and promyelocytic leukemia zinc finger genes resulting from t(11;17)(q23;q21) in a patient with acute promyelocytic leukemia. *J Clin Invest* 91: 2260–2267
99 Wells RA, Catzavelos C, Kamel-Reid S (1997) Fusion of retinoic acid receptor α to NuMA, the mitotic apparatus protein, by a variant translocation in acute promyelocytic leukaemia. *Nat Genet* 17: 109–113
100 Guidez F, Huang W, Tong J-H, Dubois C, Balitrand N, Waxman S, Michaux JL, Martiat P, Degos L, Chen Z et al (1994) Poor response to all-*trans* retinoic acid therapy in a t(11;17) PLZF/RARα patient. *Leukemia* 8: 312–317
101 Licht JD, Chomienne C, Goy A, Chen A, Scott AA, Head DR, Michaux JL, DeBlasio A, Miller W Jr, Zelenetz AD et al (1995) Clinical and molecular characterization of a rare syndrom of acute promyelocytic leukemia associated with translocation (11;17). *Blood* 85: 1083–1094
102 Kerckaert J-P, Deweindt C, Tilly H, Quief S, Lecocq G, Bastard C (1993) *LAZ3*, a novel zinc-finger encod-

ing gene, is disrupted by recurring chromosome 3q27 translocations in human lymphomas. *Nat Genet* 5: 66–70
103 Ye BH, Lista F, Lo Coco F, Knowles DM, Offit K, Chaganti RSK, Dalla-Favera R (1993) Alterations of a zinc finger-encoding gene, *BCL-6*, in diffuse large-cell lymphoma. *Science* 262: 747–750
104 Miki T, Kawamata N, Hirosawa S, Aoki N (1994) Gene involved in the 3q27 translocation associated with B-cell lymphoma, *BCL5*, encodes a Kruppel-like zinc-finger protein. *Blood* 83: 26–32
105 Baron BW, Nucifora G, McCabe N, Espinosa RI, LeBeau MM, McKeithan TW (1993) Identification of the gene associated with the recurring chromosomal translocation t(3;14)q27;q23) and t(3;22)(q27;q11) in B-cell lymphomas. *Proc Natl Acad Sci USA* 90: 5262–5266
106 Dhordain P, Albagli O, Lin JN, Ansieau S, Quief S, Leutz A, Kerckaert J-P, Evans RM, Leprince D (1997) Corepressor SMRT binds the BTB/POZ repressing domain of the LAZ3/BCL6 oncoprotein. *Proc Natl Acad Sci USA* 94: 10 762–10 767
107 Hong SH, David G, Wong CW, Dejean A, Privalsky ML (1997) SMRT corepressor interacts with PLZF and with the PML-retinoic acid receptor α (RARα) and PLZF-RARα oncoproteins associated with acute promyelocytic leukemia. *Proc Natl Acad Sci USA* 94: 9028–9033
108 Guidez F, Ivins S, Zhu J, Soderstrom M, Waxman S, Zelent A (1998) Reduced retinoic acid-sensitivities of nuclear receptor co-repressor binding to PML- and PLZF-RARα underlie molecular pathogenesis and treatment of acute promyelocytic leukemia. *Blood* 91: 2634–2642
109 Grignani F, DeMatteis S, Nervi C, Tomassoni L, Gelmetti V, Cioce M, Fanelli M, Ruthardt M, Ferrara FF, Zamir I et al (1998) Fusion proteins of the retinoic acid receptor-α recruit histone deacetylase in promyelocytic leukaemia. *Nature* 391: 815–818
110 Lin R, Nagy L, Inoue S, Shao W, Miller Jr, W, Evans R (1998) Role of histone deacetylase complex in acute promyelocytic lekaemia. *Nature* 391: 811–814
111 David G, Alland L, Hong S-H, Wong C-W, DePinho R, Dejean A (1998) Histone deacetylase associated with mSin3A mediates repression by the promyelocytic leukemia-associated PLZF protein. *Oncogene* 16: 2549–2556
112 Koken MHM, Reid A, Quignon F, Chelbi-Alix MK, Davies JM, Kabarowski JHS, Zhu J, Dong S, Chen S-J, Chen Z et al (1997) Leukemia associated retinoic acid receptor α fusion partners, PML and PLZF, heterodimerize and colocalize to nuclear bodies. *Proc Natl Acad Sci USA* 94: 10 255–10 260
113 Redner R, Rush E, Schlesinger K, Pollock S, Watkins S (1997) The t(5;17) APL fusion protein NPM-RAR does not alter PML localization. *Blood* 90: 1431
114 Ruthardt M, Orleth A, Tomassoni L, Puccetti E, Riganelli D, Alcalay M, Mannucci R, Nicoletti I, Grignani F, Fagioli M et al (1998) The acute promyelocytic leukaemia specific PML and PLZF proteins localize to adjacent and functionally distinct nuclear bodies. *Oncogene* 16: 1945–1953
115 Perez A, Kastner P, Sethi S, Lutz Y, Reibel C, Chambon P (1993) PML/RAR homodimers: distinct DNA binding properties and heteromeric interactions with RXR. *EMBO J* 12: 3171–3182
116 Licht JD, Shaknovich R, English MA, Melnick A, Li J-Y, Reddy JC, Dong S, Chen S-J, Zelent A, Waxman S (1996) Reduced and altered DNA-binding and transcriptional properties of the PLZF-retinoic acid receptor-α chimera generated in t(11;17)-associated acute promyelocytic leukemia. *Oncogene* 12: 323–336
117 Dong S, Zhu J, Reid A, Strutt P, Guidez F, Zhong H-J, Wang Z-Y, Licht J, Waxman S, Chomienne C et al (1996) Amino-terminal protein-protein interaction motif (POZ-domain) is responsible for activities of the promyelocytic leukemia zinc finger-retinoic acid receptor-α fusion protein. *Proc Natl Acad Sci USA* 93: 3624–3629
118 Chen Z, Guidez F, Rousselot P, Agadir A, Chen S-J, Wang Z-Y, Degos L, Zelent A, Waxman S, Chomienne C (1994) PLZF-RARα fusion proteins generated from the variant t(11;17)(q23;21) translocation in acute promyelocytic leukemia inhibit ligand-dependent transactivation of wild-type retinoic acid receptors. *Proc Natl Acad Sci USA* 91: 1178–1182
119 Warrell RP Jr, He LZ, Richon V, Calleja E, Pandolfi PP (1998) Therapeutic targeting of transcription in acute promyelocytic leukemia by use of an inhibitor of histone deacetylase. *J Nat Cancer Inst* 90: 1621–1625
120 Ding W, Li YP, Nobile LM, Grills G, Carrera I, Paietta E, Tallman MS, Wiernik PH, Gallagher RE (1998) Leukemic cellular retinoic acid resistance and missense mutations in the PML-RARα fusion gene after relapse of acute promyelocytic leukemia from treatment with all-*trans* retinoic acid and intensive chemotherapy [In Process Citation]. *Blood* 92: 1172–1183
121 Nagy L, Thomazy VA, Shipley GL, Fesus L, Lamph W, Heyman RA, Chandraratna RA, Davies PJ (1995) Activation of retinoid X receptors induces apoptosis in HL-60 cell lines. *Mol Cell Biol* 15: 3540–3551
122 Smeland EB, Rusten L, Jacobsen SE, Skrede B, Blomhoff R, Wang MY, Funderud S, Kvalheim G, Blomhoff HK (1994) All-*trans* retinoic acid directly inhibits granulocyte colony-stimulating factor-induced proliferation of CD34+ human hematopoietic progenitor cells. *Blood* 84: 2940–2945

Modulation of nuclear vitamin D signalling by retinoids

C. Carlberg and P. Polly

Institut für Physiologische Chemie I, Heinrich-Heine-Universität, D-40001 Düsseldorf, Germany

Introduction

The seco-steroid hormone $1\alpha,25$-dihydroxyvitamin D_3 (VD), is the most biologically active form of vitamin D_3 and an effective regulator of calcium homeostasis and bone metabolism [1]. VD has been known, for some time, to inhibit cell growth and to induce differentiation in several normal and malignant cell types [2]. Within recent years, VD has additionally been shown to induce apoptosis in human breast cancer and leukemic cell lines [3]. Similar profound effects are also known for all-*trans* retinoic acid (RA) and other synthetic retinoids, which are vitamin A derivatives [4, 5]. VD, VD analogues and retinoids have the potential to be used in the treatment of various skin diseases, including psoriasis and acne [6], and in the treatment or chemoprevention of cancers, including melanoma that are, in general, notoriously resistant to chemotherapeutic drugs *in vitro* and *in vivo* [7].

VD and retinoids are both nuclear hormones, i.e. their effects are mediated by nuclear receptors. There are two classes of retinoid receptors, RA receptors (RARs) and retinoid X receptors (RXRs), each with three subtypes and various splicing variants [8, 9], whereas only one gene for the VD receptor (VDR) has been reported [10, 11]. Recently, it has been reported that the VDR gene encodes for several splicing isoforms [12]. All-*trans* RA is a specific ligand of RARs, whereas 9-*cis* RA is a pan-agonist of all retinoid receptors [13]. The ligands induce a conformational change within the receptor, which enables them to up- or down-regulate target gene transcription. Moreover, cell-specific expression of these nuclear receptors determines their stoichiometry in the nucleus. For example, VDR, RARγ and RXRα are preferentially expressed in the skin [14]. Depending on their level of expression and their relative affinity in solution, VDR and retinoid receptors either stay as monomers or form homo- or heterodimers. Although VD and retinoids are very different in their structure (see Fig. 1), their nuclear receptors show reasonable homology and are all members of the nuclear receptor superfamily [15]. VDR and RARs modulate the expression of their target genes in response to their respective ligands by binding as heterodimeric complexes with RXR to specific DNA sequence motifs in promoter regions, referred to as response elements [16]. Response elements are composed of two core binding motifs (with the consensus sequence (A/G)G(G/T)T(C/G)A for VDR and retinoid receptors) and specificity in response element recognition preference of different dimeric complexes is largely dictated by the specific core binding motif sequence, motif spacing and orientation [16, 17, 18]. Response elements that are spaced by five or two nucleotides in a directly repeated orientation (DR5 and DR2) are specific RA response elements (RAREs)

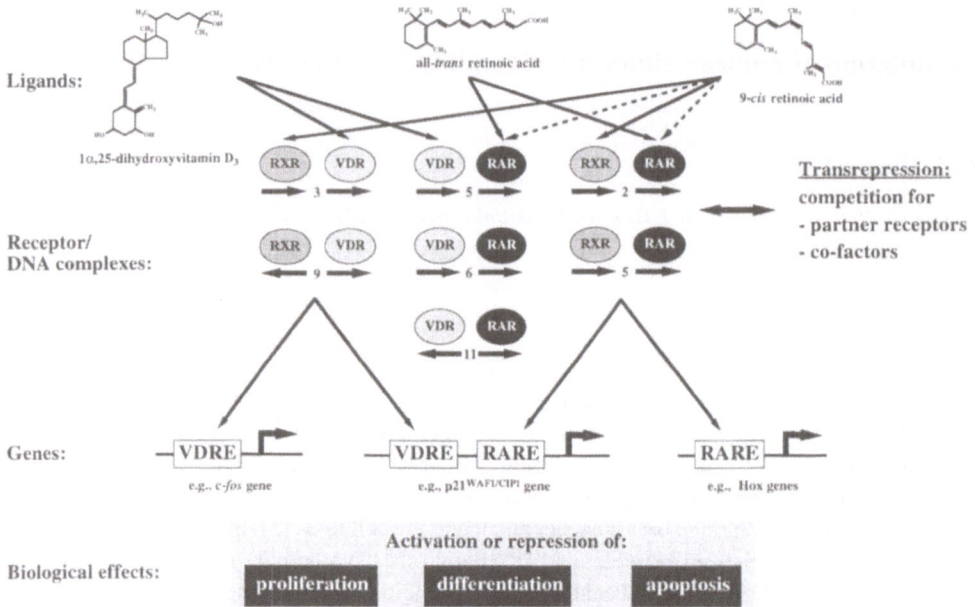

Figure 1. Interrelationship of VD and retinoid signalling. The ligands VD, all-*trans* RA and 9-*cis* RA bind and activate the nuclear receptors VDR, RAR and RXR, respectively. The latter form heterodimeric complexes on their respective response elements. These protein-DNA complexes regulate primary VD or retinoid responding genes or rarely occuring genes that contain both a VDRE and an RARE. Some of these genes are involved in the regulation (both repression and activation) of biological processes such as proliferation, differentiation, apoptosis and others. This multi-level regulation makes the prediction of the biological effect of a stimulation with VD and retinoids difficult.

and those that are spaced by three nucleotides in a directly repeated orientation (DR3) or by nine nucleotides in an inverted palindromic orientation (IP9) are specific VD response elements (VDREs) [6, 18] (Fig. 1).

The nuclear receptor superfamily

The nuclear receptor superfamily is one of the largest families of transcription factors that can be classified into several subclasses [15]. The first subclass consists of the nuclear receptors for the steroid hormones such as, estrogen, progesterone, glucocorticoids, mineralocorticoids and androgens, which are commonly referred to as the classical nuclear hormone receptors. A second subclass consists of the VDR, the RAR and the receptor for the thyroid hormone 3,5,3'-triiodothyronine (T_3), named T_3R. The natural ligands for the eight nuclear receptors of subclasses I and II were already known as nuclear hormones prior to cloning of their cognate receptors. Although the ligands are in part very different in their structure, their respective nuclear receptors are structurally related; that is they all contain a highly conserved DNA binding domain (DBD) of 66 to 70 amino acids [19], which is formed by two zinc finger structures, and a moderately conserved carboxy-terminal ligand binding domain (LBD) of approximately 250 amino acids that is composed of 11 to 12 α-helices [20]. The DBD and the LBD are separated by a hinge region that allows free rotation of the two domains relative to each other. The structures of the LBDs of RXRα [21] and RARγ [22] are known, but as yet, the LBD of VDR is unknown.

In addition to the eight established nuclear receptor gene families, the typical nuclear receptor structure has also been found in over 30 vertebrate transcription factor gene families. These transcription factors lacked a cognate ligand at the time of their discovery and were therefore referred to as "orphan" nuclear receptors [23]. Consequently, intensive screening for possible ligands for these orphan receptors started [24] and resulted in initial reports identifying 9-*cis* RA as a ligand for the retinoid X receptor (RXR) [25, 26]. The orphan members of the nuclear receptor superfamily can be subclassified into those that i) bind DNA as a monomer, ii) form homodimers and iii) form heterodimers with RXR.

VDR-RXR heterodimers

Primary stabilization of a dimeric nuclear receptor complex in solution is facilitated via dimerization of the respective receptor LBDs. Isolated DBDs are not able to dimerize in solution, but their assembly on appropriate response elements is highly cooperative. *In vitro* studies such as gel shift assays and protein-protein interaction tests like the two-hybrid system suggest that the main dimerization partner of the VDR is RXR [11]. The formation of VDR-RXR heterodimers can be obtained on all proposed natural VDREs, but in some cases (e.g. [27]) only at receptor concentrations that would be sufficient to bind any unrelated DNA sequence. Consequently, the current widely accepted model of VD signalling is based on the assumption that all transcriptionally active VDR molecules are complexed with RXR. In fact, RXR appears to have a central role in a vast variety of nuclear signalling processes, as at least half of all nuclear receptors, including many orphans, heterodimerize with RXR [28]. RXR can be activated by 9-*cis* RA, therefore this retinoid may theoretically influence all of these signalling pathways, since the formation of a heterodimer between two ligand-inducible nuclear hormone receptors bears the potential for transcriptional activation by two different ligands. However, within these different heterodimeric complexes, RXR appears to assume different conformations. When complexed with the peroxisome-proliferated activated receptor (PPAR) and nerve growth factor I-B (NGFI-B) orphan receptors, for example, it can be activated by 9-*cis* RA [29, 30], whereas when complexed with T$_3$R and RAR, for example it appears to be silent, i.e. 9-*cis* RA can not activate the dimeric receptor complex [31]. When T$_3$R-RXR or RAR-RXR heterodimers are already activated by T$_3$ or all-*trans* RA, 9-*cis* RA is able to further enhance this activation [32, 33, 34]. VDR-RXR heterodimers appear to react similarly; i.e. when VDR is not activated, 9-*cis* RA achieves only minor activation of the complex, but it is able to enhance the response after VD stimulation [35, 36]. *In vitro* experiments with a high molar excess of bacterially-expressed receptor proteins have created some confusion, as they suggest that 9-*cis* RA destabilizes the VDR-RXR heterodimer [37]. In contrast, however, *in vivo* studies on 24-hydroxylase (CYP24) gene activation supported the view of synergistic or at least additive effects of VD and 9-*cis* RA [38, 39]. Moreover, the ligand-dependent allosteric interaction between VDR and RXR has recently been demonstrated in an *in vitro* study [40].

VDR-RAR heterodimers

It was demonstrated that VDR is able to form heterodimers in solution with both types of retinoid receptors using immunoprecipitation experiments with *in vitro* translated VDR, *in vitro* translated [^{35}S]methionine-labelled RXR or RAR and an anti-VDR antibody [41]. It has been indicated that an interaction between VDR with RAR on the human osteocalcin VDRE not only results in responsiveness to VD, but also to all-*trans* RA [42]. This direct repeat-6 (DR6) type sequence was the first reported VDRE [43, 44] and was also found to be overlaid by a binding site for the AP-1 transcription factor [42], i.e. it is a complex VDRE. Functional VDR-RAR heterodimers have been observed on this above mentioned complex VDRE [45], on the complex VDRE of the rat 24-hydroxylase (CYP24) promoter (three core binding motifs, with three and six nucleotides in distance [46]) and on synthetic DR6-, DR5- and also on IP11- (inverted palindrome with 11 intervening nucleotides) type response elements [45, 47]. Moreover, VDR-RAR heterodimers have also been observed on a DR6-type VDRE from the phospholipase C-γ gene [48]. On simple response elements, i.e. response elements that are formed by only two binding sites, the DNA-binding affinity of VDR-RAR heterodimers was found to be clearly lower than that of VDR-RXR heterodimers [41], but as a component of a larger protein complex on complex VDREs, it may have a critical role.

Transrepression

In addition to gene regulation by direct interactions between the heterodimeric receptor complex and DNA response elements, protein-protein interactions that influence gene activity provide another dimension in gene regulation by nuclear hormones and their receptors, which is less well understood. Recent studies have shown that the VDR is capable of interacting with other nuclear receptors that are bound to DNA to influence gene transactivation via a protein-protein mechanism [49, 50]. The molecular ratio of VDR to RXR appears critical for a synergistic activation of VDR-RXR heterodimers by both VD and 9-*cis* RA. In the case of limiting RXR concentrations, all RXR interacting receptors create a transrepressive effect that can result in a reduction or even inversion of the synergistic effect [36, 51]. On a DR2-type RARE, such as that of the tumor necrosis factor-α type I receptor gene, but not on a DR5-type RARE, VD was shown to act via the VDR as a repressor of retinoid signalling [52]. This could result from VDR sequestering either RXR or RAR from the RXR-RAR complex that binds the DR2-type sequence. A similar squelching phenomena has already been described for RXR in T$_3$ and VD signalling [36, 53]. Therefore, the repression appears to be mediated by competitive protein-protein interactions between VDR, RAR, RXR and possibly their co-factors. Co-factors have been loosely categorized into co-activators or co-repressors, which are capable of mediating either "positive" or "negative" transactivation, respectively. Additionally, co-integrator proteins such as CREB binding protein (CBP) also form a category, functioning as mediator proteins that can integrate incoming and outgoing signals from nuclear receptors, co-activators, or co-repressors to the basal transcriptional machinery [54]. Binding of ligand to nuclear receptors results in the presentation of a short amphipathic α-helix at the carboxy-terminal LBD of the nuclear receptor, which is referred to as the activation function-2 (AF-2) domain [55, 56]. The AF-2 domain is an interface for interaction with co-factors that mediate the activation signal to the basal transcriptional machinery [21, 22]. It has been shown that; e.g. RAR is able to contact CBP and conversely, a given co-factor can interact with different nuclear receptors; e.g. by utilizing distinct domains, CBP directly interacts with the LBD of many nuclear receptors [57]. When VDR has comparable or higher affinity than RAR or RXR for heterodimeric partner proteins or for one or the other co-factors, it will compete for binding. Retinoid receptor partners or co-factors exchanging with VDR, via interactive association with the AF-2 interface of RXR or RAR, would result in transrepression of retinoid signalling by way of communication with the basal transcriptional machinery (Fig. 1). Recently, it was found that in the absence of ligands the two AF-2 domains of some heterodimeric receptor complexes (e.g. RAR-RXR heterodimers) can interfere with each other, which results in an allosteric repression [58].

From receptor complexes to biological processes

To date, approximately 200 genes, that are diverse in function, are known to be modulated in their expression or activity by VD or retinoids [59, 60]. However, only a minority of these genes are known as primary VD or retinoid responding genes and even fewer VDREs and RAREs have been identified in the promoter regions of these known genes. These well established VD responding genes, such as osteopontin, osteocalcin, calbindin and carbonic anhydrase II [1], mainly represent genes classically involved in calcium homeostasis and bone metabolism. However, few cell cycle regulated genes, such as the proto-oncogene c-*fos* [61], cyclin C [62], the CDK inhibitors p19^{INK4D} [62] and p21$^{WAF1/CIP1}$ [27] are known to be regulated by VD. Interestingly, the p21$^{WAF1/CIP1}$ gene was shown to also contain an RARE [63] and appears to be a rare example of a gene that contains both type of response elements (Fig. 1); thus suggesting that not only a VD activation of the p21$^{WAF1/CIP1}$ gene transcription can be enhanced by a RXR-selective retinoid, but also RAR-selective retinoids may show effects that are independent of the signalling from the VDR-RXR complex. The Hox genes are examples of retinoid regulated genes [64] and interestingly, the HoxA10 gene was also recently described to be a primary VD responding gene [65].

The biological profile of VD and retinoids overlaps in cellular regulation, as both types of ligands are known to regulate cellular proliferation, differentiation and apoptosis. However, it is not yet clear, which primary nuclear hormone responding genes play a key role in these processes. Moreover, it appears obvious that the effects of these genes are not all directed at the same biological processes, i.e. not all VD and retinoid responding genes promote the inhibition of proliferation and the induction of differentiation or apoptosis. In addition, most of the known VD and retinoid responding genes are activated by their respective ligands, but there are also a subgroup that do get repressed. This suggests that despite a rather advanced understanding of the function of ligand-activated receptor-response element complexes, the complex regulation of biological processes by a cascade of genes makes the prediction of the net results of a combined ligand stimulation very difficult.

Conclusion

Many different parameters have been described in the modulation of nuclear VD and retinoid signalling. The relative expression of nuclear receptors presents a major level of regulation, as heterodimerization not only increases the variety of different nuclear receptor complexes, but also links nuclear signalling pathways mediated by different hormones. One of the main investigative goals in the nuclear receptor field has been to simplify the experimental systems used, so as to uncover the core mechanisms of nuclear hormone signalling. With the use of these tools, great progress had been made during the last decade. However, in the case of VD signalling these "reductionist" approaches have resulted in only having a few VDREs presently in hand, that mediate a satisfactory response to stimulation with VD or its analogues. This suggests that both reporter systems have to get more sensitive and accurate and that VD signalling has to be investigated in more complex systems. Recent findings on nuclear matrix attachment sites in VD target genes and on histone acetyltransferase or deacetylase activity associated with co-factors emphasizes the importance of chromatin structure in the context of the VD-mediated gene regulatory system. This also includes an understanding of the role of RXR from a compulsory partner of VDR, to an assessory factor as described earlier [66], that can be advantageously chosen in the majority of VDR-containing complexes. In additon, VDR-mediated transrepression of retinoid signalling suggests a novel mechanism for the complex regulatory interaction between retinoids and VD.

Taken together, the interrelationship between VD and retinoids form receptor heterodimers to biological processes are far from being uncovered. However, the various independent "side-effects" that each of the two ligand types can have, indicate that VD analogues or retinoids with the most selective biological profile should be applied.

Acknowledgements
P. Polly is the recipient of an Alexander von Humboldt Foundation fellowship. This work was supported by the Medical Faculty of the University of Düsseldorf.

References

1 Walters MR (1992) Newly identified actions of the vitamin D endocrine system. *Endocrine Rev* 13: 719–764
2 Bouillon R, Okamura WH, Norman AW (1995) Structure-function relationships in the vitamin D endocrine system. *Endocrine Rev* 16: 200–257
3 Welsh J, Simboli-Campbell M, Narvaez CJ, Tenniswood M (1995) Role of apoptosis in the growth inhibitory effects of vitamin D in MCF-7 cells. *Adv Exp Med Biol* 375: 45–52
4 Holdener EE, Bollag W (1993) Retinoids. *Curr Opin Oncol* 5: 1059–1066
5 Gudas LJ (1994) Retinoids and vertebrate development. *J Biol Chem* 269: 15 399–15 402

6   Carlberg C, Saurat J-H (1996) Vitamin D₃-retinoids association: molecular basis and clinical application. *J Invest Dermatol Symp Proc* 1: 82–86

7   Ho VC, Sober AJ (1990) Therapy of cutaneous melanom: an update. *J Amer Acad Dermatol* 22: 159–176

8   Chambon P (1994) The retinoid signalling pathway: molecular and genetic analysis. *Semin Cell Biol* 5: 115–125

9   Giguère V (1994) Retinoic acid receptors and cellular retinoid binding proteins: complex interplay in retinoid signalling. *Endocrine Rev* 15: 61–79

10 Pike JW (1991) Vitamin D₃ receptors: structure and function in transcription. *Annu Rev Nutr* 11: 189–216

11 Carlberg C (1996) The vitamin D₃ receptor in the context of the nuclear receptor superfamily: the central role of retinoid X receptor. *Endocrine* 4: 91–105

12 Crofts LA, Hancock MS, Morrison NA, Eismann JA (1998) Multiple promoters direct the tissue-specific expression of novel N-terminal variant human vitamin D receptor gene transcripts. *Proc Natl Acad Sci USA* 95: 10529

13 Allenby G, Janocha R, Kazmer S, Speck J, Grippo JF, Levin AA (1994) Binding of 9-*cis*-retinoic acid and all-*trans*-retinoic acid to retinoic acid receptors α, β, and γ. *J Biol Chem* 269: 16 689–16 695

14 Elder JT, Aström A, Petterson U, Tavakkol A, Griffiths CEM, Krust A, Kastner P, Chambon P, Voorhees JJ (1992) Differential regulation of retinoic acid receptors and binding proteins in human skin. *J Invest Dermatol* 98: 673–679

15 Mangelsdorf DJ, Thummel C, Beato M, Herrlich P, Schütz G, Umesono K, Blumberg B, Kastner P, Mark M, Chambon P et al (1995) The nuclear receptor superfamily: the second decade. *Cell* 83: 835–839

16 Glass CK (1994) Differential recognition of target genes by nuclear receptor monomers, dimers, and heterodimers. *Endocrine Rev* 15: 391–407

17 Umesono K, Murakami KK, Thompson CC, Evans RM (1991) Direct repeats as selective response elements for the thyroid hormone, retinoic acid, and vitamin D₃ receptors. *Cell* 65: 1255–1266

18 Carlberg C (1995) Mechanisms of nuclear signalling by vitamin D3: interplay with retinoid and thyroid hormone signalling. *Eur J Biochem* 231: 517–527

19 Freedman LP, Luisi BF (1993) On the mechanism of DNA binding by nuclear hormone receptors: a structural and functional perspective. *J Cell Biochem* 51: 140–150

20 Moras D, Gronemeyer H (1998) The nuclear receptor ligand-binding domain: structure and function. *Curr Opin Cell Biol* 10: 384–391

21 Bourguet W, Ruff M, Chambon P, Gronemeyer H, Moras D (1995) Crystal structure of the ligand binding domain of the human nuclear receptor RXR-α. *Nature* 375: 377–382

22 Renaud J-P, Rochel N, Ruff V, Vivat V, Chambon P, Gronemeyer H, Moras D (1995) Crystal structure of the RAR-γ ligand-binding domain bound to all-*trans* retinoic acid. *Nature* 378: 681–689

23 O'Malley BW, Conneely OM (1992) Orphan receptors: in search of a unifying hypothesis for activation. *Mol Endocrinol* 6: 1359–1361

24 Laudet V, Adelmant G (1995) Lonesome orphans. *Curr Biol* 5: 124–127

25 Levin AA, Sturzenbecker LJ, Kazmer S, Bosakowski T, Huselton C, Allenby G, Speck J, Kratzeisen C, Rosenberger M, Lovey A et al (1992) 9-*Cis* retinoic acid stereoisomer binds and activates the nuclear receptor RXRα. *Nature* 355: 359–361

26 Heyman RA, Mangelsdorf DJ, Dyck JA, Stein RB, Eichele G, Evans RM, Thaller C (1992) 9-*cis* retinoic acid is a high affinity ligand for the retinoid X receptor. *Cell* 68: 397–406

27 Liu M, Lee M-H, Cohen M, Bommakanti M, Freedman LP (1996) Transcriptional activation of the Cdk inhibitor p21 by vitamin D₃ leads to the induced differentiation of the myelomonocytic cell line U937. *Gene Develop* 10: 142–153

28 Mangelsdorf DJ, Evans RM (1995) The RXR heterodimers and orphan receptors. *Cell* 83: 841–850

29 Kliewer SA, Umesono K, Noonan DJ, Heyman RA, Evans RM (1992) Convergence of 9-*cis* retinoic acid and peroxisome proliferator signalling pathways through heterodimer formation of their receptors. *Nature* 358: 771–774

30 Perlmann T, Jansson L (1995) A novel pathway for vitamin A signalling mediated by RXR heterodimerization with NGFI-B and NURR1. *Gene Develop* 9: 769–782

31 Forman BM, Umesono K, Chen J, Evans RM (1995) Unique response pathways are established by allosteric interactions among nuclear hormone receptor. *Cell* 81: 541–550

32 Durand B, Saunders M, Leroy P, Leid M, Chambon P (1992) All-*trans* and 9-*cis* retinoic acid induction of CRABPII transcription is mediated by RAR-RXR heterodimers bound to DR1 and DR2 repeated motifs. *Cell* 71: 73–85

33 Schräder M, Becker-André M, Carlberg C (1994) Thyroid hormone receptor monomers can function as ligand-inducible transcription factors on octamer sequences: consequences also for heterodimerization. *J Biol Chem* 269: 6444–6449

34 Schräder M, Carlberg C (1994) Thyroid hormone and retinoic acid receptors form heterodimers with retinoid X receptors on direct repeats, palindromes and inverted palindromes. *DNA Cell Biol* 13: 333–341

35 Carlberg C, Bendik I, Wyss A, Meier E, Sturzenbecker LJ, Grippo JF, Hunziker W (1993) Two nuclear signalling pathways for vitamin D. *Nature* 361: 657–660

36 Schräder M, Nayeri S, Kahlen J-P, Müller KM, Carlberg C (1995) Natural vitamin D₃ response elements

formed by inverted palindromes: polarity-directed ligand sensitivity of VDR-RXR heterodimer-mediated transactivation. *Mol Cell Biol* 15: 1154–1161

37 Cheskis B, Freedman LP (1994) Ligand modulates the conversion of DNA-bound vitamin D$_3$ receptor (VDR) homodimers into VDR-retinoid X receptor heterodimers. *Mol Cell Biol* 14: 3329–3338

38 Allegretto EA, Shevde N, Zou A, Howell SR, Boehm MF, Hollis BW, Pike JW (1995) Retinoid X receptor acts as a hormone receptor *in vivo* to induce a key metabolic enzyme for 1,25-dihydroxyvitamin D$_3$. *J Biol Chem* 270: 23 906–23 909

39 Zou A, Elgort MG, Allegretto EA (1997) Retinoid X receptor (RXR) ligands activate the human 25-hydroxyvitamin D$_3$-24-hydroxylase promoter via RXR heterodimer binding to two vitamin D-responsive elements and elict additive effects with 1,25-dihydroxyvitamin D$_3$. *J Biol Chem* 272: 19 027–19 034

40 Kahlen J-P, Carlberg C (1997) Allosteric interaction of the 1α,25-dihydroxyvitamin D$_3$ receptor and the retinoid X receptor on DNA. *Nucl Acid Res* 25: 4307–4313

41 Schräder M, Müller KM, Carlberg C (1994) Specificity and flexibility of vitamin D signalling: modulation of the activation of natural vitamin D response elements by thyroid hormone. *J Biol Chem* 269: 5501–5504

42 Schüle R, Umesono K, Mangelsdorf DJ, Bolado J, Pike JW, Evans RM (1990) Jun-Fos and receptors for vitamins A and D recognize a common response element in the human osteocalcin gene. *Cell* 61: 497–504

43 Kerner SA, Scott RA, Pike JW (1989) Sequence elements in the human osteocalcin gene confer basal activation and inducible response to hormonal vitamin D$_3$. *Proc Natl Acad Sci USA* 86: 4455–4459

44 Morrison NA, Shine J, Fragonas J-C, Verkest V, McMenemey ML, Eisman JA (1989) 1,25-dihydroxyvitamin D-responsive element and glucocorticoid repression in the osteocalcin gene. *Science* 246: 1158–1161

45 Schräder M, Bendik I, Becker-André M, Carlberg C (1993) Interaction between retinoic acid and vitamin D signalling pathways. *J Biol Chem* 268: 17 830–17 836

46 Kahlen J-P, Carlberg C (1994) Identification of a vitamin D receptor homodimer-type response element in the rat calcitriol 24-hydroxylase gene promoter. *Biochem Biophys Res Commun* 202: 1366–1372

47 Schräder M, Müller KM, Becker-André M, Carlberg C (1994) Response element selectivity for heterodimerization of vitamin D receptors with retinoic acid and retinoid X receptors. *J Mol Endocrinol* 12: 327–339

48 Xie Z, Bikle DD (1997) Cloning of the human phospholipase C-γ1 promoter and identification of a DR6-type vitamin D-responsive element. *J Biol Chem* 272: 6573–6577

49 Garcia-Villalba P, Jimenez-Lara AM, Aranda A (1996) Vitamin D interferes with transactivation of the growth hormone gene by thyroid hormone and retinoid acid. *Mol Cell Biol* 16: 318–327

50 Yen PM, Liu Y, Sugawara A, Chin WW (1996) Vitamin D receptors repress basal transcription and exert dominant negative activity on triiodothyronine-mediated transcriptional activity. *J Biol Chem* 271: 10 910–10 916

51 Macdonald PN, Dowd DR, Nakajima S, Galligan MA, Reeder MC, Haussler CA, Ozato K, Haussler MR (1993) Retinoid X receptors stimulate and 9-*cis* retinoic acid inhibits 1,25-dihydroxyvitamin D$_3$-activated expression of the rat osteocalcin gene. *Mol Cell Biol* 13: 5907–5917

52 Polly P, Carlberg C, Eisman JA, Morrison NA (1997) 1α,25-dihydroxyvitamin D$_3$ receptor as a mediator of transrepression of retinoid signalling. *J Cell Biochem* 67: 287–296

53 Lehmann JM, Zhang X-K, Graupner G, Lee M-O, Hermann T, Hoffmann B, Pfahl M (1993) Formation of retinoid X receptor homodimers leads to repression of T$_3$ response: hormonal cross talk by ligand-induced squelching. *Mol Cell Biol* 13: 7698–7707

54 Torchia J, Glass C, Rosenfeld MG (1998) Co-activators and co-repressors in the integration of transcriptional responses. *Curr Opin Cell Biol* 10: 373–383

55 Danielian PS, White R, Lees JA, Parker MG (1992) Identification of a conserved region required for hormone dependent transcriptional activation by steroid hormone receptors. *EMBO J* 11: 1025–1033

56 Durand B, Saunders M, Gaudon C, Roy B, Losson R, Chambon P (1994) Activation function 2 (AF-2) of retinoic acid receptor and 9-*cis* retinoid acid receptor: presence of a conserved autonomous constitutive activating domain and influence of the nature of the response element. *EMBO J* 13: 5370–5382

57 Kamei Y, Xu L, Heinzel T, Torchia J, Kurokawa R, Gloss B, Lin S-C, Heyman RA, Rose DW, Class CK et al (1996) A CBP integrator complex mediates transcriptional activation and AP-1 inhibition by nuclear receptors. *Cell* 85: 403–414

58 Westin S, Kurokawa R, Nolte RT, Wisely GB, McInerney EM, Rose DW, Milburn MV, Rosenfeld MG, Glass CK (1998) Interactions controlling the assembly of nuclear-receptor heterodimers and co-activators. *Nature* 395: 198–202

59 de Luca LM (1991) Retinoids and their receptors in differentiation, embryogenesis, and neoplasia. *FASEB J* 5: 2924–2933

60 DeLuca HF, Krisinger J, Darwish H (1990) The vitamin D system. *Kidney Int* 38:S2–S8

61 Candeliere GA, Jurutka PW, Haussler MR, St-Arnaud R (1996) A composite element binding the vitamin D receptor, retinoid X receptor α, and a member of the CTF/NF-1 family of transcription factors mediates the vitamin responsiveness of the c-*fos* promoter. *Mol Cell Biol* 16: 584–592

62 Danielsson C, Polly P, Schräder M, Carlberg C (1998) Cyclin C and p19^{INK4D} are primary 1α,25-dihydroxyvitamin D$_3$ responding genes. *submitted*

63 Liu M, Iavarone A, Freedman LP (1996) Transcriptional activation of the human p21$^{WAF1/CIP1}$ gene by retinoic acid receptor. *J Biol Chem* 271: 31 723–31 728

64 Langston AW, Gudas LJ (1994) Retinoic acid and homeobox gene regulation. *Curr Opin Genet Develop* 4:

550–555

65 Rots NY, Liu M, Anderson EC, Freedman LP (1998) A differential screen for ligand-regulated genes: identification of HoxA10 as a target of vitamin D_3 induction in myeloid leukemic cells. *Mol Cell Biol* 18: 1911–1918

66 Sone T, Ozono K, Pike JW (1991) A 55-kilodalton accessory factor facilitates vitamin D receptor DNA binding. *Mol Endocrinol* 5: 1578–1586

Pharmacology and molecular mechanisms of retinoid action in skin

S. Kang, G.J. Fisher and J.J. Voorhees

Department of Dermatology, University of Michigan Medical Center, 1910 Taubman Center, Ann Arbor, MI 48109, USA

Introduction

In both hypo- as well as hyper-vitaminosis A, alterations in clinical appearance and histological morphology of the skin, although non-specific, are known to occur [1, 2]. Realizing the importance of vitamin A for the normal physiology of epithelial tissue, investigative dermatologists were among the first to experiment with retinoid therapeutics [3, 4]. More than three decades ago, they introduced topical retinoic acid treatment for actinic keratosis, a premalignant skin condition characterized by cytologic atypia and a loss of normal differentiation of epidermal keratinocytes. Since then, a variety of dermatological conditions, ranging from acne to photoaged skin, have proven to be responsive to topical and/or systemic retinoid treatment [5]. In addition to the impressive clinical advances that have been made with retinoid therapy, there have also been significant advances in our understanding of the cutaneous effects of retinoids. Indeed, skin remains the only human tissue in which systematic analyses of retinoid biology have been made under *in vivo* conditions. This chapter summarizes the relevant data, emphasizing the *in vivo* studies on human skin that form the basis of prevailing knowledge about metabolism and molecular mechanisms of retinoid action.

Metabolism of retinol

All-*trans* retinol, appropriately designated vitamin A, cannot be synthesized *de novo* in the body and is therefore an essential nutrient. Through regulated metabolism, retinol is able to participate in many diverse and apparently unrelated functions in the body. For example, retinol functions in the physiology of vision when it is oxidized to retinaldehyde to generate the essential photoreactive molecule, rhodopsin [6]. In the immune system, retinol must be converted to 14-hydroxy-4,14-*retro*-retinol for proper growth and activation of lymphocytes [7, 8]. For almost all other biological functions that involve retinoids, from cellular differentiation to growth and development of organisms, sequential oxidation of retinol to retinoic acid must be achieved.

β-carotene and retinyl esters are the two major precursors of retinol in our diet. Once ingested, they are converted to retinol in the intestines, and stored in the liver after they are reconverted to retinyl esters. Retinol is mobilized from the liver when it is generated by hydrolysis of retinyl esters; it is then transported to the circulatory system where it complexes to plasma retinol binding protein. Once delivered to skin cells, retinol apparently is taken up by passive diffusion. Within cells, retinol associates with cellular retinol-binding protein (CRBP) and undergoes one of several metabolic fates. The retinoic acid status of cells is an important determinant of retinol metabolism. A cell deficient in retinoic acid oxidizes retinol to retinoic acid through the actions of retinol- and retinaldehyde-dehydrogenases. On the other hand, if a cell is sufficiently supplied with retinoic acid, retinol is esterified with fatty acids by lecithin retinol acyltransferase (LRAT) and stored intracellularly in the form of retinyl esters. The ratio of CRBP bound to retinol (holo-CRBP) to free CRBP not bound to retinol (apo-CRBP) appears to be a key factor in regulating the direction of retinol metabolism. Apo-CRBP inhibits LRAT from catalyzing the formation of retinyl esters [9], while stimulating the hydrolysis of retinyl esters [10]. When holo-CRBP predominates over apo-CRBP in cells with sufficient retinol, retinyl ester hydrolysis is reduced in favor of retinol esterification to retinyl esters.

When retinol is topically applied to human skin *in vivo*, CRBP expression is markedly induced [11]. This reaction is probably a compensatory cellular response to an excess of retinol, a mechanism to promote storage of retinol as retinyl esters. When retinol is applied to human epidermis *in vivo* [12], or added exogenously to cultured human keratinocytes [13], the level of retinyl esters increases markedly along with the increase in CRBP.

All-*trans* retinoic acid is formed from retinol by sequential oxidation, with retinaldehyde as the intermediate metabolite. Oxidation of retinol to retinaldehyde is rate limiting for retinoic acid biosynthesis. This metabolic pathway is tightly regulated; no detectable increase in the level of retinoic acid is observed in skin cells following topical application of retinol to skin [12]. Yet, retinol causes alterations in skin that closely mimic those that occur in response to retinoic acid. Histologically, these changes include thickening of the epidermis due to keratinocyte hyperproliferation, increased intercellular spaces between epidermal cells, and greater compaction of the stratum corneum. Biochemically, CRBP, cellular retinoic acid binding protein-II (CRABP-II), and retinoic acid 4-hydroxylase activity are induced [12, 14]. The ability of retinoic acid 4-hydroxylase to hydroxylate and remove retinoic acid is an important regulatory step that partly explains the failure to detect significant accumulation of retinoic acid in retinol-treated human skin. In cultured human keratinocytes, which do not express retinoic acid 4-hydroxylase activity, there is a small but detectable increase in retinoic acid after retinol has been added to the culture media [13]. This small increase represents less than a 1% conversion of added retinol to retinoic acid, however, indicating that enzymatic capacity, and not substrate availability, is the limiting factor for retinoic acid synthesis.

One distinguishing feature of the human skin response to retinol and retinoic acid is the generation of clinical erythema. Retinol causes significantly less erythema than does retinoic acid [12]. Since the erythema response of human skin to retinoic acid is dose-dependent [15], the reduced erythemogenic potential of retinol is probably due to lack of an increase in retinoic acid levels in retinol-treated skin. However, the similar biochemical and molecular responses elicited by retinol indicates that endogenously synthesized retinoic acid is much more effective than exogenously supplied retinoic acid in activating retinoid pathways in human skin.

The retinoid receptors in human skin

The discovery of retinoic acid receptors in the late 1980 s was pivotal for our understanding of the mechanisms of retinoid action, because it was the first demonstration of the existence of a retinoid-responsive transcription factor. Both retinoic acid receptors (RARs) and retinoid X receptors (RXRs) have three different family members, α, β, and γ, each encoded by different genes. All-*trans* retinoic acid (*t*-RA) and its stereoisomer 9-*cis* retinoic acid (9*c*-RA), bind to RARs with similar high affinity [16]. The off-rates for *t*-RA binding to RAR-α, RAR-β, and RAR-γ are similar [17]. However, for 9*c*-RA, the off-rates differ: it is fastest with RAR-γ and slowest with RAR-β. Because of these differences, RARs tend to prefer *t*-RA when they are in the presence of mixtures of *t*-RA and 9*c*-RA. This is especially true for RAR-γ, which demonstrates a strong preference for *t*-RA. *t*-RA does not bind to RXRs, only 9*c*-RA does [18, 19].

In human skin, transcripts for RAR-α, RAR-γ, RXR-α, and RXR-β are expressed in the epidermis [20]. mRNA for the other isoforms (RXR-γ and RAR-β) is either very low or undetectable. This pattern is also observed in cultured human keratinocytes and in dermal fibroblasts [21]. Quantitation of retinoid receptors at the protein level has been achieved utilizing keratome biopsy specimens from normal volunteers. The skin tissue obtained through this technique contains mostly epidermis with a small amount of papillary dermis. Analysis of nuclear extracts by ligand binding and immunological assays indicates that the total numbers of RARs and RXRs per cell, on average, are 1,800 and 9,400, respectively [16]. RAR-γ makes up the majority (90%) of RARs in skin. The remaining 10% of RARs are RAR-α: no RAR-β is detectable. Total RXRs are five-fold more abundant than RARs, with 90% of RXRs of the α-subtype, and the remainder of the β-subtype [16]. These relative amounts of receptor proteins parallel the relative expression of their corresponding mRNAs.

Receptor combination for retinoid response in human skin

Retinoid receptors are members of the nuclear hormone receptor superfamily. There are two important features common to this group of receptors. First, they work as paired functional units (dimers). When a dimer is made up of two identical receptors (e.g. RAR/RAR), it is called a homodimer. If the dimer consists of two different receptors, it is called a heterodimer (e.g. RAR/RXR). Second, in the presence of the relevant ligand (hormone), the dimer binds to specific sequences of DNA, called the response element, in the enhancer region of a gene promoter to regulate the transcription of hormone-responsive genes. The response element for retinoic acid (RAREs) consists of hexameric direct repeats with a consensus sequence of -AGGTCA-separated by either two (DR2) or five (DR5) nucleotide base pairs. For retinoid X response elements (RXRE), a one nucleotide (DR1) spacing separates the direct repeats. RARs and RXRs in nuclear extracts isolated from normal epidermis (endogenous receptors) and from cultured keratinocytes bind to RAREs (DR2 and DR5) and to RXRE (DR1) exclusively as heterodimers [16, 22, 23]. Neither RAR nor RXR homodimers are detected in gel shift analyses. The binding of RAR/RXR heterodimers to DNA does not require the presence of their respective ligands (i.e., *t*-RA and 9*c*-RA). However, for transactivation of RARE-containing responsive genes by the bound heterodimers, ligands are required.

Ligand requirements for retinoid receptors have been studied in keratinocytes transfected with CAT reporter genes regulated by RAREs (DR2, DR5) or RXREs (DR1). Both DR2 and DR5 types of RARE reporters are activated by *t*-RA, 9*c*-RA (via isomerization to *t*-RA), and the RAR-specific ligand, CD367, but are not affected by the synthetic RXR-specific SR-11237. Furthermore, the heterodimer requires only the presence of RAR ligand since simultaneous presence of RXR ligand does not confer additional transactivation. Ordinarily, none of the retinoids, including the RXR-specific SR-11237, activates transcription of RXRE (DR1) reporters in keratinocytes. If RXRs are overexpressed, then normally absent RXR homodimer formation occurs; under such non-physiological conditions, the dimer responds to RXR ligands. In normal epidermal keratinocytes, however, RXREs are not activated by RXR ligands because of the absence of RXR homodimers. At their normal physiological levels in human skin epidermis, RARs and RXRs bind to RAREs as heterodimers and transactivate these elements in the presence of RAR ligands.

Target genes of retinoids and activation of the signalling cascade in human skin

If the biological effects of retinoids are mediated through their receptors, then alterations in the receptors are expected to be detrimental to the normal maintenance of structure and function of skin. In fact, in receptor gene knock-out mice and in mice in which dominant negative retinoid receptor transgenes are overexpressed, normal skin development and function are lost [24–27, 28]. Furthermore, when simultaneous null mutations of two RARs are introduced, the resultant phenotypic changes are similar to those observed in vitamin A-deficient animals [29].

Several genes are known to be directly regulated by RAREs in human skin. These genes include CRABP-II [30, 31], CRBP [11], keratin 6 [32, 33], and, most recently, retinoic acid 4-hydroxylase [34]. Topical application of *t*-RA to human skin results in induction of all of these genes. CRABP-II, CRBP, and retinoic acid 4-hydroxylase are all involved in the metabolism of retinoids, thereby regulating the level of free retinoic acid. It makes teleologic sense that the promoter of these genes contains RAREs. Other genes containing functional, positive RAREs will undoubtedly be discovered in skin in the future. The protein products of one or more such genes, directly activated by *t*-RA, may then activate non-RARE-containing genes and/or engage other signalling machinery in a cascading reaction to produce the clinical features of retinoid action in skin. As most of the genes expressed within this cascade are not known to contain functional RAREs, their activation by *t*-RA may be indirect. In short, the retinoid response in human skin is initiated by a retinoid signal but may be mediated subsequently by both direct and indirect activation of genes.

A recent molecular dissection of the peeling and desquamative skin responses to topical *t*-RA demonstrates this point. The well-known clinical consequences of *t*-RA application to human skin correspond histologically to compaction of the stratum corneum and thickening of the epidermal layer [12, 15]. These histological changes are due to basal keratinocyte hyperproliferation, as indicated by a greater number of mitotic figures and enhanced expression of differentiation markers [12, 32]. That basal cell hyperproliferation is a retinoid receptor-initiated event was recently revealed in studies of transgenic mice that express mutant RXR-α receptors. In these genetically engineered mice, the expression of dominant negative (dn) RXR-α mutant receptors was targeted to the suprabasal layers of the epidermis with the keratin 10 promoter. The mutant RXR-α can dimerize with RARs, but the dnRXRα-RAR heterodimer cannot trans-activate RARE-containing genes. For that reason, overexpression of mutant receptors will tie up normal RARs in non-functional heterodimers so that functional wild-type RXRα-RAR heterodimer formation is reduced. In such a retinoid receptor-deficient state, which is present only in the suprabasal cells of the transgenic mice, topical *t*-RA application does not induce proliferation of basal cells that give rise to the epidermal hyperplasia typically seen [28]. This finding suggests that the *t*-RA-induced hyperproliferative skin response is indeed RXR-RAR mediated and that retinoids induce biological changes in suprabasal keratinocytes which, by sending secondary signals, stimulate basal keratinocytes to proliferate. More recent work suggests that heparin-binding epidermal growth factor from suprabasal keratinocytes, that is selectively up-regulated by *t*-RA via RXR/RAR heterodimers, is one such secondary signalling factor responsible for the proliferative response to retinoids [35].

Modulation of retinoid receptors by ultraviolet light

Studies on cancerous tissues have similarly demonstrated the importance of retinoid receptors for normal skin. For example, in benign skin tumors in mice, RAR transcripts are reduced [36] and they are completely lost in undifferentiated squamous cell carcinomas in man [37]. RAR-β mRNA selectively lost in premalignant oral lesions is restored by treatment with 13-*cis* RA [38]. Recently, UV irradiation of human skin was shown to cause rapid and significant reduction of retinoid receptors. This reduction was associated with almost total loss of RA-responsive gene expression in human skin *in vivo*. Both RAR-γ and RXR-α proteins, the two major receptors in skin, are reduced as early as one to two hours after UV exposure and remained reduced for a full day [34]. During this time, reduction of receptor protein levels as high as 80% was observed. The decrease in both receptor protein levels was transient, returning to pre-UV treatment levels within two days. Although the reduction in the protein levels was accompanied by a similar decrease in mRNA levels, the RAR-γ transcript lagged behind reduction of RAR-γ protein, suggesting that reduced RAR-γ expression does not fully account for the initial reduction in RAR-γ protein following UV irradiation. The reduction in the receptor mRNA levels was UV dose-dependent and was significant even at half the dose of UV that caused skin reddening (the so-called "minimal erythema dose", MED) [34].

CRABP-II and RA-4 hydroxylase are two genes that contain functional RAREs in their promoter and are thus induced by *t*-RA in human skin [31, 34]. However, if the skin is irradiated with UV two to six hours before topical application of *t*-RA, induction of the two gene transcripts is blocked [34]. This indicates that the reduction of RAR and RXR proteins by UV results in functional impairment of receptor-dependent retinoid responsiveness in human skin. Treatment of skin with *t*-RA for 24 h prior to UV irradiation reduces receptor loss and markedly accelerates receptor recovery. The mechanisms by which UV reduces retinoid receptors, and by which *t*-RA prevents the UV effect are not known. Although the reduction of retinoid receptors caused by acute UV exposure is transient, the fact that less than one MED UV can trigger it suggests that daily casual exposures to sun may lengthen the receptor recovery time and lead to a chronic state of vitamin A deficiency.

Retinoid receptor interaction with AP-1

Activator protein-1 (AP-1) is an important transcription factor that figures in the inflammation response. AP-1 is a dimeric complex of the protooncoproteins *jun* and *fos* that is induced by growth factors, cytokines, tumor promoters, and sunlight [39]. Activation of AP-1 increases the transcription of cytokines, such as interleukin-2, and certain matrix metalloproteinases [40]. In the presence of *t*-RA, RARs can inhibit the actions of AP-1. Reciprocally, elevated expression of either the *jun* or *fos* components of AP-1 can prevent activation of RAREs by RARs. This repression of gene transcription, called transrepression, is a well-known mechanism for cross-talk between retinoid receptors and AP-1 [41–43]. The molecular mechanism of transrepression described for *in vitro* systems is dependent on the presence of *t*-RA and is believed to involve direct or indirect protein-protein interactions between retinoid receptors and AP-1 components (*jun* and *fos*), and/or competition between retinoid receptors and AP-1 for a common factor (or factors) required for their activities [42, 43]. However, this phenomenon studied in the context of photoaging of human skin *in vivo* has revealed a novel mechanism.

Photoaging refers to premature skin aging caused by chronic exposure to ultraviolet irradiation from the sun [44]. A chapter in this book is devoted to this subject and interested readers should refer to it for further discussion. Here, the dermal aspect of photoaging and the mechansim of *t*-RA antagonism is presented briefly. In photoaged skin, the dermal layer, which underlies the epidermis and forms the connective tissue matrix, is markedly altered. These alterations involve an accumulation of disorganized elastin-containing fibers and reduced content of type I and type III procollagens in the extracellular matrix. This damage initiated by sun exposure involves destruction of the dermal extracellular matrix by matrix-degrading metalloproteinases (MMPs) [45]. Of the MMPs, interstitial collagenase, gelatinase B, and stromelysin 1 contain AP-1 response elements in their promoters. Ultraviolet irradiation of human skin *in vivo* up-regulates AP-1 and induces the mRNAs, proteins, and activities of these MMPs [45]. All of the *jun* and *fos* family members, with the exception of *Fos B*, are detectable in normal human skin. UV induces *cJun*, *Jun B*, and *Fra 1* proteins and UV-induced AP-1 complexes are enriched in *cJun* in human skin [46]. Elevated *cJun*, in association with constitutively expressed *cFos* produce increased AP-1 activity. The time course of *c-Jun* protein induction by UV correlates with UV-induced gene expression of AP-1 driven MMPs [45]. Collectively, these MMPs degrade the dermal matrix [47]. The ensuing dermal damage is imperfectly repaired, resulting in an invisible solar scar. Repeated exposure to sunlight over time causes an accumulation of such micro-scars, which eventually become visible as the clinical signs of photoaged skin.

If human skin is treated with *t*-RA prior to UV irradiation, UV-induction of AP-1 and MMPs is markedly inhibited [45], providing strong evidence that *t*-RA may prevent photoaging. Although *t*-RA absorption (λmax 351 nm) overlaps with the UVB range, it does not reduce

UVB-induced skin reddening, indicating that *t*-RA is not a sunscreen, at least for the chromophore(s) responsible for the sunburn reaction in skin. For *t*-RA to prevent UV-induction of AP-1 and MMP, prolonged pretreatment (>16 h) is required before UV-irradiation. Well within this time (6 h.), *t*-RA penetrates skin and fully activates retinoid receptors and the target gene CRABP-II [48], indicating that the mechanism by which *t*-RA opposes UV-induced AP-1 signalling differs from previously described transrepression. In fact, *t*-RA antagonizes UV induction of AP-1 by preventing the increase in *cJun* protein [45]. Since *t*-RA does not alter UV induction of cJun mRNA, the reduction of *cJun* protein could occur either through inhibition of its translation and/or accelerated degradation. The mechanisms by which *t*-RA blocks the UV-induced increase in *cJun* protein have yet to be elucidated.

References

1 Magro C, Crowson NA, Mihm Jr, M (1997) Cutaneous manifestations of nutritional deficiency states and gastrointestinal disease. *In*: D Elder et al (eds): *Lever's Histopathology of the Skin*, 8th ed. Lippincott-Raven, Philadelphia, 353–354
2 Peck GL, DiGiovanna JJ (1993) Retinoids. *In*: TB Fitzpatrick et al (eds): *Dermatology in General Medicine*, 4th ed. McGraw-Hill Inc., New York, 2883–2908
3 Saurat JH (1992) Side effects of systemic retinoids and their clinical management. *J Amer Acad Dermatol* 17:S23–S28
4 Lambert WE, Meyer E, DeLeeheer AP, De Bersaques J, Kint AH (1992) Pharmacokinetics and drug interactions of etretinate and acitretin. *J Amer Acad Dermatol* 27:S19–S22
5 Fritsch PO (1992) Retinoids in Psoriasis and disorders of keratinization. *J Amer Acad Dermatol* 27:S8–S14
6 Wald G (1968) The molecular basis of visual excitation. *Nature* 219: 819–822
7 Buck J, Derguini F, Levi E, Nakamiski K, Hammerling U (1991) Intra*Cell Signal* 14-hydroxy 4,14-retroretinol. *Science* 254: 1654–1656
8 Eppinger TM, Buck J, Hammerling U (1993) Growth control or terminal differentiation: endogenous production and differential activities of vitamin A metabolites in HL-60 cells. *J Exp Med* 178: 1995–2005
9 Herr F, Ong DE (1992) Differential interaction of lecithin-retinol acyltransferase with cellular retinoid-binding proteins. *Biochemistry* 31: 6748–6755
10 Boerman MHEM, Napoli JL (1991) Cholate-independent retinyl ester hydrolysis. *J Biol Chem* 263: 22 273–22 278
11 Fisher GJ, Reddy AP, Datta SC, Kang S, Yi JY, Chambon P, Voorhees JJ (1995) All-*trans* retinoic acid and all-*trans* retinol induce cellular retinol-binding protein in human skin *in vivo*. *J Invest Dermatol* 105: 80–86
12 Kang S, Duell EA, Fisher GJ, Datta SC, Wang ZQ, Reddy AP, Tavakkol A, Yi JY, Griffiths CEM, Elder JT et al (1995) Application of retinol to human skin *in vivo* induces epidermal hyperplasia and cellular retinoid binding proteins characteristic of retinoic acid but without measurable retinoic acid levels or irritation. *J Invest Dermatol* 105: 549–556
13 Kurlandsky SB, Xiao JH, Duell EA, Voorhees JJ, Fisher GJ (1994) Biological activity of all-*trans* retinol requires metabolic conversion to all-*trans* retinoic acid and is mediated through activation of nuclear retinoid receptors in human keratinocytes. *J Biol Chem* 269: 32821–32827
14 Kang S, Duell EA, Kim KJ, Voorhees JJ (1996) Liarozole inhibits human epidermal retinoic acid 4-hydroxylase activity and differentially augments human skin responses to retinoic acid and retinol *in vivo*. *J Invest Dermatol* 107: 183–187
15 Griffiths CEM, Finkel LJ, Tranfaglia MG, Hamilton TA, Voorhees JJ (1993) An *in vivo* experimental model for effects of topical retinoic acid in human skin. *Brit J Dermatol* 129: 389–394
16 Fisher GJ, Talwar HS, Xiao JH, Datta SC, Reddy AP, Gaub MP, Rochette-Egly C, Chambon P, Voorhees JJ (1994) Immunological identification and functional quantitation of retinoic acid and retinoid X receptor protein in human skin. *J Biol Chem* 269: 20 629–20 635
17 Allenby G, Bocquel MT, Saunder M, Kazmer S, Speck J, Rosenberger M, Lovey A, Kastner P, Grippo JF, Chambon P et al (1993) Retinoic acid receptors and retinoid X receptors: Interaction with endogenous retinoic acids. *Proc Natl Acad Sci USA* 90: 30–34
18 Levin AA, Sturzenbecker LJ, Kazmer S, Bosakowski T, Huselton C, Allenby G, Speck J, Kratzeisen C, Rosenberger M, Lovey A et al (1992) 9-*cis*-retinoic acid steroisomer binds and activates the nuclear receptor RXRα. *Nature* 355: 359–361
19 Heyman RA, Mangelsdorf DJ, Dyck JA, Stein RB, Eichele G, Evans RM, Thaller C (1992) 9-*cis*-retinoic acid

is a high-affinity ligand for the retinoid X receptor. *Cell* 68: 397–406

20 Elder JT, Astrom A, Pettersson U, Tavakkol A, Griffiths CE, Krust A, Kaster P, Chambon P, Voorhees JJ (1992) Differential regulation of retinoic acid receptors and binding proteins in human skin. *J Invest Dermatol* 98: 673–679

21 Elder JT, Fisher GJ, Zhang QY, Eisen D, Krust A, Kastner P, Chambon P, Voorhees JJ (1991) Retinoic acid receptor gene expression in human skin. *J Invest Dermatol* 96: 425–433

22 Xiao JH, Durand B, Chambon P, Voorhees JJ (1995) Endogenous retinoic acid receptor-retinoid X receptor heterodimers are the major functional forms regulating retinoid-responsive elements in adult human keratinocytes. *J Biol Chem* 270: 3001–3011

23 Fisher GJ, Reddy AP, Datta SC, Kang S, Yi JY, Chambon P, Voorhees JJ (1995) All-*trans* retinoic acid induces cellular retinol-binding protein *in vivo*. *J Invest Dermatol* 105: 80–86

24 Attar PS, Wertz PW, McArthur M, Imakado S, Bickenbach JR, Roop DR (1997) Inhibition of retinoid signalling in transgenic mice alters lipid processing and disrupts epidermal barrier function. *Mol Endocrinol* 11: 792–800

25 Imakado S, Bickenbach JR, Bundman DS, Rothnagel JA, Attar PS, Wang XJ, Walczak VR, Wisniewski S, Pote J, Gordon JS (1995) Targeting expression of a dominant-negative retinoic acid receptor mutant in the epidermis of transgenic mice results in loss of barrier function. *Gene Develop* 9: 317–329

26 Saitou M, Sugai S, Tanaka T, Shimouchi K, Fuchs E, Narumiyas Kakizuka A (1995) Inhibition of skin development by targeted expression of a dominant-negative retinoic acid receptor. *Nature* 374: 159–162

27 Sucov HM, Dyson E, Gumeringer CL, Price J, Chien KR, Evans RM (1994) RXR-α mutant mice establish a genetic basis for vitamin A signalling in heart morphogenesis. *Gene Develop* 8: 1007–1018

28 Feng X, Peng ZH, Di W, Li XY, Rochette-Egly C, Chambon P, Voorhees JJ, Xiao JH (1997) Suprabasal expression of a dominant-negative RXR α mutant in transgenic mouse epidermis impairs regulation of gene transcription and basal keratinocyte proliferation by RAR-selective retinoids. *Gene Develop* 11: 59–71

29 Kastner P, Mark M, Chambon P (1995) Nonsteroid nuclear receptors: what are genetic studies telling us about their role in real life? *Cell* 83: 859–869

30 Sanquer S, Gilchrest BA (1994) Characterization of human cellular retinoic acid-binding proteins-I and -II: ligand binding affinities and distribution in skin. *Arch Biochem Biophys* 311: 86–94

31 Astrom A, Tavakkol A, Pettersson U, Cromie M, Elder JT, Voorhees JJ (1991) Molecular cloning of two human cellular retinoic acid-binding proteins (CRABP). Retinoic acid-induced expression of CRABP-II but not CRABP-I in adult human skin *in vivo* and in skin fibroblasts *in vitro*. *J Biol Chem* 266: 17 662–17 666

32 Rosenthal DS, Griffiths CEM, Yuspa SH, Roop DR, Voorhees JJ (1992) Acute or chronic topical retinoic acid treatment of human skin *in vivo* alters the expression of epidermal transglutaminase, loricrin, involucrin, filaggrin, and keratins 6 and 13 but not keratins 1, 10, and 14. *J Invest Dermatol* 98: 343–350

33 Rosenthal DS, Roop DR, Huff CA, Weiss JS, Ellis CN, Hamilton T, Voorhees JJ (1990) Changes in photoaged human skin following topical application of all-*trans* retinoic acid. *J Invest Dermatol* 95: 510–515

34 Wang Z, Boudjelal M, Kang S, Voorhees JJ, Fisher GJ (1999) Ultraviolet irradiation of human skin causes functional vitamin A deficiency, preventable by all-*trans* retinoic acid pretreatmen. *Nat Medlcine* 5: 418–422

35 Xiao JH, Feng X, Di W, Peng ZH, Li LA, Voorhees JJ (1998) In all-*trans* retinoic acid (tRA)-induced epidermal hyperplasia, basal keratinocytes (Kcs) are activated by heparin-binding EGF-like growth factor (HB-EGF) delivered exclusively from suprabasal Kcs. *J Invest Dermatol (abstract)* 110: 479

36 Darwiche N, Scita G, Jones C, Rutberg S, Greenwald E, Tennenbaum T, Collins SJ, De Luca LM, Yuspa SH (1996) Loss of retinoic acid receptors in mouse skin and skin tumors is associated with activation of the ras(Ha) oncogene and high risk for premalignant progression. *Cancer Res* 56: 4942–4949

37 Darwiche N, Celli G, Tennenbaum T, Glick AB, Yuspa SH, De Luca LM (1995) Mouse skin tumor progression results in differential expression of retinoic acid and retinoid X receptors. *Cancer Res* 55: 2774–2782

38 Lotan R, Xu XC, Lippman SM, Ro JY, Lee JS, Lee JJ, Hong WK (1995) Suppression of retinoic acid receptor-β in premalignant oral lesions and its up-regulation by isotretinoin. *New Engl J Med* 332: 1405–1410

39 Herrlich P, Ponta H (1994) Mutual cross-modulation of steroid/retinoic acid receptor and AP-1 transcription factor activities: a novel property with practical implications. *Trends Endocrinol Metab* 5: 341–346

40 Angel P, Karin M (1991) The role of Jun, Fos and AP-1 complex in cell proliferation and transformation. *Biochim Biophys Acta* 1072: 129–157

41 Chen JY, Penco S, Ostrowski J, Balaguer P, Pons M, Starrett JE, Reczek P, Chambon P (1995) RAR-specific agonist/antagonists which dissociate transactivation and AP-1 transrepression inhibit anchorage-independent cell proliferation. *EMBO J* 14: 1187–1197

42 Saatcioglu F, Claret FX, Karin M (1994) Negative transcriptional regulation by nuclear receptors. *Semin Cancer Biol* 5: 347–359

43 Pfahl M (1993) Nuclear receptor/AP-1 interaction. *Endocrine Rev* 14: 651–658

44 Kang S, Fisher GJ, Voorhees JJ (1997) Photoaging and topical tretinoin: therapy, pathogenesis, and prevention. *Arch Dermatol* 133: 1280–1284

45 Fisher GJ, Datta SC, Talwar HT, Wang ZQ, Varani J, Kang S, Voorhees JJ (1996) Molecular basis of sun-induced premature skin ageing and retinoid antagonism. *Nature (London)* 379: 335–339

46 Fisher GJ, Talwar H, Lin J, Lin P, McPhillips F, Wang ZQ, Li XY, Wan Y, Kang S, Voorhees JJ (1998) Retinoic acid inhibits induction of c-JUN protein by ultraviolet radiation that occurs subsequent to activation of mito-

gen-activated protein kinase pathways in human skin *in vivo*. *J Clin Invest* 101: 1432–1440

47 Fisher GJ, Wang ZQ, Datta SC, Varani J, Kang S, Voorhees JJ (1997) Pathophysiology of premature skin aging induced by ultraviolet light. *New Engl J Med* 337: 1419–1428

48 Tavakkol A, Zouboulis CC, Duell EA, Voorhees JJ (1994) A retinoic acid-inducible skin-specific gene (RIS-1/psoriasin): molecular cloning and analysis of gene expression in human skin *in vivo* and cultured skin cells *in vitro*. *Mol Biol Rep* 20: 75–83

Synthetic retinoids and their usefulness in biology and medicine

M.I. Dawson[1], X. Zhang[2], P.D. Hobbs[3] and L. Jong[3]

[1] Molecular Medicine Research Institute, 325 East Middlefield Road, Mountain View, CA 94043, USA
[2] The Burnham Institute, 10901 N. Torrey Pines Road, La Jolla, CA92037, USA
[3] SRI International, 333 Ravenswood Avenue, Menlo Park, CA94025, USA

Introduction

The natural retinoids—all-*trans*-retinoic acid (*trans*-RA) and its 9-*cis* isomer (9-*cis*-RA)—and their synthetic analogs have important functions in regulating cell proliferation, differentiation, and morphogenesis. For these reasons, retinoids are successfully used as therapeutic agents to treat several proliferative diseases, such as acne, psoriasis, skin cancer, and acute promyeloblastic leukemia, and to prevent the recurrence of head-and-neck and certain skin cancers [1–3]. Because of their ability to modulate cell differentiation and proliferation, retinoids at therapeutically effective doses may have such toxic side effects on normal cells that their use is limited [4]. A continuing goal of numerous drug development programs has been to identify therapeutically effective retinoids that have reduced systemic side effects [5]. Such an approach is possible because retinoid activities are regulated through their receptor proteins.

Mechanism of retinoid action

Retinoids exert their pleiotropic effects through binding to their nuclear receptor proteins, which function as gene transcription factors (reviewed in [6]). These nuclear receptors are members of the steroid/thyroid hormone superfamily of receptors. Other family members include the estrogen, progesterone, glucocorticoid, androgen, vitamin D, thyroid hormone, and orphan receptors. The latter receptors, which include the peroxisome proliferator-activated receptor (PPARs), chicken ovalbumin upstream promoter transcription factors (COUP-TFs) I and II, liver X receptor, and nur77, were designated as orphans because, on discovery, their natural ligands were unknown. Based on their amino acid sequences or homologies, the retinoid receptors are most closely related to the vitamin D and thyroid hormone receptors. Two classes of retinoid receptors—the retinoic acid receptors (RARs) and the retinoid X receptors (RXRs)—exist. *trans*-RA is a natural ligand for the RARs, and 9-*cis*-RA is a natural ligand for both the RARs and RXRs [7, 8]. Other, as yet unidentified, natural ligands for these two receptor classes may also exist. Each retinoid receptor class has three subtypes, which are designated α, β, and γ (for brief overviews see [9–11]). Due to alternate gene splicing these subtypes have different isoforms, which are designated by numerical subscripts. Because discoveries in this field were made by several different groups, receptor terminology is not standardized. This chapter uses the most common nomenclature—class, subtype, and isoform—to designate these receptor types.

The retinoid receptors have six major domains, A through F. The major functional domains are the DNA-binding domain (DBD, or domain C) and the retinoid or ligand-binding domain (LBD or domain E), which are separated by a hinge region (domain D). Sub-domains include those for dimerization and activation functions (AFs) 1 and 2. The homology between the DBDs and LBDs of the RAR and RXR subtypes is shown in Table 1. Receptor subtype isoforms have the closest homology. The subtypes in each class have close homology in their DBDs but discrete differences in their LBDs. The classes also differ in their DNA-binding domains. The cell type and differentiation state determine which retinoid receptors are expressed and translated to protein [6]. For example, in skin the predominant receptor is RARγ, whereas RARα is only found at low levels, and RARβ message and protein have not been detected [12]. Fully transformed cancer cells may lose their ability to express RARβ [13]. Retinoids are able to induce the synthesis of RARβ in estrogen-dependent breast cancer cell lines [14]. Introduction of the gene for RARβ into estrogen-independent human breast cancer cell lines, such as MDA-MB-231, restores its sensitivity to growth inhibition by trans-RA and other retinoids [14, 15].

Table 1. Amino acid identity comparison between the DNA-binding domain (DBD) and ligand-binding domain (LBD) sequences of the retinoid receptor classes and subtypes[a]

| Retinoid receptor | Amino acid identity | | | |
| | DBD | | LBD | |
	Sequence	%	Sequence	%
hRARα	88–153	100	98–462	100
hRARβ	81–146	97	200–448	82
hRARγ	90–155	97	200–454	76
hRXRα	135–200	100 (62)[b]	225–462	100 (27)[b]
mRXRβ	82–147	92	171–410	88
mRXRγ	139–204	95	229–463	86

[a]Taken from Mangelsdorf et al. [6].
[b]Numbers in parentheses refer to % identify with hRARα.

The retinoid receptors function as dimers. Thus, the RAR subtypes heterodimerize with the RXR subtypes. The RXRs also can homodimerize or heterodimerize with the vitamin D, thyroid hormone, and the orphan receptors. The receptor dimers act either directly or indirectly to regulate gene function. The dimers bind directly to retinoid response elements (RAREs and RXREs). These response elements (REs) are specific sequences in the promoter regions of retinoid-responsive genes, which typically consist of two conserved sequences of six nucleotide bases that are separated by discrete numbers of bases. For example, RXR-RAR dimers bind to RAREs which are direct repeats of AGGTCA separated by five (synthetic DR-5 sequence) or two (synthetic DR-2) nucleotides or are palindromic, inverted, or more complex in structure [6].

RXR-RXR homodimers bind to the DR-1 RE and the more complex RXREs [16]. In the presence of a retinoid, the dimer complex undergoes a conformational change, to which the RARE or RXRE responds to activate or repress gene transcription to form messenger RNA. Variations in the bases present in a consensus sequence and their neighboring nucleotides can influence gene responses to retinoids [17]. The receptor binding downstream in the response element (at the 3'-end) is the dominant receptor, which controls the transcriptional response on complexation to a retinoid. Because the upstream (or 5'-binding receptor) in the RXR-RAR heterodimer can also be occupied by a retinoid, RXR is not a silent partner [18]. Chambon and coworkers reported that both RXR subtype-selective 2-(4-carboxyphenyl)-2-(5,6,7,8-tetrahydro-5,5,8,8-tetramethyl-2-naphthalenyl)-1,3-dioxolane ammonium salt (SR11237) [19] and RARα-selective 4-[(5,6,7,8-tetrahydro-5,5,8,8-tetramethyl-2-naphthalenyl)aminocarbonyl]benzoic acid (Am80) [20] bound to the RXR-RAR heterodimer to induce gene transcription [21]. These results were confirmed by Botling et al. [22]. Therefore, the structures of the retinoids bound will determine whether one or both members of the dimeric receptor complex are occupied by ligand. Using RAR-selective (E)-4-[2-(5,6,7,8-tetrahydro-5,5,8,8-tetramethyl-2-naphthalenyl)propenyl]benzoic acid (TTNPB), Ro13-7410 [23] and RXR-selective 6-[1-(3,5,5,8,8-pentamethyl-5,6,7,8-tetrahydro-2-naphthalenyl)cyclopropyl]-3-pyridinecarboxylic acid (LGD100268) [24], Westin et al. [25] demonstrated that the concentration of both ligands could affect the binding of the steroid receptor coactivator protein NCoA-1/SRC-1 to the RXR-RAR complex. RARs and RXRs can also antagonize the effects of each other [26, 27].

The retinoid receptors can also function indirectly by modulating the activity of other transcription factors. The activator protein-1 (AP-1) complex is a dimer consisting of members of the Jun-Fos family of proteins that binds to AP-1 sites in the promoter regions of many genes, including those involved in the remodeling processes necessary for cell proliferation and differentiation. Jun (Jun D, Jun B, and c-jun) and Fos (c-fos, fra-1, fra-2) family members also bind to the cyclic adenosine monophosphate (cAMP) response element (CRE) to which the CRE-binding protein (CREB) and cAMP-responsive element modulator (CREM) bind. Jun and CREB can also heterodimerize. CREB-binding protein (CBP) binds to Jun in the AP-1 dimer to stimulate gene transcription from AP-1 sites. In the presence of retinoids, retinoid receptors bind to CBP and inhibit its ability to activate the AP-1 complex [28]. CBP can also function as a coactivator or cointegrator of RAR and RXR signalling [29].

Activity is also influenced by families of intermediary proteins that bind to AF-2 domains of the receptors and function as coactivators or corepressors of gene transcription [30, 31] Coactivators of retinoid-induced gene transcription include SRC-1/NCoA-1, TIF2/GRIP-1/NCoA-2, and PCIP/ACTR/AIB1. Corepressor proteins include N-CoR, SMRT, and TRIP [32–36].

Recent reports also indicate that some synthetic retinoids act through other signalling pathways that do not involve the retinoid receptors. For example, N-(4-hydroxyphenyl) all-trans-retinamide (4-HPR), which is currently undergoing clinical trials for prevention of breast cancer and which is reported to prevent ovarian cancer [37], appears to function through an anti-oxidant pathway to induce cancer cell apoptosis [38]. 6-[3-(1-Adamantyl)-4-hydroxyphenyl]-2-naphthalenecarboxylic acid (AHPN, CD437) is able to induce cell cycle arrest and apoptosis in cancer cell lines that lack functional retinoid receptors and do not respond to growth inhibition by the natural retinoids [39]. These results were confirmed in an ovarian cancer cell line by Chao et al. [40], and in lung cancer cell lines by Sun et al. [41], and Li et al. [42].

Retinoid activities can also be modulated by the cellular retinoic acid-binding proteins (CRABPs) I and II. These proteins bind the natural retinoid *trans*-RA and potentiate its metabolic oxidation at the 4-position. (*E*)-4-[2-(3,5-Di-*t*-butylphenyl)propenyl]benzoic acid (CD55), which does not bind to CRABP and cannot be metabolized at its position corresponding to the 4-position of *trans*-RA, is far more potent than *trans*-RA in inducing the differentiation of HL-60 leukemia cells [43].

The diversity arising from the six retinoid receptor subtypes, their multiple heterodimeric partners, the variations in retinoid response elements and their adjoining nucleotides, and the intermediary proteins allows multiple signalling pathways for retinoids. These multiplicities permit different cell types to respond differently to retinoids and allow the separation of retinoid functions. Using the natural retinoid receptor ligands to study retinoid signalling pathways can be problematic because their olefinic (C=C) bonds readily isomerize [44]. For example, within the cell, 9-*cis*-RA can be converted to *trans*-RA, 13-*cis*-RA, and other bond isomers or to metabolites that have different receptor affinities and pharmacokinetics. The identification of selective retinoids provides unique opportunities for (1) simplifying mechanistic studies on retinoid signalling pathways [45] and (2) identifying more effective retinoids for treating proliferative diseases.

Both pharmaceutical and academic research groups have been generous in sharing retinoids for academic research, provided that the appropriate material transfer agreements are implemented.

Retinoid selectivity

Both class- and subtype-selective retinoids have been reported. Many of these compounds were first designed as therapeutic candidates having greater stability than *trans*-RA. After reports of the three RAR subtypes in 1987–1988 [45–49], retinoids began to be evaluated for gene transcriptional activation activities in cotransfection assays using gene constructs for the receptor subtypes, and receptor class and subtype selectivity was finally observed. The first synthetic RAR class-selective retinoid reported was TTNPB (Ro13-7410) [23]. Fully aromatic analogs of TTNPB, 6-(5,6,7,8-tetrahydro-5,5,8,8-tetramethyl-2-naphthalenyl)-2-naphthalenecarboxylic acid (TTNN, or SR3957) [51, 52] and 4-(5,6,7,8-tetrahydro-5,5,8,8-tetramethyl-2-anthracenyl)benzoic acid (TTAB, CD367, or SR3961) [52–54], were synthesized next. TTNPB, with only one internal olefinic bond, is far more stable to isomerization than the natural retinoids, whereas TTNN and TTAB are fully aromatic. Because only TTNN, of the three, showed selectivity for RARβ and RARγ, it provided the initial lead for the development of RARβ- and RARγ-selective retinoids. TTNPB, TTNN, and TTAB are all derived from the first reported retinoidal benzoic acid, (*E*)-4-[2-methyl-4-(2,6,6-trimethylcyclohexenyl)-1,3-butadienyl]benzoic acid (SR3987), which was originally prepared by the Dawson group [55, 56].

The first RARα-selective retinoids identified were the Shudo group's 4-[(5,6,7,8-tetrahydro-5,5,8,8-tetramethyl-2-naphthalenyl)carbamoyl]benzoic acid (Am80) and 4-[(5,6,7,8-tetrahydro-5,5,8,8-tetramethyl-2-naphthalenyl)carboxamido]benzoic acid (Am580), which are analogs of TTNPB in which the internal propenyl bridge was replaced by amide groups [20]. Of the two, Am580, with an EC_{50} of 3.4×10^{-10} M, is two-fold more potent than Am80 at inducing HL-60 leukemia cell differentiation.

RARγ subtype-selective retinoids were reported next. These compounds were derived from TTNN. The first leads in this series were 6-[(5,6,7,8-tetrahydro-5,5,8,8-tetramethyl-2-naphthalenyl)carbonyl]-2-naphthalenecarboxylic acid, which activated both RARβ and γ, and the related secondary alcohol [56–59], which activated RARγ five times more potently than the parent ketone in the cotransfection assay. This result indicated that a hydrogen donor in the bridge region of the retinoid skeleton imparted RARγ-selectivity [59]. Based on these leads, the Dawson group and the Bristol-Myers Squibb group converted the ketone to the oxime, which showed high RARγ selectivity [40, 60–62]. Another retinoid reported as RARγ-selective is AHPN [58], which is also derived from TTNN, and yet another retinoid reported by the Dawson group, RARβ,γ-selective (*E*)-4-[2-(3-*t*-butyl-4-methoxyphenyl)propenyl]benzoic acid (SR3968) [57]. (*E*)-4-[2-(3-Hexyl-5,6,7,8-tetrahydro-5,5,8,8-tetramethyl-2-naphthalenyl)-ethenyl]benzoic acid (Ro 44-4753) is also RARγ-selective [63].

Identification of RARβ-selective retinoids has been challenging. Three that have been reported are the AHPN analog 6-[3-(1-methylcyclohexyl)-4-methoxyphenyl]-2-naphthalenecarboxylic acid (CD2019) [58, 64], 4-[2,2-difluoro-1-(5,6,7,8-tetrahydro-5,5,8,8-tetramethyl-2-naphthalenyl)ethenyl]benzoic acid (BMS188970) [60], and 6-[4-3-(2-methyl-2-propyl)-4-(2-methyl)propyloxyphenyl]-2-naphthalenecarboxylic acid (Ro 48-2249) [63].

After the report that 9-*cis*-RA was the natural ligand for the RXRs [7, 8], several synthetic retinoids were found to show minimal transcriptional activation of RXRα. One of these was 4-[(5,6,7,8-tetrahydro-5,5,8,8-tetramethyl-2-naphthalenyl)carbonyl]benzoic acid (SR11004) [65]. Analogs with more polar or less polar modifications in the bridge linking the two aromatic rings were next synthesized to determine what groups enhanced RXR activity. Cotransfection assays indicated that increasing lipophilic substitution in the bridge region increased both RXR class-selectivity and activity. This discovery led to the first reported RXR-selective retinoids — the thioketal of 4-(5,6,7,8-tetrahydro-5,5,8,8-tetramethyl-2-naphthalenyl)carbonylbenzoic acid (SR11203) and 4-[2-methyl-1-(5,6,7,8-tetrahydro-5,5,8,8-tetramethyl-2-naphthalenyl)propenyl]benzoic acid (SR11217) [19, 65]. Selectivity was further enhanced by using a cyclopropyl ring as the bridge (SR11246) [65] or 3-methyl group on the tetrahydronaphthalene ring (LDG1069, or SR11247) [24, 65]. Characteristic of most of these analogs are the benzoic or heterocyclic carboxylic acid terminus that originated with the Dawson group [65] and the 5,6,7,8-tetrahydronathphalene ring that originated with the Loeliger group [23]. Retinoids with high RXR subtype-selectivity have not yet been reported.

Receptor binding affinity and transcriptional activation activity are the two parameters used to establish retinoid selectivity for a particular receptor. Selectivity based on preferential binding affinity to one of the retinoid receptor classes or subtypes can be less reliable than that based on activation of a particular retinoid receptor for gene transcription from a defined retinoid response element. Selective retinoids have been identified by both methods, and combination of the two provides a means of identifying both selective transcriptional agonists and antagonists. An overview of the literature on selective retinoids indicates that the criteria used to identify selectivity should be carefully understood by researchers before mechanistic studies using such retinoids as selective probes are undertaken. Consulting the original literature is highly recommended. Such an approach should save valuable time and resources.

Selectivity based on receptor binding affinity

Much of the research on receptor selectivity has been conducted by pharmaceutical companies, such as Allergan, Bristol-Myers Squibb, CIRD-Galderma, Eisai, Hoffmann-La Roche, and Ligand Pharmaceuticals, in the search for improved therapeutic agents. After protection of the retinoid receptor cotransfection assays by patents, several companies engaged in developing selective retinoids as drugs no longer report selectivity data based on the results of RAR or RXR transcriptional activation activity in these cotransfection assays. Consequently, binding affinity and cell culture assays for selectivity are now being reported.

K_D, K_d, apparent K_d, K_i, and IC_{50} values are used to report binding affinity data. Usually, K_D and K_d values refer to saturation constants determined from Scatchard analysis of saturation binding data from experiments in which the concentration of the radiolabeled ligand, alone or in the presence of the non-labeled version of the same ligand, is increased. IC_{50} values refer to the concentration of another ligand that competitively displaces 50% of the radiolabeled ligand from its binding site. Once K_d and IC_{50} values are determined, the K_i value is calculated from the Cheng-Prusoff equation as follows [67]:

$$K_i = IC_{50} / (1 + [\text{concentration of tritiated retinoid standard}])$$

At low concentrations of radiolabeled ligand, the IC_{50} value can be the same as or approach the K_i value. Many of the K_d or K_D values reported in the retinoid literature are actually these K_i values. Apparrent K_d values have been determined by using the Clark equation on K_i values [68]. Tritiated standards for RAR binding studies include commercially available [11,12-$^3H_2$]trans-RA and [11,12-$^3H_2$]9-cis-RA, [$^3H_2$]AHPN, and [$^3H_2$]Am80; those for RXR are [11,12-$^3H_2$]9-cis-RA and [^3H]LDG1069. Our and other groups have preferred using tritiated aromatic retinoids as standards because of their higher stability compared to the tritiated poly-olefinic natural retinoids. These standards have different affinities for the receptors, as illustrated in Table 2. If binding to a natural retinoid is also reported, extrapolations for comparison purposes can be made between different data sets.

Receptor sources for binding studies also differ. Nuclear extracts from COS-7 [57], or HeLa cells provide sources of over-expressed recombinant receptors that permit binding studies to be conducted in the presence of endogenous levels of heterodimeric partners and intermediary proteins. Purified recombinant receptors from *Escherichia coli* or baculovirus/Sf12 insect expression systems have also been used [83, 84]. To permit purification by affinity chromatography, these receptors or their ligand-binding, plus D and/or E, domains are expressed as chimeric proteins with maltose- or glutathione-*S*-transferase-binding protein. In some cases a polyhistidine tag is attached to the receptor to allow purification on a nickel-chelated support. Variations in receptor structure and source can produce different K_i values. Only compounds in a series

Table 2. Retinoid receptor binding affinities

Structure	Name/code number	RARα	RARβ	RARγ	RXRα	RXRβ	RXRγ	Reported as	Receptor source[a]	Tritiated standard	Ref.[b]
2-1	trans-RA	0.37	0.37	0.22	nb[b]	nb	nb	K_d (dissociation)	Baculovirus	trans-RA	Allegretto et al. (1993)
		15.5	4.5	3.0				K_d	COS-7	TTAB	Bernard et al. (1992)
		2.3	0.4	0.3				k_d (apparent)	E. coli	trans-RA	Yu et al. (1996)
		15	13	18	>10000	>10000	>10000	K_d	Baculovirus	trans- or 9-cis-RA	Hembree et al. (1996)
		1.2	0.8					K_i	MBP-RAR-DE	Am80	Eyrolles et al. (1994a)
		14.4	16.6	6	>10000			IC_{50}	E. coli	trans-RA	Crettaz et al. (1990)
		7	5	3				IC_{50}	nr	trans-RA	Bollag et al. (1995)
		16	7					K_i	COS-7	TTAB	Charpentier et al. (1995)
		3.3	4.1	0.04	330	54	120	K_i	E. coli	Am80 (RARα,β)/ -cis-RA (RARγ, RXR)	Umemiya et al. (1997)
2-2	9-cis-RA	0.31	0.20	0.78	1.62	2.36	2.29	K_d (dissociation)	Baculovirus	cis-RA	Allegretto et al. (1993)
		7	7	17	32	12	4	K_d	Baculovirus	trans- or 9-cis-RA	Hembree et al. (1996)
		19	28	11	45		11	IC_{50}	E. coli	trans- or 9-cis-RA	Dawson et al.
		1.2		15	70			K_d	E. coli	trans-RA	Lamour et al. (1996)
			17		12	3.8		IC_{50}	nr	trans-RA	Bollag et al. (1995)
		13		1.0				K_i	E. coli	9-cis-RA	Umemiya et al. (1997)
2-3	TTNPB/Ro13-7410	28.2	23.7	2				IC_{50}	E. coli	trans-RA	Crettaz et al. (1990)
	TTNPB	30	3		>1000	>1000	>1000	K_d	Baculovirus	trans- or 9-cis-RA	Agarwal et al. (1996)
											Hembree et al. (1996)
		4	5	5				IC_{50}	E. coli	trans-RA	Lamour et al. (1996)
		21	22	15				K_i	COS-7	TTAB	Charpentier et al. (1995)
		36						IC_{50}	E. coli	trans-RA	Dawson et al.
2-4	TTNN/SR3957	115	5.4	40				IC_{50}	E. coli	trans-RA	Dawson et al.
	TTNN	580	13					IC_{50}	E. coli	trans-RA	Charpentier et al. (1995)
	Ro19-0645	480	50.3	190				IC_{50}	E. coli	trans-RA	Crettaz et al. (1990)
	Ro19-0645	460	26					IC_{50}	E. coli	trans-RA	Keidel et al. (1994)
2-5	CD367	3	4	1.5				K_d	COS-7	TTAB	Delescluse et al. (1991)
	TTAB/SR3961	5.3	3	2				K_i	COS-7	TTAB	Charpentier et al. (1995)
	Ro40-6976	1.3	7.7	2.5				IC_{50}	E. coli	trans-RA	Dawson et al.
		10.3	5.9					IC_{50}	E. coli	trans-RA	Crettaz et al. (1990)
2-6	UAB8	14	16	10				IC_{50}	E. coli	trans-RA	Muccio et al. (1998)

Table 2. (Continued)

Structure	Name/code number	RARα	RARβ	RARγ	RXRα	RXRβ	RXRγ	Reported as	Receptor source[a]	Tritiated standard	Ref.[b]
2-7	Am80	62	280	816				K_d	COS-7	TTAB	Bernard et al. (1992)
		22	280	1720				K_D	COS-7	TTAB	Delescluse et al. (1991)
		88	>1000	>1000				IC_{50}	E. coli	trans-RA	Dawson et al.
		3.9	30	nb[c]	nb	nb	nb	K_i	E. coli	Am80	Kagechika et al. (1997)
		6.5	30					K_i	E. coli	Am80	Umemiya et al. (1997)
		~8	~12					IC_{50}	E. coli	Am80	Eyrolles et al. (1994)
2-8	Am580	8	131	450				K_d	COS-7	TTAB	Bernard et al. (1992)
	Ro40-6055	6	130	827	>1000			K_D	COS-7	TTAB	Delescluse et al. (1991)
		23	>1000	390				IC_{50}	E. coli	trans- or 9-cis-RA	Dawson et al.
		0.95	21					K_i	E. coli	Am80	Kagechika et al. (1997)
		68.6	3320					IC_{50}	E. coli	trans-RA	Crettaz et al. (1990)
		39	870	2700	10000			IC_{50}	E. coli	trans-RA	Apfel et al. (1992)
		39	870	4800				IC_{50}	nr	trans-RA	Toma et al. (1998)
2-9	Am555S	~2	~40					K_i	E. coli	Am80	Hashimoto et al. (1996)
2-10	AGN 193836	8.4	17374	>30000				K_D	Baculovirus	trans-RA	Teng et al. (1996)
2-11	CD2019	920	26	160				K_d	COS-7	TTAB	Bernard et al. (1992)
2-12	CD417	6500	36	426				K_d	COS-7	TTAB	Bernard et al. (1992)
2-13	Ro48-2249	1000	28	39				IC_{50}	nr	trans-RA	Toma et al. (1998)
2-14	Ro44-4753	2700	3000	210				IC_{50}	nr	trans-RA	Toma et al. (1998)
2-15	SR11254	>1000	>1000	20				IC_{50}	Baculovirus	trans-RA	Dawson et al.
2-16	BMS185354	700	50	3.3				K_d (apparent)	E. coli	trans-RA	Yu et al. (1996)
2-17	AGN 190299	1250	67	57				K_d	Baculovirus	trans- or 9-cis-RA	Hembree et al. (1996)
2-18	CD437	6500	2480	77				K_d	COS-7	TTAB	Bernard et al. (1992)
	AHPN/SR11248	440	680	25				IC_{50}	Baculovirus	trans-RA	Chao et al. (1997)

Table 2. (Continued)

Retinoid		Receptor binding (nM)						Reported as	Assay conditions		Ref.[b]
Structure	Name/code number	RARα	RARβ	RARγ	RXRα	RXRβ	RXRγ		Receptor source[a]	Triticated standard	
2-19	LDG1069	>1000	>1000	>1000	14			K_d	Baculovirus	trans- or 9-cis-RA	Boehm et al. (1994)
	SR11247	100	>1000 (47%)[d]	3800	82	5.9	8.3	IC_{50}	Baculovirus	trans- or 9-cis-RA	Dawson et al.
	LDG1069	180	50	130	16			K_i	E. coli	Am80 (RARα,β) /9-cis-RA (RARγ, RXR)	Umemiya et al. (1997)
	Ro26-4456	9500	9900	5100	190			IC_{50}	nr	trans-RA	Bollag et al. (1995)
		9500	9900	5100	110			IC_{50}	nr	trans-RA	Toma et al. (1998)
2-20	SR11217	>10000	>10000	>10000	388	121	915	K_d	Baculovirus	trans- or 9-cis-RA	Hembree et al. (1996)
		>1000		>1000	370			IC_{50}	Baculovirus	trans- or 9-cis-RA	Dawson et al.
2-21	SR11237	>10000	>10000	5556	72	110	61	K_d	Baculovirus	trans- or 9-cis-RA	Hembree et al. (1996)
				>1000 (0)	180			IC_{50}	Baculovirus	trans- or 9-cis-RA	Dawson et al.
2-22	AGN 191701	>10000	>10000	>10000	308	387	301	K_d	Baculovirus	trans- or 9-cis-RA	Hembree et al. (1996)
		nr	9000	712				K_d	Baculovirus	trans-RA	Elmazar et al. (1992)
2-23	LDG100268	>1000	>1000	>1000	3.2	6.2	9.7	K_i	Baculovirus	trans-RA or 9-cis-RA	Lala et al. (1996)
2-24	LDG100568	306	400	437	2	2	4	K_i	Baculovirus	trans-RA (RAR)/ LG-D1069 (RXR)	Canan Koch et al. (1996)
2-25	LGD100754	1791	2587	6094	8	9	14	K_i	Baculovirus	trans-RA (RAR/LGD)1069 (RXR)	Canan Koch et al. (1996)
nr[c]	Ro25-7386	>10000	>10000	>10000	110			IC_{50}	nr	trans-RA	Toma et al. (1998)

[a] COS-7, nuclear extracts obtained from cells transfected with a RAR or RXR gene; E. coli, extracts of transfected cells expressing recombinant human RAR or RAR-DEF; baculovirus, baculovirus/Sf21 expression system; MBP-RAR-DE, maltose-binding protein–RAR–DE domain chimeric receptor.
[b] Agarwal et al. (1996) [74]; Allegretto et al. (1993) [9]; Apfel et al. (1992) [77]; Bernard et al. (1992) [58]; Boehm et al. (1994) [24]; Bollag et al. (1995) [3];Canan Koch et al. (1996) [82]; Chao et al. (1997) [40]; Charpentier et al. (1995) [71]; Crettaz et al. (1990) [80]; Dawson et al. [unpublished]; Delescluse et al. (1991) [54]; Elmazar et al. (1992) [81]; Eyrolles et al. (1994) [69]; Hashimoto et al. (1996) [79]; Hembree et al. (1996) [27]; Kagechika et al. (1997) [76]; Keidel et al. (1994) [10]; Lala et al. (1996) [36]; Lamour et al. (1996) [73]; Muccio et al. (1996) [75]; Teng et al. (1996) [80]; Toma et al. (1998) [78]; Umemiya et al. (1997) [72]; Yu et al. (1996) [60]. [c] nr, not reported. [d] Number in parentheses refers to % binding at cited concentration. [e] nb, no binding.

Table 2. (Continued)

2-1

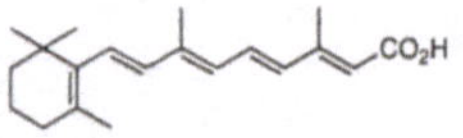

2-2

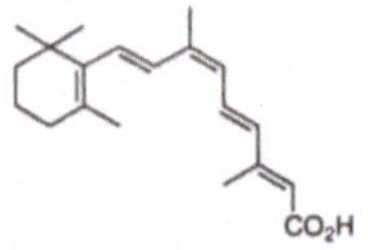

2-3

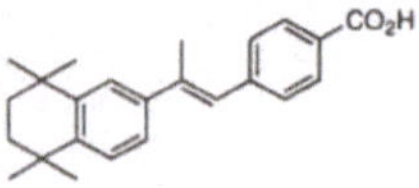

2-4

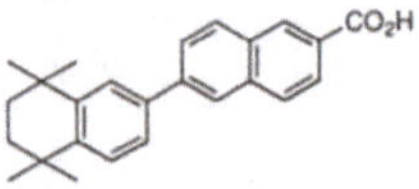

2-5

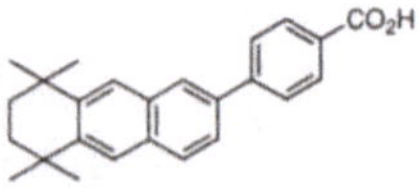

2-6

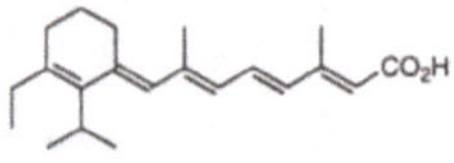

2-7

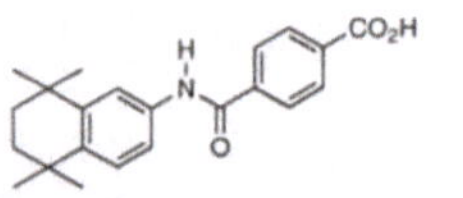

2-8

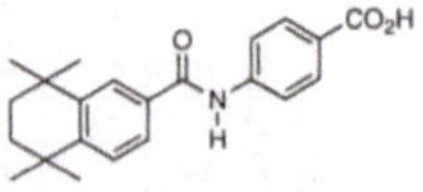

2-9

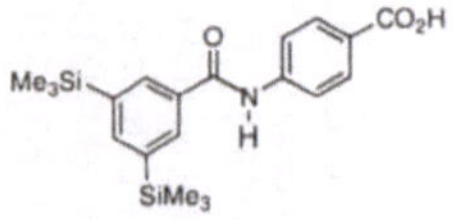

2-10

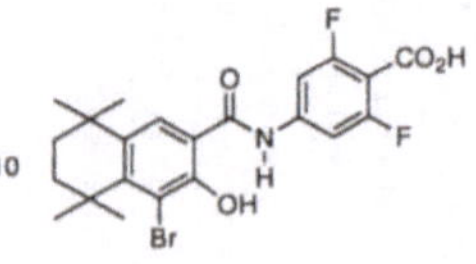

2-11

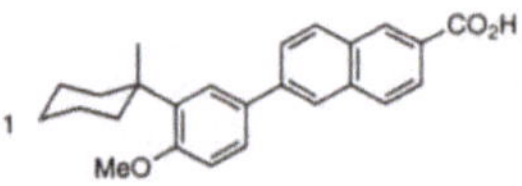

2-12

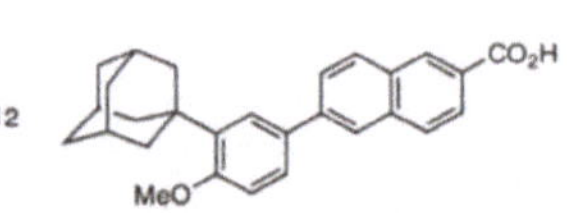

2-13

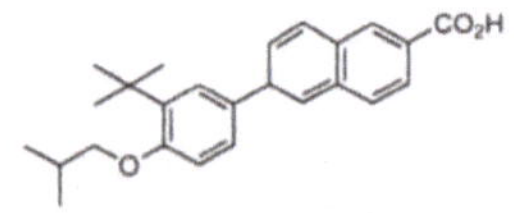

2-14

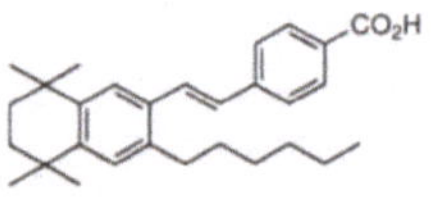

2-15

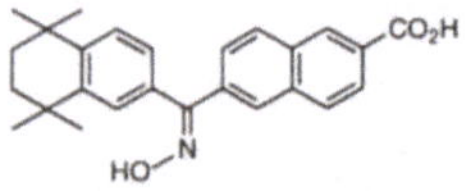

2-16

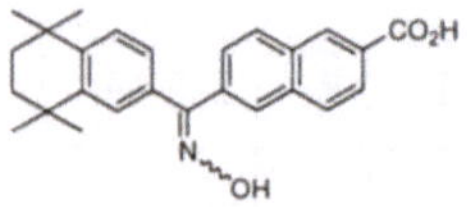

2-17

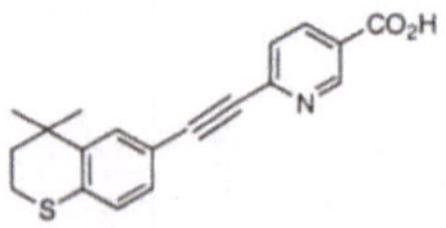

2-18

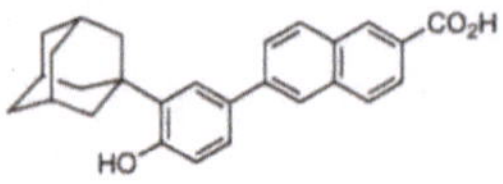

2-19

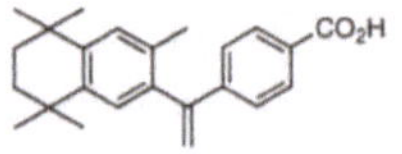

2-20

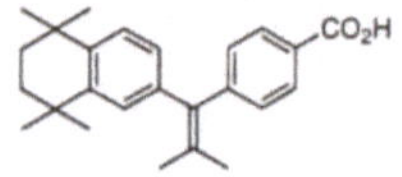

2-21

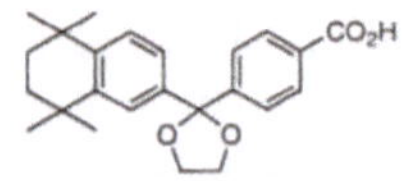

2-22

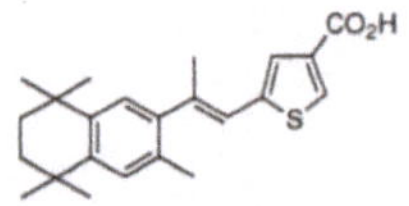

2-23

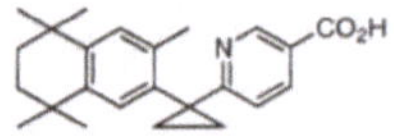

2-24

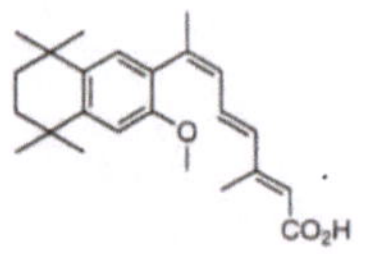

2-25

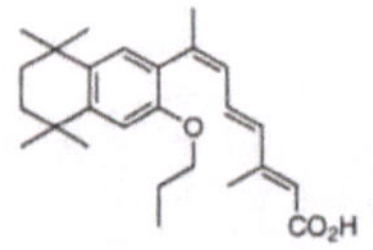

should be compared unless the series contains internal standards to establish consistency with other experimental methods.

In Table 2 are reported binding affinities to RARα, β, and γ and RXRα, β, and γ that have been determined for various class- and subtype-selective retinoids. As this table indicates, binding affinities that have been reported by different groups can vary widely. However, values in the same data set do reflect homology differences in the LBDs of the various subtypes. Generally, we have found that IC_{50} or K_i values for RARβ and RARγ are closer to each other than to those of RARα, reflecting the greater homology between their LBDs [6]. Interestingly, we observed that while some retinoids have a very high K_i or IC_{50} value that could indicate low or no appreciable binding to RARβ and RARγ, these retinoids can have the ability to transcriptionally activate either RARβ or RARγ. This dichotomy may be caused by the higher lipophilicity of the RARβ,γ-selective retinoids compared to the RARα-selective retinoids or the more lipophilic character of the RARβ and RARγ LBDs. Using a detergent in the binding assay may circumvent this problem.

Selectivity based on receptor transcriptional activation

The transient cotransfection assay was first reported by the Chambon [85], Evans [86], and Pfahl [87] groups to identify retinoids that selectively activated one RAR subtype to induce gene transcription from a retinoid response element. For example, in this assay, human COS-1, COS-7, CV-1, or HeLa cell lines are cotransfected with three gene constructs: (1) a retinoid response element linked to a reporter, (2) a retinoid receptor, and (3) an internal standard. The reporter construct contains a retinoid response element, such as the natural β2RARE sequence found in the promoter region of the RARβ gene, which responds to RARs, or a synthetic response element, such as the palindromic thyroid hormone response element (TREpal), which responds to both RARs and RXRs. The response element is coupled to a promoter sequence, such as *tk* from the thymidine kinase gene, and a reporter, such as the gene for chloramphenicol acetyl transferase (CAT), secreted alkaline phosphatase (SeAP), or luciferase (LUC). The β-galactosidase gene is used as a standard to determine cotransfection efficiency. After cotransfection, messenger RNAs for the retinoid receptor and β-galactosidase are expressed and translated to protein. The retinoid receptor binds as a dimer to its retinoid response element in the reporter construct. In the presence of retinoid, the receptor complex is activated to induce the promoter to activate reporter gene transcription. Reporter message is then translated to protein. Interestingly, the level of reporter enzyme activity correlates with the ability of a retinoid to activate a particular receptor on its response element. Constitutive activity (promoter activity in the absence of added retinoid) is subtracted, and retinoid-induced activation is normalized based on the transfection efficiency, determined by measuring the β-galactosidase level. Test retinoid activity is expressed relative to the transactivation activity achieved for a standard retinoid on the same receptor. We have expressed retinoid activity relative to that of either 1 μM *trans*-RA for activating the RARs or 1 μM 9-*cis*-RA for activating the RXRs, with each of these standards designated as 100%. Generally, the CAT reporter assay is more reliable for determining EC_{50} values from dose-response curves when fewer replicates are performed. The luciferase

Table 3. Retinoid receptor transcriptional activation activities of receptor-selective retinoid agonists compared to all-*trans*- and 9-*cis*-retinoic acids

Retinoid			Receptor activation EC_{50}, nM (activation at 1 μM, %)						Transfection assay type					Ref.[e]
Structure	Name/code number	Receptor selectivity	RARα	RARβ	RARγ	RXRα	RXRβ	RXRγ	Receptor type[a]	Response element[b]	Reporter[c] vector	Cell line	EC_{50} response[d]	
3-1	*trans*-RA	RAR	54 (82)	4.5 (78)	2 (72)	530 (53)			wt	(TREpal)$_2$	CAT	CV-1	50% of 1μM *trans*- or 9-*cis*-RA	Dawson et al. (1995)
			27 (100)	12 (97)	8 (98)	430 (86)			wt	(TREpal)$_2$	CAT	CV-1	50% of 10μM *trans*- or 9-*cis*-RA	Jong et al. (1993)
			352	82	10	916	1492	1130	wt	TREpal (RAR/CRBPII (RXR))	LUC	CV-1	50% maximal	Allegretto et al. (1993)
			17	23	12				wt	mLamB1	CAT	HeLa	50% of 1μM *trans*-RA	Ostrowski et al. (1998)
			21	3.6	2.5				wt	(TREpal)$_2$	CAT	HeLa	50% maximal	Delescluse et al. (1991)
			20	18	5				wt	mLamB1	CAT	HeLa	50% of 1μM *trans*-RA	Reczek et al. (1995)
			7	1	0.7				wt[f]	TREpal	LUC	CV-1	50% maximal	Johnson et al. (1995)
			6.7	2.8	3.5	26			nr[f]	nr	nr	nr	50% maximal	Bollag et al. (1995), Toma et al. (1998)
3-2	9-*cis*-RA	RAR/RXR	23 (100)	2.6 (100)	4.3 (100)	6 (100)			wt	(TREpal)$_2$	CAT	CV-1	50% of 1μM *trans*- or 9-*cis*-RA	Dawson et al. (1995)
			45 (110)	22 (86)	12 (127)	13 (100)			wt	(TREpal)$_2$	CAT	CV-1	50% of 10μM *trans*- or 9-*cis*-RA	Jong et al. (1993)
			102	3.3	6.0	13			chimeric ER	ERE	CAT	HeLa	50% of max. *trans*- or 9-*cis*-RA	Beard et al. (1995)
			191	51	45	253	221	147	wt	TREpal (RAR/CRBPII (RXR))	LUC	CV-1	50% maximal	Allegretto et al. (1993)
			13	14	18	5			nr	nr	nr	nr	50% maximal	
			13	14	18	2.5	>5000		nr	nr	nr	nr	50% maximal	Toma et al. (1998)
3-3	TTNPB/Ro13-7410 TTNPB	RAR	2.1		2				chimeric ER	βRARE	SeAP	COS	50% maximal	Keidel et al. (1994)
			21	4	2.4	>1000			chimeric ER	ERE	CAT	HeLa	50% maximal	Jiang et al. (1995)
			85 (80)	6 (69)	24 (112)	>10000 (13)			wt	(TREpal)$_2$	CAT	CV-1	50% of 10μM *trans*- or 9-*cis*-RA	Jong et al. (1993)
			30	3	22 (75)	>5000			wt	TREpal (RAR/CRBPII (RXR))	LUC	CV-1	50% of 1μM *trans*-RA on RARα	Nagy et al. (1995)
			110 (85)	26 (86)			>5000	>5000	chimeric ER	ERE	CAT	nr	50% maximal	Lehmann et al. (1991)
3-4	UAB8	RAR	33	20	2	>2000			wt	(TREpal)$_2$	CAT	CV-1	50% of 1μM *trans*-RA	Muccio et al. (1998)
3-5	CD367 TTAB/SR3961	RAR	0.2	0.37	0.25				wt	(TREpal)$_2$	CAT	HeLa	50% maximal	Delescluse et al. (1991)
			6 (119)	0.9 (151)	0.9 (86)				ER chimeric	ERE	CAT	CV-1	50% of 1μM *trans*-RA on RARα	Lehmann et al. (1991)
3-6	SR11365	RAR	27 (95)	23 (56)	19 (48)	>1000 (6)			wt	(TREpal)$_2$	CAT	CV-1	50% of 1μM *trans*- or 9-*cis*-RA	Dawson et al.
3-7	SR11256	RAR	0.2 (127)	4 (109)	2 (99)	2300 (27)			wt	(TREpal)$_2$	CAT	CV-1	50% of 1μM *trans*- or 9-*cis*-RA	Dawson et al.
3-8	Am580 Ro40-6055	RARα	0.36	25	28				wt	(TREpal)$_2$	CAT	HeLa	50% maximal	Delescluse et al. (1991)
			3.2	200	400				wt	mLamB1	CAT	HeLa	50% of 1μM *trans*-RA	Ostrowski et al. (1998)
			17 (70)	30 (113)	>1000 (28)	>1000 (10)			wt	(TREpal)$_2$	CAT	CV-1	50% of 1μM *trans*- or 9-*cis*-RA	Dawson et al.
			3	25	40	10000			nr	nr	nr	nr	50% maximal	Bollag et al. (1995), Toma et al. (1998)
3-9	Am80	RARα	1.5	6.9	149				wt	(TREpal)$_2$	CAT	HeLa	50% maximal	Delescluse et al. (1991)
			45	235	591	>5000	>5000	>5000	wt	TREpal (RAR/CRBPII (RXR))	LUC	CV-1	50% maximal	Nagy et al. (1995)
			21 (81)	76 (80)	>1000 (33)	>5000 (16)			wt	(TREpal)$_2$	CAT	CV-1	50% of 1μM *trans*- or 9-*cis*-RA	Dawson et al.
3-10	AGN 193836	RARα	~600 (58)						chimeric ER	ERE	LUC	CV-1	50% of 1μM *trans*-RA	Teng et al. (1996)
3-11	CD2019	RARβ	20	3.8	47				wt	(TREpal)$_2$	CAT	HeLa	50% maximal (vs. *trans*-RA)	Bernard et al. (1992)
3-12	BMS188970	RARβ	250 (70)	11 (115)	60 (65)				chimeric Gal4	Gal4	CAT	HeLa	50% of 1μM *trans*-RA	Yu et al. (1996)
3-13	Ro48-2249	RARβ	2000	39	160	10000			nr	nr	nr	nr	50% maximal	Bollag et al. (1995), Toma et al. (1998)
3-14	TTNN/SR3957 Ro19-0645	RARβ,γ	>1000 (23)	42 (82)	60 (95)				chimeric ER	EREβ	CAT	CV-1	50% of 1μM *trans*-RA on RARα	Lehmann et al. (1991)
			67						wt		SeAP	COS	50% maximal	Keidel et al. (1994)
3-15	BMS185282	RARγ	>1000 (0)	600 (55)	100 (90)				chimeric Gal4	Gal4	CAT	HeLa	50% of 1μM *trans*-RA	Yu et al. (1996)
			>1000 (5)	290 (75)	50 (98)				wt	mLamB1	CAT	HeLa	50% of 1μM *trans*-RA	Reczek et al. (1995)
			na	na	450				chimeric Gal4	Gal4	CAT	HeLa	50% of 1μM *trans*-RA	Chen et al. (1995)

Table 3. (Continued)

Retinoid			Receptor activation EC$_{50}$ nM (activation at 1 µM, %)						Transfection assay type					Ref.[c]
Structure	Name/code number	Receptor selectivity	RARα	RARβ	RARγ	RXRα	RXRβ	RXRγ	Receptor type	Response element[b]	Reporter vector[c]	Cell line	EC$_{50}$ response[d]	
3-16	BMS185283	RARγ	>1000 (0)	>1000 (10)	800 (55)				chimeric Gal4	Gal4	CAT	HeLa	50% of 1 µM trans-RA	Yu et al. (1996)
			>10000 (10)	>1000 (24)	350 (65)				wt	mLamB1	CAT	HeLa	50% of 1 µM trans-RA	Reczek et al. (1995)
			na	363	58				chimeric Gal4	Gal4	CAT	HeLa	50% of 1 µM trans-RA	Chen et al. (1995)
3-17	BMS185354	RARγ	>1000 (5)	200 (105)	30 (100)				chimeric Gal4	Gal4	CAT	HeLa	50% of 1 µM trans-RA	Yu et al. (1996)
			na	450	30				chimeric Gal4	Gal4	CAT	HeLa	50% of 1 µM trans-RA	Chen et al. (1995)
3-18	SR11254	RARγ	>10000 (13)	1600 (46)	22 (84)	>10000 (13)			wt	(TREpal)$_2$	CAT	CV-1	50% of 1 µM trans- or 9-cis-RA	Chao et al. (1993, 1995,1997)
			>10000 (3)	500 (65)	<100 (168)	>10000 (0)			chimeric ER	ERE	CAT	CV-1	50% of 1 µM trans- or 9-cis-RA	Dawson et al.
3-19	Ro44-4753	RARγ	1000	88	15	10000			nr	nr	nr	nr	50% maximal	Bollag et al. (1995), Toma et al. (1998)
3-20	CD437 AHPN/SR11248	RARγ	140	28	7.3				wt	(TREpal)$_2$	CAT	HeLa	50% maximal	Bernard et al. (1992)
			>1000 (21)	63 (78)	69 (88)				wt	(TREpal)$_2$	CAT	CV-1	50% of 1 µM trans-RA	Dawson et al.
3-21	LGD100568	RAR/RXR	40 (35)	8 (54)	17 (66)	5 (51)	8 (102)	9 (68)	wt	TREpal (RAR)/CRBPII or (CPRE) (RXR)	LUC	CV-1	50% maximal	Canan Koch et al. (1996)
3-22	SR11217	RXR	>1000 (0)	980 (21)	>1000 (0)	86 (77)			wt	(TREpal)$_2$	CAT	CV-1	50% of 1 µM 9-cis-RA	Dawson et al. (1995)
			>10000	>10000	>10000	12			chimeric ER	ERE	CAT	HeLa	50% maximal trans- or 9-cis-RA	Jiang et al. (1995)
3-23	LDG1069	RXR	>10000	>10000	>10000	33	24	25	wt	(TREpal)$_2$(RAR)/(CRBPII) or (CPRE)$_3$ (RXR)	LUC	CV-1	50% maximal	Boehm et al. (1994)
	SR11247		>10000 (10)	>10000 (2)	1000 (54)	42 (95)			chimeric ER	ERE	CAT	CV-1	50% of 10µM trans-RA	Dawson et al.
	SR11247		>1000 (17)	112 (67)	122 (9)	4			wt	(TREpal)$_2$	CAT	CV-1	50% of 1 µM trans- or 9-cis-RA	Dawson et al. (1995)
	Ro26-4456		10000	1000	1000	5			nr	nr	nr	nr	50% maximal	Bollag et al. (1995)
			685	520	930				nr	nr	nr	nr	50% maximal	Toma et al. (1998)
3-24	SR11246	RXR	>1000 (6)	>1000 (12)	>1000 (9)	55 (98)			wt	(TREpal)$_2$	CAT	CV-1	50% of 1 µM trans- or 9-cis-RA	Dawson et al. (1995)
			>1000 (2)	>1000 (33)	>1000 (15)	17 (102)			wt	(TREpal)$_2$	CAT	CV-1	50% of 1 µM trans- or 9-cis-RA	Dawson et al.
3-25	SR11345	RXR	>1000 (0)	>1000 (-6)	>1000 (-12)	21 (107)			wt	(TREpal)$_2$	CAT	CV-1	50% of 1 µM trans- or 9-cis-RA	Dawson et al.
3-26	LDG100268	RXR	>10000	>10000	>10000	4	3	4	wt	(TREpal)$_2$(RAR)/(CRBPII) or (CPRE)$_3$ (RXR)	LUC	CV-	50% maximal	Boehm et al. (1995)
3-27	AGN 191701	RXR	>5000	1990	979	215	180	105	wt	(TREpal)$_2$ (RAR)/CRBPII (RXR)	LUC	CV-	50% maximal	Nagy et al. (1995)
			>10000	>10000	>10000	201			chimeric ER	ERE	CAT	HeL	50% of maximal (trans-RA)	Beard et al. (1995)
3-28	AGN 192849	RXR	>1000 (6)	>1000 (4)	>1000 (0)	54 (91)	57 (100)	42 (85)	wt	(TREpal)$_2$ (RAR)/CRBPII or (CPRE)$_3$ (RXR)	LUC	CV-	50% maximal	Beard et al. (1996)
3-29	HX600	RXR	>1000 (0)	>1000 (0)	>1000 (-10)	>1000 (55)	>1000 (40)	200 (91)	wt	DR-5 (RAR)/DR-1 (RXR)	CAT	HeL	50% of 1 µM 9-cis-RA	Umemiya et al. (1997)
nr[e]	Ro25-7386	RXR	>10000	>10000	>10000	1.4			nr	nr	nr	nr	50% maximal	Toma et al. (1998)

Table 3 (Continued)

3-1

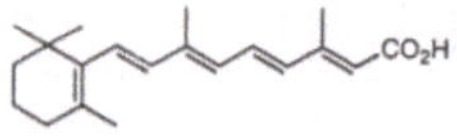

3-2

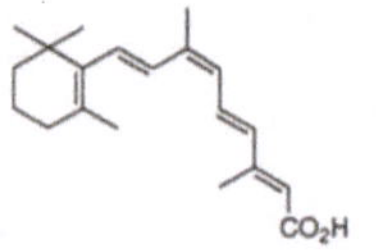

3-3

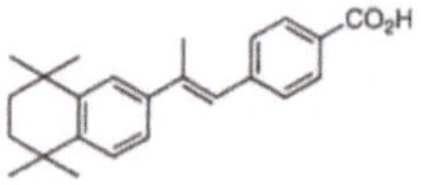

3-4

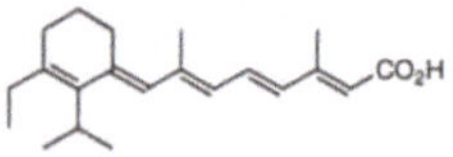

3-5

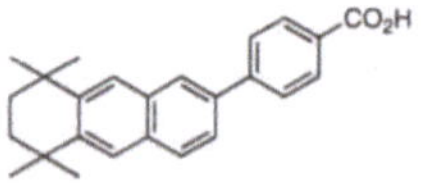

3-6

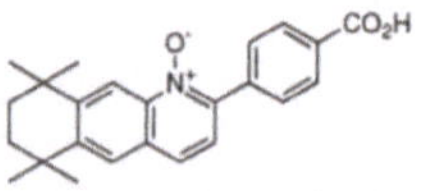

3-7

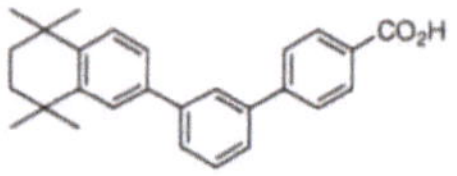

3-8

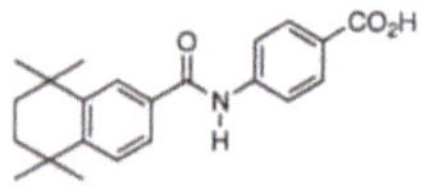

3-9

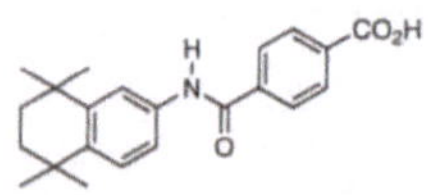

3-10

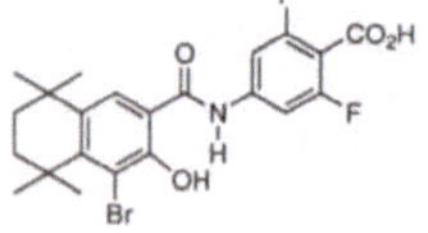

3-11

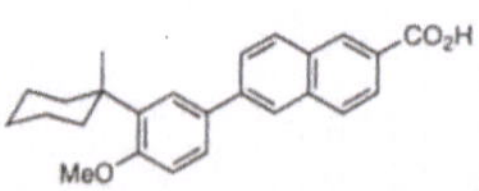

3-12

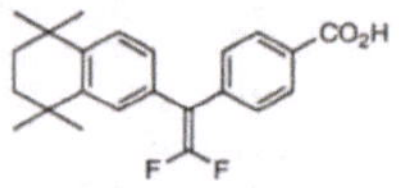

3-13

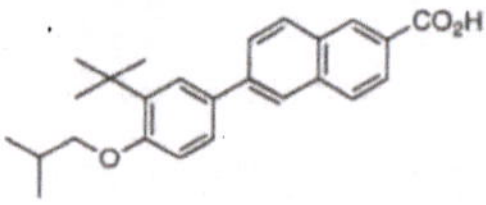

3-14

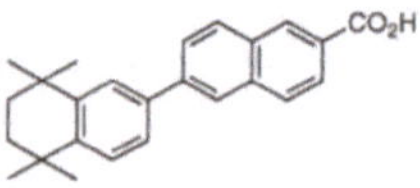

3-15

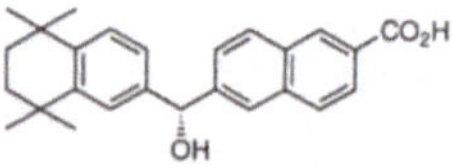

3-16

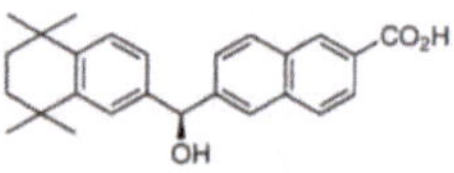

3-17

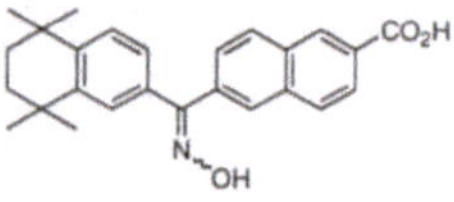

3-18

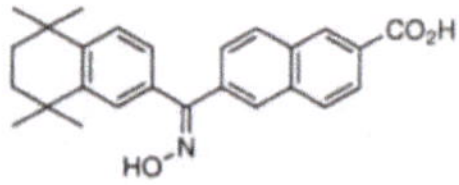

3-19

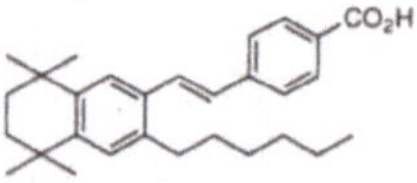

3-20

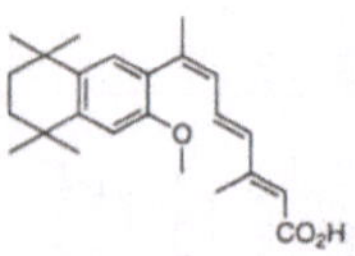

3-21

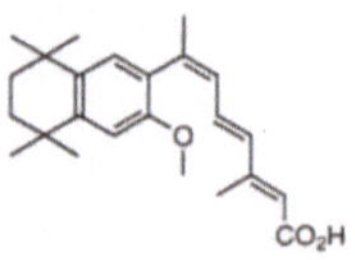

3-22

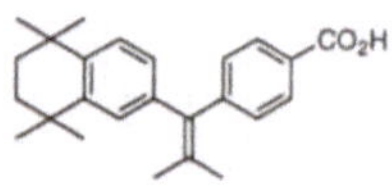

3-23

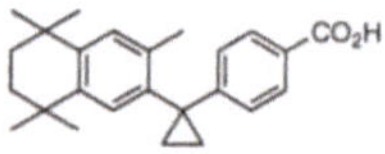

3-24

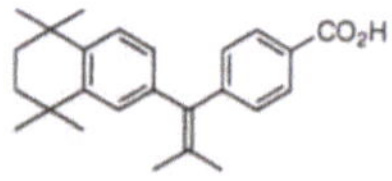

3-25

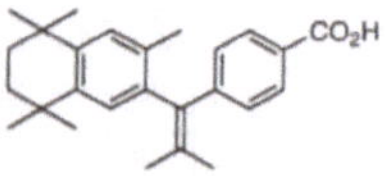

3-26

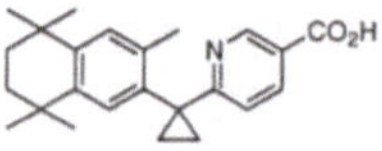

3-27

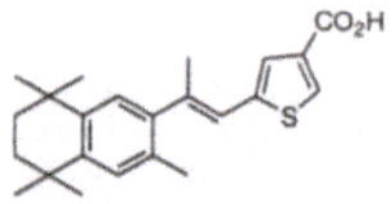

3-28

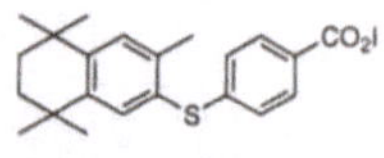

3-29

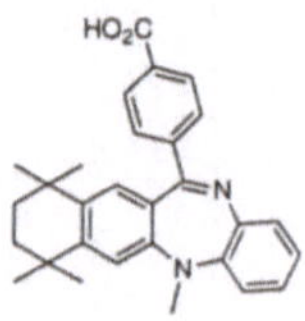

Table 3. Footnotes

[a] Transfected receptor: wt, wild-type or native human or mouse RAR or RXR; chimeric receptor composed of ER (A/B) or Gal4 DNA-binding domain linked to RAR (D/E/F) or RXR (D) domain.
[b] Response element linked to reporter: (TREpal)$_2$, two copies of palindromic TRE, which is activated by both RARs and RXRs; CRBPII, cellular retinol-binding protein II (DR-1) activated by RXRs or RXRα,γ; mLamB1, complex RARE of mouse laminin B1 gene activated by RARs; CPRE, chick ovalbumin upstream promoter response element activated by RXRβ; ERE, *Xenopus* vitellogenin A2 estrogen response element activated by estrogen receptor; Gal4, galactose-inducible yeast transcription factor.
[c] CAT, chloramphenicol acetyltransferase gene; LUC, luciferase gene; SeAP, secreted alkaline phosphatase gene.
[d] EC$_{50}$ as 50% maximal is retinoid concentration giving one-half of its maximum response, which may not be its response at 1 or 10 μM; 50% of 1 μM *trans*-RA is retinoid concentration giving 50% of the response of 1 μM *trans*-RA, which is designated as 100% for every RAR subtype or for only RARα, with RARβ and RARβ then expressed as percent of RARα at 1 μM *trans*-RA; 50% of 1 μM 9-*cis*-RA is retinoid concentration giving 50% of the response of 1 μM 9-*cis*-RA on RAR or RXR, which is designated as 100%.
[e] Allegretto et al. (1993) [9]; Beard et al. (1995) [93]; Beard et al. (1996) [98]; Bernard et al. (1992) [58]; Boehm et al. (1994) [24]; Boehm et al. (1995) [97]; Bollag et al. (1995) [3]; Canan Koch et al. (1996) [82]; Chao et al. (1993, 1995, 1997) [40, 61, 62]; Chen et al. (1995) [96]; Dawson et al. (1995) [65]; Dawson et al. [unpublished]; Delescluse et al. (1991) [54]; Jiang et al. (1995) [94]; Johnson et al. (1995) [92]; Jong et al. (1993) [89]; Keidel et al. (1994) [10]; Lehmann et al. (1991) [52]; Muccio et al. (1998) [75]; Nagy et al. (1995) [95]; Ostrowski et al. (1998) [90]; Reczek et al. (1995) [91]; Teng et al. (1996) [80]; Toma et al. (1998) [78]; Umemiya et al. (1997) [99]; Yu et al. (1996) [59].
[f] nr, not reported.

assay is more amenable to automation in 96-well plates and is reliable provided that sufficient replicates are run. We found that on the (TREpal)$_2$-*tk*-CAT, RARγ shows the highest constitutive activity.

In the presence of certain retinoid antagonists, which are termed inverse agonists, transcriptional activity can drop to below constitutive levels to produce a negative response. Interestingly, treatment of human keratinocytes with the inverse agonist 4-[2-(5,6-dihydro-5,5-dimethyl-8-*p*-tolyl-2-naphthalenyl)ethynyl]benzoic acid (AGN 193109) and RAR agonist TTNPB (AGN 193840) resulted in mutual repression of their ability to attenuate the expression of the serum-induced differentiation marker MRP-8 [88]. Whether constitutive activity arises from a non-liganded RAR or an RAR complexed to some as yet unidentified ligand or coactivator remains to be established.

Identifying candidates for mechanistic studies based on the ability of retinoids to activate the retinoid receptors should be approached knowledgeably because of differences in cotransfection assays (see Tab. 3). For example, RE, receptor type, reporter, cell type, and method of expressing activation have been varied.

In collaboration with the Pfahl group, we first reported that retinoid transcriptional activity depended on the RE used in the reporter construct [100, 101]. This finding has since been confirmed by another group [17]. Whether natural or synthetic, these REs are often repeated multiple times to enhance the transcriptional response. If the tertiary structure of this sequence differs from that found in natural gene sequences, the transcriptional response can be altered. Retinoid receptors are overexpressed in the cotransfection assay compared to their endogenous levels. Responses may differ because overexpressed receptor levels exceed the endogenous levels of their heterodimeric partners, coactivators, and corepressors. In the cotransfection assay, the overexpressed RAR or RXR is thought to bind the RE as a homodimer. To circumvent retinoid receptor homodimer formation, chimeric receptors are used in cotransfection assays. Gene sequences coding for the LBD of a retinoid receptor are linked to the DBD of another hormone receptor, such as the estrogen or glucocorticoid receptor, and the reporter construct contains either an estrogen or glucocorticoid RE. In the presence of added retinoid, the retinoid LBD (domain E) activates the steroid DBD (domain C) of the chimeric receptor bound to its steroid RE to induce reporter gene synthesis. The influences of the natural retinoid REs and the levels of heterodimeric partners and intermediary proteins on the receptor response to a particular ligand are thereby avoided, but they are replaced by the effects of modulators of the steroid response. Alternatively, the retinoid receptor D, E, and F domains can be coupled to the Gal4 DBD, which activates the Gal4 element, which only responds to the RXR-RAR dimer [102, 103].

Both potency and efficacy must be considered in determining retinoid receptor selectivity. Activity parameters that we generally report are (1) the concentration of retinoid required to give 50% of the transcriptional activation activity shown by the comparison standard at 1 μM (EC_{50} or AC_{50} value) on the same receptor, and (2) the percent retinoid activity at 1 μM compared to that of the 1 μM standard. Selectivity determined by comparison of EC_{50} or AC_{50} values alone may not reflect the relative receptor activities at other, particularly higher, concentrations. For example, in Figure 1A are dose-response curves for activation of each of the RAR subtypes by Am580 and Am80 on the $(TREpal)_2$, and in Figure 1B dose-response curves for SR11254 and AHPN. At 10 nM, Am580 is more RARα selective than Am80 on this RE, but at 100 nM, Am80 is the more selective. At 10 nM, SR11254 is more RARγ selective than AHPN, while selectivity is reversed at 1000 nM. AHPN is a potent activator of RARβ, perhaps

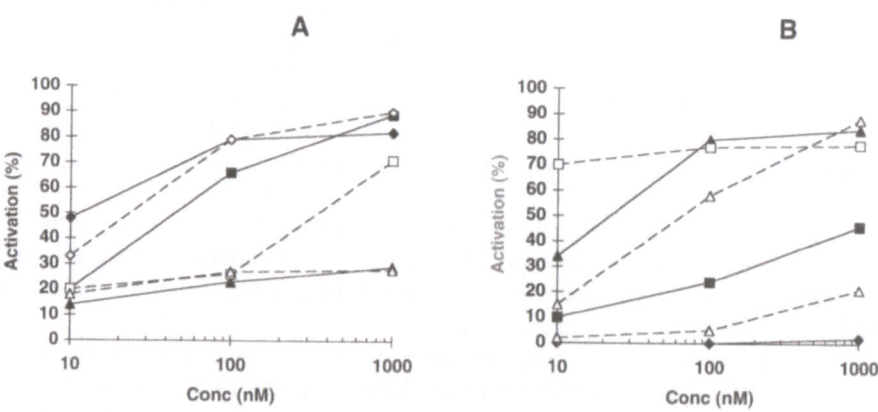

Figure 1. Concentration-dependent transcriptional activation of the retinoic acid receptor subtypes α, β, and γ on the $(TREpal)_2$-*tk*-CAT reporter construct in CV-1 cells by (A) RARα-selective retinoids Am580 (RARα, ◆; RARβ, ■; RARγ, ▲) and Am80 (RARα, ◇; RARβ, □; RARγ, △), and (B) RARγ-selective retinoids SR11254 (RARα, ◆; RARβ, ■; RARγ, ▲) and AHPN (RARα, ◇; RARβ, □; RARγ, △).

due to its ability to induce the synthesis of nur77 [42]. This orphan receptor heterodimerizes with RXR on the βRARE and induces RARβ synthesis in the presence of a RXR-selective ligand. Because reported EC_{50} or AC_{50} values may refer to (1) 50% of the maximal response of the test retinoid or (2) its concentration giving 50% of the response of a 1 or 10 μM retinoid standard, such as trans-RA or 9-cis-RA, and the activation induced by the 1 or 10 μM standard retinoid on every RAR subtype may be treated as 100% activation or only that induced on RARα, with the activations by the retinoid on RARβ and RARγ expressed as percentages of that on RARα, EC_{50} values reported by different groups vary, as can be seen in Table 3.

If the objective is to determine the role of a particular receptor subtype on cell differentiation, using EC_{50} or AC_{50} values alone as a measure of receptor selectivity may give reliable results because the retinoid concentration required to induce differentiation can be in the nanomolar range, which would approximate the value for the 50% response. However, if the objective is to determine the role of a particular retinoid receptor in inhibiting cancer cell growth, responses at 1 μM or the actual dose-response curves are important because higher retinoid concentrations (submicromolar to micromolar range) are required for growth inhibition. At higher concentrations, selectivity may be lost. For example, 1 μM Am580 is not RARα-selective on the $(TREpal)_2$, and 1 μM AHPN activates RARβ in addition to RARγ (see Figure 1). Endogenous REs and dimeric partner levels may also change selectivity patterns. Because of these complexities, we try to use more than one subtype-selective retinoid so that correlation analyses of selectivity with biological activity can be performed. Cotransfection assays have been conducted at retinoid concentrations as high as 10 μM. In this range, retinoids, particularly the more lipophilic ones, can have general toxic effects that are probably not mediated through interactions with retinoid receptors. Therefore, cotransfection and other experiments using high retinoid concentrations may not reflect actual selectivities.

In Table 3 are listed the structures and transcriptional activation activities of retinoids reported as receptor-selective based on their preferential ability to induce one retinoid receptor complex to activate gene transcription from a particular RE compared to the activities of the standards. These retinoid agonists are commonly used to identify the retinoid receptor having the most influence on a particular retinoid-signalling pathway. Four retinoids—TTNPB [23], TTNN [51], TTAB [53, 54], SR11256 [104], and SR11365 [105]—are listed as RAR class selective because they only activate the RARs. Of these, TTNPB, TTAB, and the N-oxide SR11365 are the most transcriptionally selective across all RAR subtypes, whereas TTNN preferentially activates RARβ and RARγ. At higher concentrations SR11256, 4-[3-(5,6,7,8-tetrahydro-5,5,8,8-tetramethyl-2-naphthalenyl)phenyl]benzoic acid, can also activate the RXRs. The parent 4-(6,7,8,9-tetrahydro-6,6,9,9-tetramethylnaphtho[2,3-b]pyridin-2-yl)benzoic acid from which SR11365 is derived was first reported by the Shudo group [106].

Three RARα-selective retinoids, Am580 [20], Am80, and AGN193836 [80] are listed. RARα subtype-selective retinoids were identified first and appear to be characterized by a group capable of hydrogen-bonding to serine 232 of helix 3 in the RARα LBD [90]. All are aromatic amides and should not be stored in media for long periods of time, because we found that degradation can occur under such conditions. 4-[(3-Bromo-3-hydroxy-5,6,7,8-tetrahydro-5,5,8,8-tetramethyl-2-naphthalenyl)carboxamido]-2,6-difluorobenzoic acid (AGN 193836) and its analogs, reported by Chandraratna and coworkers [80], are very selective but have lower potency in the cotransfection assay. 4-[3,5-Bis(trimethylsilyl)phenylcarboxamido]benzoic acid (Am555S) is reported to be an RARα-selective inducer of HL-60 leukemia cell differentiation [79] with an EC_{50} value $(3.0 \times 10^{-9}$ M) almost nine-fold higher than that of Am580.

Identification of RARβ-selective retinoids has been the most challenging. Three retinoids with RARβ subtype selectivity, CD2019 [58, 64], BMS188970 [60], and Ro48-2249 [63] are listed. Because the aromatic methyl ether group of CD2019 could undergo cleavage in cell or organ culture or *in vivo* to give a retinoid capable of activating RARγ, RARβ selectivity should be confirmed using another RARβ-selective retinoid. Our group has not yet identified a potent RARβ agonist, although we have identified several retinoids that cause the activation of both RARβ and RXRα in the cotransfection assay. These compounds may, in fact, only be RXR-selective and form an activated RXR-nur77 complex on the β2RARE to induce RARβ synthesis [107].

Several RARγ-subtype selective retinoids have been reported. Mutational analysis by Ostrowski et al. [90] indicates that these retinoids, such as the oxime of 6-(5,6,7,8-tetrahydro-5,5,8,8-tetramethyl-2-naphthalenyl)-2-naphthalenecarboxylic acid (SR11254 is the *E*-isomer and BMS185354 is the *E/Z* mixture), bind selectively because of their apparent ability to hydrogen-bond to methionine 272 in helix 5 of the RARγ LBD. We found that SR11254 is relatively stable on storage in dimethyl sulfoxide but does hydrolyze if stored for long periods in aqueous medium. CD437 (AHPN) [58, 64] is widely used as a probe to establish RARγ selectivity. As Figure 1B indicates, RARγ-selective activation by AHPN on the (TREpal)₂ is concentration-dependent and RARβ is potently activated at 1 μM. Because AHPN has the ability to inhibit cell growth and induce apoptosis through a non-retinoid receptor-mediated pathway, its use as an RARγ probe should be confirmed using other RARγ-selective retinoids having dissimilar structure, for example, RARγ-selective SR11254. Since AHPN has a phenolic hydroxyl group that is susceptible to oxidation, solutions should be stored under inert gas. Ro 44-4753 is also RARγ-selective [63].

Due to the ability of RXR to heterodimerize with other nuclear receptors, including the orphans, several groups have focussed on identifying RXR-selective retinoids for use either in mechanistic studies to establish retinoid signalling pathways or in drug development to determine if their complexation in the heterodimer complex will enhance the activation produced by the ligand of the heterodimeric partner. For example, the use of combination therapy for diabetes could potentiate the effects of other receptor-active drugs, such as the thiazolidinediones which are the ligands for the RXR heterodimeric partner PPARγ. 4-[1-(5,6,7,8-Tetrahydro-3,5,5,8,8-pentamethyl-2-naphthalenyl)ethenyl]benzoic acid (LGD1069) was reported as RXR-selective on the CRBPII DR-1 RXRE [24]. We found that weak activation of the RARs on the (TREpal)₂ also occurs with this retinoid (designated SR11247) [65], and Bollag et al. [63] also report that this retinoid (designated Ro26-4456) is RXR-selective but that weak activation of the RARs also occurs. It remains to be established whether LGD1069 exerts its *in vivo* anti-cancer effects by functioning as a RAR/RXR panagonist, or by activating a RXR-nur77 dimer [107] or an RXR-orphan receptor dimer. The cyclopropane-bridged retinoids LGD100268 [97] and 4-[1-(5,6,7,8-tetrahydro-5,5,8,8-tetramethyl-2-naphthalenyl)cyclopropyl]benzoic acid (SR11246) [65] are more RXR-selective. The isobutylidene-bridged RXR agonist 4-[2-methyl-1-(3,5,5,8,8-pentamethyl-5,6,7,8-tetrahydro-2-naphthalenyl)propenyl]benzoic acid

(SR11345) enhanced the ability of RAR-selective SR11365 to inhibit breast cancer cell growth but had little inhibitory effect alone [105]. Ro25-7386 (structure not reported) was more RXRα selective than LGD1069 (designated Ro26-4456) [78].

Selectivity based on bioassays

The HL-60 cell differentiation assay has been extensively used by Shudo, Hashimoto, and their co-workers to search for RAR agonists and synergists. RAR- and RARα-selective compounds induce differentiation as measured by nitrotetrazolium blue staining of these cells. Preferential selectivity for RARα or RARβ is determined by binding affinity. This differentiation assay has proved useful for identifying retinoid antagonists and synergists. The antagonists are unable to induce cell differentiation, but they competitively block the activity of RARα-selective Am80 (see discussion below). The synergists have minimal or no differentiating activity but potentiate the activity of Am80. LGD1069 is such a synergist [108].

Umemiya et al. [72, 99] reported that the RXR synergists 4-[5*H*-2,3(2,5-dimethyl-2,5-hexano)-5-methyldibenzo[*b,e*][1,4]diazepin-11-yl]benzoic acid (HX600) [99], which also binds to RARβ, and 4-[2,3-(2,5-dimethyl-2,5-hexano)dibenzo[*b,f*][1,4]thiazepin-11-yl]benzoic acid (HX630) [72], which also binds to RARα and β, were inactive alone but enhanced the ability of RARα-selective Am80 to induce HL-60 differentiation. HX600 behaved as a RXR agonist on the Gal4 response element. At higher concentrations, HX600 antagonized *trans*-RA-induced HL-60 differentiation. HX630 was a more potent synergist than HX600.

Other synergists are 4-(5,6,7,8-tetrahydro-5,5,8,8-tetramethyl-2-naphthalenylamino)benzoic acid (DA010) and related low-molecular-weight alkyl diaryl amines, such as the methyl analog DA011, which only weakly induced differentiation alone but potently enhanced that induced by Am80 [109]. The most potent synergists in this series were the *n*-propyl, *i*-propyl, cyclopropyl, and cyclopropylmethyl tertiary amines. The first three synergists were inactive alone at concentrations below 1 μM. In the 4-(5,6,7,8-tetrahydro-3,5,5,8,8-pentamethyl-2-naphthalenylamino)benzoic acid series, the ethyl and *n*-propyl analogs were the most potent and were also inactive alone below 1 μM. TZ245 and TZ335, the thiazolidinedione derivatives of RXR-selective LDG1069 and 4-[(5,6,7,8-tetrahydro-5,5,8,8-tetramethyl-2-naphthalenyl)-methylamino]benzoic acid (DA011), respectively, were unable to induce differentiation at 1 μM but effectively enhanced Am80 activity [108]. TZ335 was the more potent of the two, having synergistic activity comparable to that of LDG1069. On the basis of their activity, these compounds appear to be selective for RXR.

Retinoid antagonists

Retinoid antagonist activity is indicated if a retinoid binds to a retinoid receptor but is not able to induce the conformational change in the receptor LBD AF-2 region of helix 12 necessary to release a corepressor and/or bind a coactivator protein [90, 110] so that gene transcription on a retinoid RE is activated. Thus, the conformation of the receptor does not undergo the appropriate changes that induce gene transcription by the ligand-receptor dimer complex on a retinoid RE. Because the recombinant receptors necessary to conduct binding studies may not be available, some groups have demonstrated antagonist activity by inhibiting agonist activity in competition experiments, such as blocking receptor activation in the cotransfection assay or inhibiting HL-60 leukemia cell differentiation [72, 99]. Generally, an antagonist concentration that is 100-fold or more higher than that of an agonist having a similar receptor binding affinity is necessary to achieve inhibition of agonist activity in cell culture. Retinoid transcriptional antagonists are listed in Table 4.

Several RAR class-selective antagonists have been reported. The potent antagonists first reported by groups at Allergan [92] and Bristol-Myers Squibb [117] are characterized by an 8-aryl-substituted 5,6-dihydronaphthalene ring and are analogs of TTNPB. These antagonists are related to (2E,4E,6Z,8E)-3-methyl-7-(p-tolyl)-9-(2,6,6-trimethylcyclohexenyl)-2,4,6,8-nonatetraenoic acid (SR11302), which reduced the constitutive activity of the ER-RARα and ER-RARγ chimeric receptors on the ERE, an indication of antagonist activity [118]. (E)-4-{2-[4-(3-Methylbutylthio)phenyl]propenyl}benzoic acid (SR3986) did not activate the chimeric ER-RARs on the ERE but did bind to the RARs [52]. Although SR3986 preferentially binds to RARγ, the difference in affinities is insufficient to confer subtype selectivity in the sub- and micromolar range, and so SR3986 should be considered an RAR panantagonist. TTAB analogs SR11330 and SR11335, which have a 2,2,2-trifluoro-1-hydroxyethyl and 2,2,2-trifluoro-1-methoxyethyl group, respectively, at the 4-position of their tetrahydroanthracene rings are reported to antagonize $trans$-RA activation of the ER-RARs on the ERE [111]. SR11335 shows low binding affinity (IC_{26} = 1000 nM) to recombinant RARβ, which was the only subtype evaluated.

Several antagonists have been reported by Shudo and coworkers. Analogs of TTAB having an i-propyl or benzyl group at the 1-position of the 5,6,7,8-tetrahydro-5,5,8,8-tetramethylnaphth[2,3-d]imidazol-2-yl ring antagonize the ability of Am80 to bind to RARα and RARβ in HL-60 leukemia cell extracts and inhibit the differentiation of these cells induced by $trans$-RA [69]. The antagonist 4-(13H-10,11,12,13-tetrahydro-10,10,13,13,15-pentamethyldinaphtho[2,3-b][1,2-e][1,4]diazepinyl)benzoic acid (LE135) binds to RARα and RARβ but not to RARγ. LE135 antagonizes HL-60 cell differentiation by Am80. Its benzo analog 4-(5H-7,8,9,10-tetrahydro-5,7,7,10,10-pentamethylbenzo[e]naphtho[2,3-b][1,4]diazepin-13-yl)benzoic acid (LE540) also binds to the RXRs and antagonizes HL-60 cell differentiation by Am80 [69, 99]. Their receptor binding affinity is lower than that found for the TTNPB analog antagonists. 4-[3-(4-Diamantyl)-4-methoxyphenylcarboxamido]benzoic acid (TD550) is reported to antagonize HL-60 differentiation by $trans$-RA or Am80 [112].

(E)-7-[3-(1-Adamantyl)-4-methoxyphenyl]-3-methyl-2,4,6-octatrienoic acid (CD2366) is reported to be a RAR antagonist with four- and eight-fold higher affinity for RARα than for RARβ and RARγ, respectively [81]. Because this compound has the 3-(1-adamantyl)-4-methoxyphenyl terminus found in the methyl ether of AHPN, it may have other effects on cells. Its conjugated trienic bond system could also isomerize or be metabolized in cell culture.

Hoffmann-La Roche's (E)-6-[1-(4-carboxyphenyl)propen-2-yl]-4,4-dimethyl-7-heptyloxy-3,4-dihydrobenzo[2H]thiopyran-1,1-dioxide (Ro41-5253) was one of the first RAR subtype-selective antagonists reported. Ro41-5253 was originally synthesized by the Klaus group

Table 4. Retinoid receptor antagonists that bind to the receptors but are unable to activate them to induce gene transcription from retinoid response elements

Retinoid		Receptor selectivity	Receptor binding affinity (nM)						Binding assay conditions		
Structure	Name/code number		RARα	RARβ	RARγ	RXRα	RXRβ	RXRγ	Reported as	Receptor source [a]	Tritiated standard
4-1	SR3986	RAR	100	100	38				IC$_{50}$	Baculovirus	trans-RA
4-2	CD2366	RAR	9.2	38	72				K$_d$	COS-7	TTAB
4-3	R = i-C$_3$H$_7$ R = CH$_2$Ph	RAR RAR	2 000 >10 000 (45)	3 000 4 000					IC$_{50}$ IC$_{50}$	HL-60 HL-60	Am80 Am80
4-4	SR11330 (R = CH(OH)CF$_3$) SR11335 (R = CH(OMe)CF$_3$)	RAR RAR							IC$_{50}$ IC$_{50}$	E. coli E. coli	trans-RA trans-RA
4-5	AGN 192870 (R = H) AGN 193109 (R = Me)	RAR RAR	147 16 2	33 7 2	42 7 3	nb^g nb	nb nb	nb nb	K$_d$ K$_d$ K$_d$	Baculovirus Baculovirus Baculovirus	trans- or 9-cis-RA trans- or 9-cis-RA trans- or 9-cis-RA
4-6	TD550	RAR	~100	~500		nb	nb	nb	IC$_{50}$	HL-60	Am80
4-7	ER 27191	RAR									
4-8	CD2366	RAR	9.2	38	72				K$_d$	COS-7	TTAB
4-9	BMS185411	RARα	2.5	40	412				K$_d$		trans-RA
4-10	Ro41-5253	RARα	60	2 400	3 300				IC$_{50}$	Baculovirus	trans-RA
4-11	AGN 194301	RARα	2.8	320	7258				K$_d$	Baculovirus	trans-RA
4-12	LE135 (R = H, H)	RARαβ	1 400	220	nb	nb	nb	nb	K$_d$	E. coli	Am80
4-13	AGN 193644	RARβ,γ	240	4	11				K$_d$	Baculovirus	trans-RA
4-14	CD2665	RARβ,γ									
4-15	SR11253	RARγ	1000	2 400	44				IC$_{50}$	Baculovirus	trans-RA
4-16	LE540	RAR/RXR	1 300	500	490	3800	1700	1500	K$_d$	E. coli	Am80
4-17	LGD100754	RXR	1 791	2 587	6 094	8 (3)^h	9 (10)	14 (12)	K$_i$	Baculovirus	trans-RA/LDG1069 (9-cis-RA)

[a] COS-1, nuclear extracts containing receptor expressed in transfected COS-1 cells; HL-60, cell extracts; Baculovirus, recombinant receptors from the baculovirus/Sf21 expression system.

[b] wt, wild-type; chimeric ER, RAR DEF or RXR D domain fused to ER DNA-binding domain

[c] CAT, chloramphenicol acetyl transferase gene; LUC, luciferase gene; SeAP, secreted alkaline phosphatase gene.

[d] TREpal, palindromic TRE; CRBPII, cellular retinol-binding protein II DR-1 RXRE for RXRα,γ; CPRE, chicken ovalbumin upstream promoter response element for RXRβ; ERE, estrogen response element; mLamB1, mouse laminin B1 gene RARE.

[e] 50% maximal, concentration giving half of maximal response; 50% of 1 μM trans-RA, concentration giving 50% of response of 1 μM trans-RA.

[f] Apfel et al. (1992) [77]; Canan Koch et al. (1996) [82]; Chao et al. (1993, 1995, 1997) [40, 61, 62]; Elmazar et al. (1997) [81]; Eyrolles et al. (1994) [69]; Johnson et al. (1995) [92]; Kaneko et al. (1991) [112]; Keidel et al. (1994) [10]; Klein et al. (1996) [88]; Lee et al. (1996) [111]; Lehman et al. (1991) [52]; Meister et al. (1998) [116]; Ostrowski et al. (1998) [90]; Standeven et al. (1997) [115]; Teng et al. (114); Ueno et al. (1998) [113]; Umemiya et al. (1997) [99].

[g] nb, no binding.

[h] Numbers in parentheses refer to binding versus 9-cis-RA.

[i] Numbers in parentheses refer to receptor activation at 1 μM retinoid.

[j] na, no activity.

Table 4. (Continued)

| Structure | Receptor transactivation activity ED$_{50}$, nM (activation at 1 μM, %) | | | | | | Cotransfection assay conditions | | | | | Ref.[f] |
	RARα	RARβ	RARγ	RXRα	RXRβ	RXRγ	Receptor vector[b]	Response element[c]	Reporter[d]	Cell line	EC$_{50}$ response[e]	
4-1	>10 000 (7)[i]	>10 000 (-7)	>10 000 (13)	>10 000 (0)			chimeric ER	ERE	CAT	CV-1	50% of 10 μM trans-RA	Lehmann et al. (1991)
4-2							wt	CRBPI/2 (DR-1)	CAT	HeLa	50% maximal	Elmazar et al. (1997)
4-3												Eyrolles et al. (1994a) Eyrolles et al. (1994a)
4-4	>10 000 (15) >10 000 (5)	>10 000 (5) >10 000 (0)	>10 000 (4) >10 000 (10)	>10 000 (-10) >10 000 (10)			chimeric ER chimeric ER	ERE ERE	CAT CAT		50% of 1 μM trans-RA 50% of 1 μM trans-RA	Lee et al. (1996) Lee et al. (1996)
4-5	>1000 (0) na[j]	>1000 (50) na	>1000 (-85) na				chimeric ER chimeric ER wt	ERE ERE TREpal	CAT CAT LUC	CV-1 CV-1 CV-1	50% maximal 50% maximal 50% maximal	Klein et al. (1996) Klein et al. (1996) Johnson et al. (1995)
4-6												Kaneko et al. (1991)
4-7												Ueno et al. (1998)
4-8												Elmazar et al. (1997)
4-9	>1000 (na)	288	28				wt	mLamB1	CAT	HeLa	50% of 1 μM trans-RA	Ostrowski et al. (1998)
4-10	>1000 (22)						chimeric ER	ERE	SeAP	COS-1	50% maximal	Apfel et al. (1992) Keidel et al. (1994)
4-11	>1000 (na)	>1000 (na)	>1000 (na)	>1000 (na)			RAR-P-GR	MTV-4-(R5G)	LUC	CV-1	50% maximal	Teng et al. (1997)
4-12												Umemiya et al. (1997a)
4-13												Standeven et al. (1997)
4-14												Meister et al. (1998)
4-15	>10 100 (1)[i]	>10 000 (-3)	>10 000 (-56)				chimeric ER	ERE	CAT	CV-1	50% of 1 μM trans-RA on RARα	Chao et al. (1993, 1995, 1997)
4-16												Umemiya et al. (1997b)
4-17	>10 000 (4)	>10 000 (10)	192 (27)	>10 000 (2)	>10 000 (13)	>10 000 (4)	wt	TREpal (RAR)/CRBPII or CRPE (RXR)	LUC	CV-1	50% maximal	Canan Koch et al. (1996)

Table 4 (Continued)

[77] and is one of the most RARα-selective and potent retinoid antagonists identified, although the reported dose-response curve indicated that at 1 μM it weakly activated ER-RARα on the

EREvit-*tk*-SeAP reporter construct. Structure-activity studies indicate that the 3'-heptyloxy group confers antagonist activity and binding affinity to Ro41-5253 while the sulfone group induces subtype selectivity. Allergan's 4-[8-bromo-2,2-dimethyl-4-(4-methylphenyl)-2*H*-dihydrobenzopyranyl-6-carboxamido]-2,6-difluorobenzoic acid (AGN 194301) is more RARα selective [114]. The RARα-selective antagonist 4-[5,6-dihydro-5,5-dimethyl-8-(4-methylphenyl)carboxamido]benzoic acid (BMS185411) preferentially antagonizes binding of agonists to RARα but also binds well to RARβ [91].

2-(6-Carboxynaphthalen-2-yl)-2-(5,6,7,8-tetrahydro-5,5,8,8-tetramethyl-2-naphthalenyl)-1,3-dithiolane (SR11253) actually functions as a partial agonist that shows selectivity for RARγ. However, at low concentrations, since SR11253 binds to but is unable to activate gene transcription from the (TREpal)$_2$, SR11253 can be used as an antagonist [40, 61, 62]. No RARβ-selective antagonists have been reported yet. 4-[2-(5,5-Dimethyl-5,6-dihydro-8-(4-methylphenyl)-2-naphthalenyl)azo]benzoic acid (AGN 193644) is an antagonist that selectively binds to both RARβ and RARγ [115]. The incorporation of the azo bridge in a TTNPB analog was first reported by Kagechika et al. [43].

(2*E*,4*E*,6*Z*)-3-Methyl-7-(3-propyloxy-5,6,7,8-tetrahydro-5,5,8,8-tetramethyl-2-naphthalenyl)-2,4,6-octatrienoic acid (LGD100754) was reported by Canan Koch et al. [22] as a potent RXR-selective homodimer antagonist. This retinoid could also weakly activate RARγ on the TREpal. However, in the context of the RXRα-PPARγ heterodimer on the PPAR RE, LGD100754 behaved as an agonist. Interestingly, the 6*E* isomer of this propyl ether does not have affinity to the RARs. Its related methyl ether, (2*E*,4*E*,6*Z*)-3-methyl-7-(3-methoxy-5,6,7,8-tetrahydro-5,5,8,8-tetramethyl-2-naphthalenyl)-2,4,6-octatrienoic acid (LGD100568), is somewhat less RXR-selective but has three- to four-fold higher RXR affinity. Because the tetraene side chain of the methyl ether analog could isomerize to an analog of *trans*-RA, lack of class specificity could result on isomerization. No RXR subtype-selective antagonist has been reported.

Retinoids able to inhibit transcriptional activation from AP-1 sites have been reported. The first report in this series identified several retinoids that were able to inhibit tumor promoter-induced activation of the -73ColCAT reporter construct, in which the AP-1 site of the collagenase promoter is linked to the CAT reporter, after cotransfection with a RAR vector but were unable to induce reporter gene transcription from the (TREpal)$_2$ in the presence of the same cotransfected RAR [118]. These retinoids were also reported to inhibit cancer cell growth in cell culture. We subsequently found that other retinoid antagonists, such as Ro41-5253 and SR11253, inhibited AP-1 activity in the presence of endogenous retinoid receptor levels in -73ColCAT-transfected HeLa cells. Ro41-5253 inhibited anchorage-dependent but not anchorage-independent growth of the MCF-7 human breast cancer cell line [84]. Chen et al. reported several RAR-specific agonists/antagonists that dissociate transcriptional activities from AP-1 transrepression and inhibit anchorage-independent cell proliferation [117].

Use of selective retinoids as mechanistic probes

For the reasons enumerated above, we and our collaborators recommend using several different class- or subtype-selective retinoids to more accurately make correlations to deduce the importance of a particular receptor class or subtype in a signalling pathway. Thus, we established that RARα-selective agonists were more effective than a RARα antagonist at inhibiting MCF-7 breast cancer cell anchorage-independent growth; however, both RARα-selective agonists Am580 and Am80 and antagonist Ro41-5253 inhibited anchorage-dependent growth after 6 days [84]. RARγ-selective retinoids were far less potent. This MCF-7 breast cancer cell line requires estrogen for growth and has all three RAR subtypes. Subsequently, susceptibility to growth inhibition by retinoids was found to correlate with RARα levels in several breast cancer cell lines, regardless of their ER status [119]. Toma et al. also found that the RARα antagonist Ro41-5253 induced breast cancer cell apoptosis in addition to inhibiting cell proliferation [121].

The MDA-MB-231 human breast cancer cell line grows independently of estrogen and RARγ predominates. MDA-MB-231 cells are resistant to growth inhibition by both natural and RARα- and RARγ-selective retinoids. However, the Fontana group established that AHPN inhibits growth through a retinoid receptor-independent pathway that involved cell cycle arrest and induction of p21(WAF1/Cip1), jun, and apoptosis [39, 42]. Many cancer cell lines lose their ability to express RARβ, and loss of this receptor is associated with maligant progression [13]. The Fontana group demonstrated that MDA-MB-231 cells transfected with the RARβ gene expressed RARβ and were sensitive to growth inhibition by retinoids [15]. Subsequently, Zhang and collaborators showed that the RXR-nur77 complex in the presence of an RXR-selective retinoid could activate the β2RARE, induce RARβ synthesis, and inhibit cancer cell growth [105, 107].

Whether receptor subtypes have distinct roles in mediating the retinoid response appears to depend on the system used. Using mutational deletions in mice, Chambon and coworkers established that both retinoid receptor classes are essential for life, but the subtypes may be redundant [121, 122]. For example, in contrast to the results of Dawson et al. [184] and Fitzgerald et al. [119], Toma et al. [78] reported a redundant receptor response for the induction of apoptosis in MCF-7 cells, which possess RARα, RARγ, and RXR, after treatment with 10^{-8} to 10^{-6} M RARα-selective Am580 (Ro40-6055), RARβ-selective Ro48-2249, RARγ-selective Ro44-4783, or RXR-selective Ro25-7386 for 2 to 6 days. The levels of apoptosis induction by the four retinoids were similarly dose- and time-dependent. In contrast, RAR/RXR-selective 9-*cis*-RA was more potent than *trans*-RA at 4 days of treatment, but after 6 days the results were similar. Delipidized serum was used by Dawson et al. [84], whereas Toma et al. [78] used non-treated serum, which may have enhanced retinoid efficacy. Other studies have indicated that the RAR subtypes may have different functions [123]. Further investigation is required to determine whether differences in response to selective retinoids are reflective of differences in the levels of RAR subtype proteins present in the cell line or differences in the actual functions of the subtypes.

Selective retinoids are used to investigate what retinoid receptor classes or subtypes are most important in a particular retinoid response. The following examples illustrate the roles of receptor class-selective retinoids. Charpentier et al. [71] showed that RAR class-specific TTNPB and TTAB were potent inhibitors of F9 murine teratocarcinoma cell differentiation. Nagy et al. [95] reported that RAR-selective Am80 and RXR-selective (E)-5-[2-(5,6,7,8-tetrahydro-3,5,5,8,8-pentamethyl-2-naphthalenyl)propenyl]-3-thiophenecarboxylic acid (AGN 191701) had different signalling pathways in HL-60 cells. Am80 induced HL-60 cell differentiation, whereas AGN 191701 in the presence of RAR-selective Am80 induced apoptosis. RAR-specific TTNPB inhibited chondrogenesis, whereas RXR-selective SR11217 and SR11237 were inactive [94]. RAR antagonists SR11335 and SR11330 inhibited the differentiation of HL-60 cells induced by *trans*-RA, a finding that also substantiates the RAR signalling pathway for differentiation [111]. Neither RAR-selective TTNPB nor RXR-selective LG100268 alone was capable of inducing HL-60 cell apoptosis, but the combination was an effective inducer [97]. Beard et al. [98] reported that RXR-selective retinoids had a role in transglutaminase (TGase) induction in HL-60 cells, whereas RAR-specific TTNPB was inactive. TGase increases during cell differentiation. RXR-selective LG100268 at 10 nM showed half the activity of 1 μM *trans*-RA in activating TGase. Using RAR-selective TTAB and RXR-selective CD2425, Joseph et al. [124] established that both receptor classes were required to induce TGase and apoptosis in RPMI human myeloma cells. The RXR antagonist LGD100754 inhibited TGase expression in HL-60 cells treated with RXR-selective agonist LGD1069 (targretin) [82]. Thus, RXR-selective retinoids have an important role in TGase induction.

The RXR-RXR antagonist LDG100754 and RAR-selective TTNPB activated RXR-RAR heterodimers on DR-2 and DR-5 RAREs, whereas LDG100268 did not; however, only TTNPB dissociated the corepressor SMRT from the RXR-RAR complex [36]. The synergy between RXR-selective retinoids and RAR-selective retinoids was ably demonstrated by Wu et al. [105]. RXR-selective SR11345 at 1.0 μM, which only inhibited MDA-MB-231 breast cancer cell growth by less than 5%, enhanced the inhibitory activity of RAR-selective SR11365 from below 5% to 55%. In the ZR-75-1 human breast cancer cell line, inhibition by SR11365 rose from 40% to 70%, after addition of SR11345, which alone was inactive.

Subtype-selective retinoids are used to determine the role of the α, β, and γ receptor subtypes. RARα-selective Am580 was more potent than Am80 in inhibiting interleukin-6 (IL-6) expression induced by IL-1α in MC3T3-E1 murine osteogenic fibroblasts [76]. Both compounds were potent inducers of HL-60 cell differentiation (EC_{50} = 0.34 and 0.79 nM, respectively). Their thiazolidinedione derivatives TZ181 and TZ185, respectively, were less potent [108]. The RAR antagonist LE540 enhanced IL-6 levels. These results indicate that the RAR agonists attenuate IL-6 expression. The RARα antagonist Ro41-5253 inhibited the teratogenic effects of Am580 in rat limb bud in culture, showing that RARα has a role in morphogenesis [125]. Ro41-5253 also counteracted the ability of *trans*-RA to induce both HL-60 cell differentiation and B-lymphocyte polyclonal activation [77] and to inhibit tumor cell-induced angiogenesis [126], results that confirm the importance of RARα in cell differentiation and proliferation.

Because the skin contains RARα and RARγ proteins, with the latter predominating (1:6) [12], research is devoted to establishing the predominant signalling pathway in the skin in order to develop more effective dermatological agents [127]. Yu et al. [59] reported that RARα-selective retinoids had greater potency than RARγ-selective retinoids at reducing utricule size in rhino mouse skin. Unfortunately, because this experiment was performed at millimolar dose levels, interpretation of results is difficult. Standeven et al. [115] found that RARα was not involved in skin toxicity in the female hairless mouse skin irritation assay because skin irritation induced by the RAR panagonist TTNPB was blocked by the RAR panantagonist AGN 193109 but not by an RARα-selective antagonist (AGN 194307). Chen et al. [96] and Reczek et al. [91] established that the RARβ,γ-selective (+)- or (S)-diarylmethanol BMS185283 had 0.4% of the potency of *trans*-RA in the IS3 rabbit repeat skin irritation test and 2% of its potency in utricule reduction in the rhino mouse, a result indicating that the synthetic analog had a better therapeutic index than that of *trans*-RA and that both effects were mediated by RARγ. The RARγ-selective *E/Z*-oxime BMS185354 was more potent in both assays, a result indicating a major role for RARγ in the skin.

Lu et al. [128] extended the research of the Fontana group [39] by demonstrating that an AHPN analog, 6-[4-(1-adamantyl)-1,3-dioxaindanyl]-2-naphthalenecarboxylic acid (MX-3350-1)[129], inhibited NCI-H292 lung cancer growth in a nu/nu mouse xenograft model. At 2 μM, MX-3350-1 also inhibited the growth of several cancer cell lines that were refractory to the effects of *trans*-RA. MX-3350-1 activated RARγ on three RARE-CAT constructs (CRPB1, ApoA-1, and DR-5).

RARγ-selective retinoids CD437 and CD2325 have been reported to inhibit the growth and induce the apoptosis of LAN-5 neuroblastoma cells in culture more effectively than RARα-selective Am580 (CD336), RARβ-selective CD2019, RARγ-selective CD666, or *trans*-RA [116]. As indicated in Table 5, differences in growth inhibition by the subtype-selective retinoids at 1 μM were small. The three subtype-selective agonists and the RARβ,γ-selective antagonist CD2665 were weak inhibitors, a finding which could suggest receptor redundancy. An explanation for the higher inhibitory potency of *trans*-RA could be its ability to isomerize in culture to the 9-*cis* isomer that would also activate the RXRs. Interestingly, two RARγ-selective retinoids—CD437 and its analog CD2325—were more potent inhibitors than *trans*-RA. They induced cell apoptosis, whereas *trans*-RA induced cell differentiation. The apoptotic

Table 5. Binding affinity of retinoids and their effect at 1.0 μM on LAN-5 neuroblastoma cell growth on day 7[a]

Retinoid[b]	K_d (nM)			Cell growth (% Control)
	RARα	RARβ	RARγ	
trans-RA	16	7	3	36
Am580 (CD336)	8	131	450	72
CD2019	110	26	160	63
CD437 (AHPN)	6500	2480	77	21
CD666	2240	2300	68	62
CD2325	1144	1245	53	21
CD2665[c]	-	++	++	61

[a]From Meister et al. [116].
[b]No 50% transcriptional activation for RXRα at 1 μM retinoid.
[c]Transcriptional antagonist of RARβ,γ on a RARE.

activity of CD437 and CD2325 may indicate a mechanism of action independent of RARγ for AHPN (CD437) [39].

Selective retinoids as therapeutic agents

Both *trans*-RA (tretinoin) and its prodrug 13-*cis* isomer (isotretinoin, roaccutane, accutane) are effective for the treatment of acne [130] and some skin cancers [2]. *trans*-RA has demonstrated efficacy against acute promyeloblastic leukemia [131]. Although relapse occurs with time [132], 13-*cis*-RA is reported to inhibit the appearance of second primary tumors in head-and-neck cancer patients [1]. Because of the possibility of double-bond isomerization in these retinoids, efficacy due to an RAR- and/or RXR-mediated pathway remains to be established. (*E*)-3,7-Dimethyl-9-(4-methoxy-2,3,6-trimethylphenyl)-2,4,6,8-nonatetraenoic acid (acitretin, or Ro10-1670) is used in Europe for treatment of psoriasis and several cancers [133]. This retinoid is also capable of undergoing bond isomerization. The efficacy of tretinoin, isotretinoin, and acitretin in inhibiting cancer cell growth is enhanced in the presence of cytokines [134] The positive responses of these three retinoids alone and in combination with interferon-α in the prevention and treatment of preneoplastic diseases (actinic keratosis, epidermodysplasia verruciformis, oral leukoplakia, laryngeal dysplasia, bronchial squamous metaplasia, bronchial atypia, cervical dysplasia, vulval dystrophy, and myelodysplastic syndrome) and neoplastic diseases (basal cell carcinoma, skin squamous cell carcinoma, keratoacanthoma, cutaneous T cell lymphoma, T cell lymphoma, bladder papilloma, acute promyelocytic leukemia, juvenile chronic myeloid leukemia) has been reviewed by Bollag [3]. 4-HPR (fenretinide) is undergoing clinical trials for breast cancer prevention in patients at risk [37]. A 40% reduction in contralateral breast cancer was observed in premenopausal women. A combination trial with tamoxifen is underway. The acyclic retinoid (*E*)-3,3,7,11,15-tetramethyl-2,4,6,10,14-hexadecapentaenoic acid (polyprenoic acid) prevented the appearance of second primary tumors in surgically-treated liver cancer patients [135].

The RAR/RXR panagonist 9-*cis*-RA (ALRT1057, LDG1057, panretin), RXR-selective LDG1069 (targretin, bexarotene), and RARα-selective Am80 are also undergoing clinical trials in cancer patients. A Phase I trial of 9-*cis*-RA in patients with advanced non-small cell lung, breast, colorectal, non-melanoma skin, head-and-neck, or ovarian cancer, indicated that toxic effects were typical of retinoids but were milder [136]. Based on the decrease in plasma concentration with time, 9-*cis*-RA at high doses appeared to induce its own metabolism. No tumor responses were observed [137, 138]. 9-*cis*-RA induced complete remission in 33% of relapsed and 80% of new acute promyelocytic leukemia patients [139, 140]. A combination trial with interferon α is ongoing. Clinical trial results indicate that LGD1069 has efficacy against cutaneous T cell lymphoma and Kaposi's sarcoma but did produce changes in liver function and leukopenia, hypertriglyceridemia, and hypercalcemia in some patients [139]. A combination trial with the antiestrogen tamoxifen is ongoing. Am80 has been successfully used to treat acute promyelocytic leukemia relapse, which was refractory to *trans*-RA [141]. Complete remission occurred in 14 of 24 patients. Toxic effects were characteristic of hypervitaminosis A but deemed milder than those for *trans*-RA [142]. Information about retinoids undergoing clinical trials in the U.S., from which some of the above has been obtained, can be found at the U.S. National Cancer Institute website address (http://cancernet.nci.nih.gov.).

Orally administered 4-[3,5-bis(trimethylsilyl)phenylcarboxamido]benzoic acid (Am555S, or TAC-101) [143] reduced liver metastasis in experimental models in mice by decreasing liver weight and nodule number, thereby prolonging life [144]. These models included AZ-521 human gastric adenocarcinoma in male BALB/c nu/nu mice and Co-3 human colon adenocarcinoma in male KSN nude mice.

Several drugs have recently become available for treatment of proliferative skin diseases. In addition to topical *trans*-RA and oral 13-*cis*-RA [145], the methyl ether of AHPN (adapalene, or CD417) is now available for the topical treatment of acne. Ethyl 6-[2-(2*H*-3,4-dihydro-4,4-dimethylbenzothiopyran-6-yl)ethynyl]-3-pyridinecarboxylate (tazarotene or AGN190168) is successful in treatment of psoriasis [146, 147], as is acitretin [133]. The parent acid (AGN190299) of tazarotene is RAR-selective with preferential binding affinity to RARβ and RARγ (Tab. 2). Topical Am80 has been reported as efficacious in clinical studies on psoriasis and palmoplantar pustulosis patients [148, 149]. 9-*cis*-RA is undergoing trials in psoriasis patients.

An ongoing goal in pharmaceutical development has been the reduction of retinoid toxicity to facilitate patient compliance. This strategy has been most effectively achieved using the pro-drug approach, as exemplified by 13-*cis*-RA, which will isomerize *in vivo* to *trans*- and 9-*cis*-RAs. Ethyl esters, such as tazarotene, are hydrolyzed to release the parent acid. Toxicity is also reduced by conversion of retinoids to glucuronide derivatives, an approach that is being explored by the groups of Olson and Curley using the natural retinoids [150, 151]. Abou-Issa et al. (1997) [152] extended this approach to the glucuronide of 4-HPR, which was found to be more potent and less toxic than 4-HPR in the rat mammary tumor model. The glucuronamide [153] also proved effective and more stable.

Despite substantial effort, no synthetic RAR or RXR class- or RAR subtype-specific retinoid is yet available for treatment of cancer, although two are available for treatment of proliferative skin diseases. This situation indicates to the authors that standard drug discovery must be coupled with studies on drug mechanism at the molecular level if synthetic retinoids are going to have a significant impact in cancer therapy. By thoroughly understanding the molecular mechanisms by which these selective retinoids function, we and other groups plan to identify more effective retinoids for therapeutic applications.

Acknowledgements
Research by the Dawson group reported in this chapter was supported by United States National Cancer Institute Grant P01 CA51993 (M.I.D., X.Z.), which is gratefully acknowledged. Transcriptional analyses were performed by the group of Dr. Xiao-kun Zhang at the Burnham Institute, La Jolla, CA. The authors thank Drs. Marcus Boehm, Rosh Chandraratna, Thomas Curley, Hiroyu Kagechika, James Olson, Magnus Pfahl, and Peter Reczek for providing reprints, preprints, and other information for this chapter, and Ms. Anne Miller and Ms. Mary Williamson of SRI for their assistance.

References

1   Hong WK, Lippman SM, Itri LM, Karp DD, Lee JS, Byers RM, Schantz SP, Kramer AM, Lotan R, Peters LJ et al (1990) Prevention of second primary tumors with isotretinoin in squamous-cell carcinoma of the head and neck. *New Engl J Med* 323: 795–801
2   Hong WK, Itri LM (1994) Retinoids and human cancer. *In*: MB Sporn, AB Roberts, DS Goodman (eds): *The retinoids. Biology, chemistry, and medicine*. Raven Press, New York, 579–630
3   Bollag W (1995) Retinoids and interferon-α in the prevention and treatment of preneoplastic and neoplastic diseases: a review. *In*: F Patel (ed.): *Retinoids Today and Tomorrow*, issue 40. Mediscript, London, 26–31
4   Armstrong RB, Ashenfelter KO, Eckoff C, Levin AA, Shapiro SS (1994) General and reproductive toxicology of retinoids. *In*: MB Sporn, AB Roberts, DS Goodman (eds): *The retinoids. Biology, chemistry, and medicine*. Raven Press, New York, 545–572
5   Armstrong RB, Kim HJ, Grippo JF, Levin AA (1992) Retinoids for the future: Investigational approaches for the identification of new compounds. *J Amer Acad Dermatol* 27, S38–S42
6   Mangelsdorf DJ, Umesono K, Evans RM (1994) The retinoid receptors. *In*: MB Sporn, AB Roberts, DS Goodman (eds): *The retinoids. Biology, chemistry, and medicine*. Raven Press, New York, 319–350
7   Heyman RA, Mangelsdorf DJ, Dyck JA, Stein RB, Eichele G, Evans RM, Thaller C (1992) 9-*Cis* retinoic acid is a high affinity ligand for the retinoid X receptor. *Cell* 68: 397–406
8   Levin AA, Sturzenbecker LJ, Kazmer S, Bosakowski T, Huselton C, Allenby G, Speck J, Kratzeisen C, Rosenberger M, Lovey A et al (1992) 9-*Cis* retinoic acid stereoisomer binds and activates the nuclear receptor RXRα. *Nature* 355: 359–361
9   Allegretto EZ, McClurg MR, Lazarchik SB, Clemm DL, Kerner SA, Elgort MG, Boehm MF, White SK, Pike JW, Heyman RA (1993) Transactivation properties of retinoic acid and retinoid X receptors in mammalian cells and yeast. *J Biol Chem* 268: 26 625–26 633
10 Keidel S, LeMotte P, Apfel C (1994) Different agonist- and antagonist-induced conformational changes in retinoic acid receptors analyzed by protease mapping. *Mol Cell Biol* 14: 287–298
11 Ostrowski J, Hammer L, Roalsvig T, Pokornowski K, Reczek PR (1995) The *N*-terminal portion of domain E of retinoic acid receptors α and β is essential for the recognition of retinoic acid and various analogs. *Proc Natl Acad Sci USA* 92: 1812–1816
12 Fisher GJ, Talwar HS, Xiao J-H, Datta SC, Reddy AP, Gaub M-P, Rochette-Egly C, Chambon P, Voorhees JJ (1994) Immunological identification and functional quantitation of retinoic acid and retinoid X receptor proteins in human skin. *J Biol Chem* 269: 20 629–20 635
13 Geisen C, Denk C, Gremm B, Baust C, Karger A, Bollag W, Schwarz E (1997) High-level expression of the retinoic acid receptor β gene in normal cells of the uterine cervix is regulated by the retinoic acid receptor α and is abnormally down-regulated in cervical carcinoma cells. *Cancer Res* 57: 1460–1467
14 Liu Y, Lee M-O, Wang H-G, Li Y, Hashimoto Y, Klaus M, Reed JC, Zhang X (1996) Retinoic acid receptor β mediates the growth inhibitory effect of retinoic acid by promoting apoptosis in human breast cancer cells. *Mol Cell Biol* 16: 1138–1149
15 Li X-S, Shao Z-M, Sheikh MS, Eiseman JL, Sentz D, Jetten AM, Chen JC, Dawson MI, Fontana J (1995) Retinoic acid nuclear receptor β (RARβ) inhibits breast carcinoma growth and tumorigenicity. *J Cell Physiol* 16: 449–458
16 Mangelsdorf DJ, Ong ES, Dyck JA, Evans RM (1990) Nuclear receptor that identifies a novel retinoic acid response pathway. *Nature* 345: 224–229
17 Nakshatri H, Bhat-Nakshatri P (1998) Multiple parameters determine the specificity of transcriptional response by nuclear receptors HNF-4, ARP-1, PPAR, RAR and RXR through common response elements. *Nucl Acid Res* 26: 2491–2499
18 Kurokawa R, DiRenzo J, Boehm M, Sugarman J, Gloss B, Rosenfeld MG, Heyman RA, Glass CK (1994) Regulation of retinoid signalling by receptor polarity and allosteric control of ligand binding. *Nature* 371: 528–531
19 Lehmann JM, Jong L, Fanjul A, Cameron JF, Lu XP, Haefner P, Dawson MI, Pfahl M (1992) Retinoids selective for retinoid X receptor response pathways. *Science* 258: 1944–1946
20 Kagechika H, Kawachi E, Hashimoto Y, Himi T, Shudo K (1988) Retinobenzoic acids 1 Structure-activity relationships of aromatic amides with retinoidal activity. *J Med Chem* 31: 2182–2192 [published erratum appears in *J Med Chem* 32: 2583 (1989)]
21 Roy B, Taneja R, Chambon P (1995) Synergistic activation of retinoic acid (RA)-responsive genes and induction of embryonal carcinoma cell differentiation by an RA receptor α (RARα)-, RARβ-, or RARγ-selective ligand in combination with a retinoid X receptor-specific ligand. *Mol Cell Biol* 15: 6481–6487
22 Botling J, Castro DS, Oberg F, Nilsson K, Perlmann T (1997) Retinoic acid receptor/retinoid X receptor heterodimers can be activated through both subunits providing a basis for synergistic transactivation and cellular differentiation. *J Biol Chem* 272: 9443–9449
23 Loeliger P, Bollag W, Mayer H (1980) Arotinoids, a new class of highly active compounds. *Eur J Med Chem-Chim Ther* 15: 9–15
24 Boehm MF, Zhang L, Badea BA, White SK, Mais DE, Berger E, Suto CM, Goldman ME, Heyman RA (1994) Synthesis and structure-activity relationships of novel retinoid X receptor-selective retinoids. *J Med Chem* 37:

2930–2941
25 Westin S, Kurokawa R, Nolte RT, Wisely GB, McInerney EM, Rose DW, Milburn MV, Rosenfeld MG, Glass CK (1998) Interactions controlling the assembly of nuclear-receptor heterodimers and co-activators. *Nature* 395: 199–202
26 Husmann M, Lehmann J, Hoffman B, Hermann T, Tzukerman M, Pfahl M (1991) Antagonism between retinoic acid receptors. *Mol Cell Biol* 11: 4097–4103
27 Hembree JR, Agarwal C, Beard RL, Chandraratna RAS, Eckert RL (1996) Retinoid X receptor-specific retinoids inhibit the ability of retinoic acid receptor-specific retinoids to increase the level of insulin-like growth factor binding protein-3 in human ectocervical epithelial cells. *Cancer Res* 56: 1794–1799
28 Kamei Y, Xu L, Heizel T, Torchia J, Kruokawa R, Gloss B, Lin S-C, Heyman RA, Rose DW, Glass CK et al (1996) A CBP integrator complex mediates transcriptional activation and AP-1 inhibition by nuclear receptors. *Cell* 85: 403–414
29 Chakravarti D, LaMorte VJ, Nelson MC, Nakajima T, Schulman IG, Juguilon H, Montminy M, Evans RM (1996) Role of CBP/P300 in nuclear receptor signalling. *Nature* 383: 99–103
30 Huang N, vom Baur E, Garnier J-M, Lerouge T, Vonesch J-L, Lutz Y, Chambon P, Losson R (1998) Two distinct nuclear receptor interaction domains in NSD1, a novel SET protein that exhibits characteristics of both corepressors and coactivators. *EMBO J* 17: 3398–3412
31 Glass CK, Rose DW, Rosenfeld MG (1997) Nuclear receptor coactivators. *Curr Opin Cell Biol* 9: 222–232
32 Burris TP, Nawaz Z, Tsai M-J, O'Malley BW (1995) A nuclear hormone receptor-associated protein that inhibits transactivation by the thyroid hormone and retinoic acid receptors. *Proc Natl Acad Sci USA* 92: 9525–9529
33 Chen JD, Evans RM (1995) A transcriptional co-repressor that interacts with nuclear hormone receptors. *Nature* 377: 454–457
34 Hörlein AJ, Näär AM, Heinzel T, Torchia J, Gloss B, Kurokawa R, Ryan A, Kamei Y, Söderström M, Glass CK et al (1995) Ligand-independent repression by the thyroid hormone receptor mediated by a nuclear receptor co-repressor. *Nature* 377: 397–404
35 Kurokawa R, Söderström M, Hörlein A, Halachmi S, Brown M, Rosenfeld MG, Glass CK (1995) Polarity-specific activities of retinoid acid receptors determined by a co-repressor. *Nature* 377: 451–453
36 Lala DS, Mukherjee R, Schulman IG, Canan Koch SS, Dardashti LJ, Nadzan AM, Croston GE, Evans RM, Heyman RA (1996) Activation of specific RXR heterodimers by an antagonist of RXR homodimers. *Nature* 383: 450–453
37 Formelli F, Barua AB, Olson JA (1996) Bioactivities of N-(4-hydroxyphenyl)retinamide and retinoyl β-glucuronide. *FASEB J* 10: 1014–1024
38 Oridate N, Suzuki S, Higuchi M, Mitchell MF, Hong WK, Lotan R (1997) Involvement of reactive oxygen species in N-(4-hydroxyphenyl)retinamide-induced apoptosis in cervical carcinoma cells. *J Nat Cancer Inst* 89: 1191–1198
39 Shao ZM, Dawson MI, Li X-S, Rishi AK, Sheikh MS, Han Q-X, Ordonez JV, Shroot B, Fontana JA (1995) p53 independent G_0/G_1 arrest and apoptosis induced by a novel retinoid in human breast cancer cells. *Oncogene* 11: 493–504
40 Chao W, Hobbs PD, Jong L, Zhang X, Zheng Y, Wu Q, Shroot B, Dawson MI (1997) Effects of receptor class- and subtype-selective retinoids and an apoptosis-inducing retinoid on the adherent growth of the NIH:OVCAR-3 ovarian cancer cell line in culture. *Cancer Lett* 113: 1–7
41 Sun S-Y, Yue P, Dawson MI, Shroot B, Michel S, Lamph WW, Heyman RA, Teng M, Chandraratna RAS, Shudo K et al (1997) Differential effects of synthetic nuclear retinoid receptor-selective retinoids on the growth of human non-small cell lung carcinoma cells. *Cancer Res* 57: 4931–4939
42 Li Y, Lin B, Agadir A, Liu R, Dawson MI, Reed JC, Fontana JA, Bost F, Hobbs PD, Zheng Y et al (1998) Molecular determinants of AHPN (CD437)-induced growth arrest and apoptosis in human lung cancer cell lines. *Mol Cell Biol* 18: 4719–4731
43 Kagechika H, Himi T, Namikawa K, Kawachi E, Hashimoto Y, Shudo K (1989) Retinobenzoic acids. 3. Structure-activity relationships of retinoidal azobenzene-4-carboxylic acids and stilbene-4-carboxylic acids. *J Med Chem* 32: 1098–1108
44 Allenby G, Bocquel M-T, Saunders M, Kazmer S, Speck J, Rosenberger M, Lovey A, Kastner P, Grippo JF, Chambon P et al (1993) Retinoic acid receptors and retinoid X receptors: interactions with endogenous retinoic acids. *Proc Natl Acad Sci USA* 90: 30–34
45 Levin AA (1995) Receptors as tools for understanding the toxicity of retinoids. *Toxicol Lett* 82/83: 91–97
46 Giguere V, Ong ES, Segui P, Evans RM (1987) Identification of a receptor for the morphogen retinoic acid. *Nature* 330: 624–629
47 Krust A, Kastner P, Petrovich M, Zelent A, Chambon P (1989) A third human retinoic acid receptor, hRAR-γ. *Proc Natl Acad Sci USA* 86: 5310–5314
48 Petkovich M, Brand NJ, Krust A, Chambon P (1987) A human retinoic acid receptor which belongs to the family of nuclear receptors. *Nature* 330: 444–450
49 Benbrook D, Lernhardt E, Pfahl M (1988) A new retinoic acid receptor identified from a hepatocellular carcinoma. *Nature* 333: 669–672
50 Brand N, Petkovich M, Krust A, Chambon P, de Thé H, Marchio A, Tiollais P, Dejean A (1988) Identification

of a second human retinoic acid receptor. *Nature* 332: 850–853

51 Dawson MI, Chan RL-S, Derdzinski K, Hobbs PD, Chao W and Schiff LJ (1983) Synthesis and pharmaco-logical activity of 6-[(E)-2-(2,6,6-trimethyl-1-cyclohexen-1-yl)ethen-1-yl]- and 6-(1,2,3,4-tetrahy-dro-1,1,4,4-tetramethyl-6-naphthyl)-2-naphthalenecarboxylic acids. *J Med Chem* 26: 1653–1656

52 Lehmann JM, Dawson MI, Hobbs PD, Husmann M, Pfahl M (1991) Identification of retinoids with nuclear receptor subtype-selective activities. *Cancer Res* 51: 4804–4809

53 Dawson MI, Hobbs PD, Derdzinski KA, Chao W-R, Frenking G, Loew GH, Jetten AM, Napoli JL, Williams JB, Sani BP et al (1989) Effect of structural modifications in the C7–C11 region of the retinoid skeleton on biological activity in a series of aromatic retinoids. *J Med Chem* 32: 1504–1517

54 Delescluse C, Cavey MT, Martin B, Bernard BA, Reichert U, Maignan J, Darmon M, Shroot B (1991) Selective high affinity retinoic acid receptor α or β-γ ligands. *Mol Pharmacol* 40: 556–562

55 Sporn MB, Newton DL (1979) Chemoprevention of cancer with retinoids. *Fed Proc* 38: 2528–2534

56 Newton DL, Henderson WR, Sporn MB (1980) Structure-activity relationships of retinoids in hamster tra-cheal organ culture. *Cancer Res* 40: 3413–3425

57 Graupner G, Malle G, Maignan J, Lang G, Pruniéras M, Pfahl M (1991) 6'-Substituted naphthalene-2-car-boxylic acid analogs, a new class of retinoic acid subtype-specific ligands. *Biochem Biophys Res Commun* 179: 1554–1561

58 Bernard BA, Bernardon J-M, Delescluse C, Martin B, Lenoir M-C, Maignan J, Charpentier B, Pilgrim WR, Reichert U, Shroot B (1992) Identification of synthetic retinoids with selectivity for human nuclear retinoic acid receptor γ. *Biochem Biophys Res Commun* 186: 977–983

59 Yu K-L, Ostrowski J, Chen S, Tramposch KM, Reczek PR, Mansuri MM and Starrett JE Jr (1996) Structural modifications of 6-naphthalene-2-carboxylate retinoids. *Bioorg Med Chem Lett* 6: 2865–2870

60 Yu KL, Spinazze P, Ostrowski J, Currier SJ, Pack EJ, Hammer L, Roalsvig T, Honeyman JA, Tortolani DR, Reczek PR et al (1996) Retinoic acid receptor β,γ-selective ligands: Synthesis and biological activity of 6-sub-stituted 2-naphthoic acid retinoids. *J Med Chem* 39: 2411–2421

61 Chao W-R, Jong L, Costa E, Lehmann J, Pfahl M, Dawson MI (1993) Effects of retinoic acid receptor-spe-cific and retinoid X receptor-specific retinoids on the induction of ornithine decarboxylase in mouse epider-mis treated with the tumor promoter 12-O-tetradecanoylphorbol-13-acetate. American Association of Cancer Research Special Conference, Mechanism of Action of Retinoids, Vitamin D, and Steroid Hormones. Banff, Alberta, Canada

62 Chao W, Rudd CJ, Costa E, Jong L, Hobbs PD, Lehmann JM, Pfahl M, Lombardo A, Ely KA, Quick T et al (1995) Effects of receptor-selective retinoids on epidermal cells in the presence of the tumor promoter 12-O-tetradecanoylphorbol-13-acetate. American Association of Cancer Research Special Conference, Mechanism of Action of Retinoids, Vitamin D and Steroid Hormones. Whistler, British Columbia, Canada

63 Bollag W, Apfel C, LeMotte P (1995) Retinoids: Achievements, prospectives and future goals. *In:* S Waxman (ed.): *Challenges in Modern Medicine, Vol. 10. Proceedings of the 6th Conference on Differentiation Therapy.* Ares Serono Symposia Publications, Rome, 285–302

64 Reichert U, Bernandon JM, Charpentier B, Nedoncelle P, Martin B, Bernard BA, Asselineau D, Michel S, Lenoir MC, Delescluse C et al (1993) Synthetic retinoids: Receptor selectivity and biological activity. *In:* BA Bernard, B Shroot (eds): *From Molecular Biology to Therapeutics. (Pharmacology of the Skin, Vol. 5).* Karger, Basel, 117–127

65 Dawson MI, Jong L, Hobbs PD, Cameron JF, Chao W, Pfahl M, Lee M-O, Shroot B and Pfahl M (1995) Conformational effects on retinoid receptor selectivity. 2. Effects of retinoid bridging group on retinoid X receptor activity and selectivity. *J Med Chem* 38: 3368–3383

66 Dawson MI, Chan R, Hobbs PD, Chao W and Schiff LJ (1983) Aromatic retinoic acid analogues. 2. Synthesis and pharmacological activity. *J Med Chem* 26: 1282–1293

67 Cheng Y-C, Prusoff WH (1973) Relationship between the inhibition constant (K_i) and the concentration of inhibitor which causes 50% inhibition (IC_{50}) of an enzymatic reaction. *Biochem Pharmacol* 22: 3099–3108

68 Clark AJ (1933) *The Mode of Action of Drugs on Cells.* Arnold, London

69 Eyrolles L, Kagechika H, Kawachi E, Fukasawa H, Iijima T, Matsushima Y, Hashimoto Y, Shudo K (1994) Retinobenzoic acids. 6. Retinoid antagonists with a heterocyclic ring. *J Med Chem* 37: 1508–1517

70 Crettaz M, Baron A, Siegenthaler G, Hunziker W (1990) Ligand specificities of recombinant retinoic acid receptors RAR α and RAR β. *Biochem J* 272: 391–397

71 Charpentier B, Bernardon J-M, Eustache J, Millois C, Martin B, Michel S, Shroot B (1995) Synthesis, struc-ture-affinity relationships, and biological activities of ligands binding to retinoic acid receptor subtypes. *J Med Chem* 38: 4993–5006

72 Umemiya H, Kagechika H, Fukasawa H, Kawachi E, Ebisawa M, Hashimoto Y, Eisenmann G, Erb C, Pornon A, Chambon P et al (1997) Action mechanism of retinoid-synergistic dibenzodiazepines. *Biochem Biophys Res Commun* 233: 121–125

73 Lamour FPY, Lardelli P, Apfel CM (1996) Analysis of the ligand-binding domain of human retinoic acid receptor α by site-directed mutagenesis. *Mol Cell Biol* 16: 5386–5392

74 Agarwal C, Chandraratna RAS, Teng M, Nagpal S, Rorke EA, Eckert RL (1996) Differential regulation of human ectocervical epithelial cell line proliferation and differentiation by retinoid X receptor- and retinoic acid receptor-specific retinoids. *Cell Growth Differ* 7: 521–530

75 Muccio DD, Brouillette WJ, Breitman TR, Taimi M, Emanuel PD, Zhang X, Chen G, Sani BP, Venepally P, Reddy L et al (1998) Conformationally defined retinoic acid analogues. 4. Potential new agents for acute promyelocytic and juvenile myelomonocytic leukemias. *J Med Chem* 41: 1679–1687

76 Kagechika H, Kawachi E, Fukasawa H, Saito G, Iwanami N, Umemiya H, Hashimoto Y, Shudo K (1997) Inhibition of IL-1-induced IL-6 production by synthetic retinoids. *Biochem Biophys Res Commun* 231: 243–248

77 Apfel C, Bauer F, Crettaz M, Forni L, Kamber M, Kaufmann F, LeMotte P, Pirson W, Klaus M (1992) A retinoic acid receptor α antagonist selectively counteracts retinoic acid effects. *Proc Natl Acad Sci USA* 89: 7129–7133

78 Toma S, Isnardi L, Riccardi L and Bollag W (1998) Induction of apoptosis in MCF-7 breast carcinoma cell line by RAR and RXR selective retinoids. *Anticancer Res* 18: 935–942

79 Hashimoto Y, Kagechika H, Kawachi E, Fukasawa H, Saito G, Shudo K (1996) Evaluation of differentiation-inducing activity of retinoids on human leukemia cell lines HL-60 and NB4. *Biol Pharm Bull* 19: 1322–1328

80 Teng M, Duong TT, Klein ES, Pino ME, Chandraratna RAS (1996) Identification of a retinoic acid receptor α subtype specific agonist. *J Med Chem* 39: 3035–3038

81 Elmazar MMA, Rühl R, Reichert U, Shroot B, Nau H (1997) RARα-mediated teratogenicity in mice is potentiated by an RXR agonist and reduced by an RAR antagonist: Dissection of retinoid receptor-induced pathways. *Toxicol Appl Pharmacol* 146: 21–28

82 Canan Koch SS, Dardashti LJ, Hebert JJ, White SK, Croston GE, Flatten KS, Heyman RA, Nadzan AM (1996) Identification of the first retinoid X receptor homodimer antagonist. *J Med Chem* 39: 3229–3234

83 Lombardo A, Costa E, Chao WR, Toll L, Hobbs PD, Jong L, Lee M-O, Pfahl M, Ely KR, Dawson MI (1994) Recombinant human retinoic acid receptor β: Binding of synthetic retinoids and transcriptional activation. *J Biol Chem* 269: 21 490–21 497

84 Dawson MI, Chao W-R, Pine P, Jong L, Hobbs PD, Rudd CK, Quick TC, Niles RM, Zhang X, Lombardo A et al (1995) Correlation of retinoid binding affinity to retinoic acid receptor α with retinoid inhibition of growth of estrogen receptor-positive MCF-7 mammary carcinoma cells. *Cancer Res* 55: 4446–4451

85 Vasios GW, Gold JD, Petkovich M, Chambon P, Gudas LJ (1989) A retinoic acid-responsive element is present in the 5′ flanking region of the laminin B1 gene. *Proc Natl Acad Sci USA* 86: 9099–9103

86 Umesono K, Giguere V, Glass CK, Rosenfeld MG, Evans RM (1988) Retinoic acid and thyroid hormone induce gene expression through a common responsive element. *Nature* 336: 262–265

87 Pfahl M, Tzukerman M, Zhang X, Lehmann JM, Hermann T, Wills KN, Graupner G (1990) Nuclear retinoic acid receptors: Cloning, analysis, and function. *In*: L Packer (ed.): *Methods in Enzymology, Vol. 189. Retinoids. Part A. Molecular and Metabolic Aspects.* Academic Press, San Diego, 256–270

88 Klein ES, Pino ME, Johnson AT, Davies PJA, Nagpal S, Thacher SM, Krasinski G, Chandraratna RAS (1996) Identification and functional separation of retinoic acid receptor neutral antagonists and inverse agonists. *J Biol Chem* 271: 22 692–22 696

89 Jong L, Lehmann JM, Hobbs PD, Harlev E, Huffman JC, Pfahl M, Dawson MI (1993) Conformational effects on retinoid receptor selectivity. 1. Effect of 9-double bond geometry on retinoid X receptor activity. *J Med Chem* 36: 2605–2613

90 Ostrowski J, Roalsvig T, Hammer L, Marinier A, Starrett JE Jr, Yu K-L, Reczek PR (1998) Serine 232 and methionine 272 define the ligand binding pocket in retinoic acid receptor subtypes. *J Biol Chem* 273: 3490–3495

91 Reczek PR, Ostrowski J, Yu KL, Chen S, Hammer L, Roalsvig T, Starrett JE Jr, Driscoll JP, Whiting G, Spinazze PG et al (1995) Role of retinoic acid receptor γ in the Rhino mouse and rabbit irritation models of retinoid activity. *Skin Pharmacol* 8: 292–299

92 Johnson AT, Klein ES, Gillett SJ, Wang L, Song TK, Pino ME, Chandraratna RAS (1995) Synthesis and characterization of a highly potent and effective antagonist of retinoic acid receptors. *J Med Chem* 38: 4764–4767

93 Beard RL, Chandraratna RAS, Colon DF, Gillett SJ, Henry E, Marler DK, Song T, Denys L, Garst ME, Arefieg T et al (1995) Synthesis and structure-activity relationships of stilbene retinoid analogs substituted with heteroaromatic carboxylic acids. *J Med Chem* 38: 2820–2829

94 Jiang H, Penner JD, Beard RL, Chandraratna RAS, Kochhar DM (1995) Diminished teratogenicity of retinoid X receptor-selective synthetic retinoids. *Biochem Pharmacol* 50: 669–676

95 Nagy L, Thomázy VA, Shipley GL, Fésüs L, Lamph W, Heyman RA, Chandraratna RAS, Davies PJA (1995) Activation of retinoid X receptors induces apoptosis in HL-60 cell lines. *Mol Cell Biol* 15: 3540–3551

96 Chen S, Ostrowski J, Whiting G, Roalsvig T, Hammer L, Currier SJ, Honeyman JA, Kwasniewski B, Yu K-L, Sterzycki R et al (1995) Retinoic acid receptor γ mediates topical retinoid efficacy and irritation in animal models. *J Invest Dermatol* 104: 779–783

97 Boehm MF, Zhang L, Zhi L, McClurg MR, Berger E, Wagoner M, Mais DE, Suto CM, Davies PJA, Heyman RA et al (1995) Design and synthesis of potent retinoid X receptor-selective ligands that induce apoptosis in leukemia cells. *J Med Chem* 38: 3146–3155

98 Beard RL, Colon DF, Song TK, Davies PJA, Kochhar DM, Chandraratna RAS (1996) Synthesis and structure-activity relationships of retinoid X receptor selective diaryl sulfide analogs of retinoic acid. *J Med Chem* 39: 3556–3563

99 Umemiya H, Fukasawa H, Ebisawa M, Eyrolles L, Kawachi E, Eisenmann G, Gronemeyer H, Hashimoto Y, Shudo K, Kagechika H (1997) Regulation of retinoidal actions by diazepinylbenzoic acids. Retinoid syner

gists which activate the RXR-RAR heterodimers. *J Med Chem* 40: 4222–4234
100 Pfahl M (1995) Retinoid receptor action and how to separate desirable from undesirable effects. AACR Special Conference, Mechanism of Action of Retinoids, Vitamin D, and Steroid Hormones. Whistler, British Columbia, Canada
101 La Vista Picard N, Hobbs PD, Pfahl M, Dawson MI, Pfahl M (1996) The receptor-DNA complex determines the retinoid response: A mechanism for the diversification of the ligand signal. *Mol Cell Biol* 16: 4137–4146
102 Nagpal S, Friant S, Nakshatri H, Chambon P (1993) RARs and RXRs: Evidence for two automous transactivation functions (AF-1 and AF-2) and heterodimerization *in vivo*. *EMBO J* 12: 2349–2360
103 Forman BM, Umesono K, Chen J, Evans RM (1995) Unique response pathways are established by allosteric interactions among nuclear hormone receptors. *Cell* 81: 541–550
104 Kizaki M, Dawson MI, Heyman R, Elstner E, Morosetti R, Pakkala S, Chen D-L, Ueno H, Chao W, Morikawa M et al (1996) Effects of novel retinoid X receptor-selective ligands on myeloid leukemia differentiation and proliferation *in vitro*. *Blood* 87: 1977–1984
105 Wu Q, Dawson MI, Zheng Y, Hobbs PD, Agadir A, Jong L, Li Y, Liu R, Lin B, Zhang X-K (1997) Inhibition of *trans*-retinoic acid-resistant human breast cancer cell growth by retinoid X receptor-selective retinoids. *Mol Cell Biol* 17: 6598–6608
106 Eyrolles L, Kawachi E, Kagechika H, Hashimoto Y, Shudo K (1994) Synthesis and biological activity of carboxyphenylquinolines and related compounds as new potent retinoids. Retinobenzoic acids. VII. *Chem Pharm Bull* 42: 2575–2581
107 Wu Q, Li Y, Liu R, Agadir A, Lee M-O, Liu Y, Zhang X (1997) Modulation of retinoic acid sensitivity in lung cancer cells through dynamic balance of orphan receptors nur77 and COUP-TF and their heterodimerization. *EMBO J* 16: 1656–1669
108 Ebisawa M, Kawachi E, Fukasawa H, Hashimoto Y, Itai A, Shudo K, Kagechika H (1998) Novel thiazolidinedione derivatives with retinoid synergistic activity. *Biol Pharm Bull* 21: 547–549
109 Ohta K, Tsuji M, Kawachi E, Fukasawa H, Hashimoto Y, Shudo K, Kagechika H (1998) Potent retinoid synergists with a diphenylamine skeleton. *Biol Pharm Bull* 21: 544–546
110 Driscoll JE, Seachord CL, Lupisella JA, Darveau RP, Reczek PR (1996) Ligand-induced conformational changes in the human retinoic acid receptor detected using monoclonal antibodies. *J Biol Chem* 271: 22 969–22 975
111 Lee MO, Dawson MI, Picard N, Hobbs PD, Pfahl M (1996) A novel class of retinoid antagonists and their mechanism of action. *J Biol Chem* 271: 11 897–11 903
112 Kaneko S, Kagechika H, Kawachi E, Hashimoto Y, Shudo K (1991) Retinoid antagonists. *Med Chem Res* 220–225
113 Ueno H, Kizaki M, Matsushita H, Muto A, Yamato K, Nishihara T, Hida T, Yoshimura H, Koeffler HP, Ikeda Y (1998) A novel retinoic acid receptor (RAR)-selective antagonist inhibits differentiation and apoptosis of HL-60 cells: Implications of RARα-mediated signals in myeloid leukemic cells. *Leuk Res* 22: 517–525
114 Teng M, Duong TT, Johnson AT, Klein ES, Wang L, Khalifa B, Chandraratna RAS (1997) Identification of highly potent retinoic acid receptor α-selective antagonists. *J Med Chem* 40: 2445–2451
115 Standeven AM, Teng M, Chandraratna RAS (1997) Lack of involvement of retinoic acid receptor α in retinoid-induced skin irritation in hairless mice. *Toxicol Lett* 92: 231–240
116 Meister B, Fink F-M, Hittmair A, Marth C, Widschwendter M (1998) Antiproliferative activity and apoptosis induced by retinoic acid receptor-γ selectively binding retinoids in neuroblastoma. *Anticancer Res* 18: 1777–1786
117 Chen J-Y, Penco S, Ostrowski J, Balaguer P, Pons M, Starrett JE Jr, Reczek P, Chambon P, Gronemeyer H (1995) RAR-specific agonist/antagonists which dissociate transactivation and AP-1 transrepression inhibit anchorage-independent cell proliferation. *EMBO J* 14: 1187–1197
118 Fanjul A, Dawson MI, Hobbs PD, Jong L, Cameron JF, Harlev E, Graupner G, Lu X-P, Pfahl M (1994) A new class of retinoids with selective inhibition of AP-1 inhibits proliferation. *Nature* 372: 107–111
119 Joseph B, Lefebvre O, Mereau-Richard C, Danze PM, Belin-Plancot MT, Formstecher P (1998) Evidence for the involvement of both retinoic acid receptor- and retinoic X receptor-dependent signalling pathways in the induction of tissue transglutaminase and apoptosis in the human myeloma cell line RMPI 8226. *Blood* 91: 2423–2432
120 Fitzgerald P, Teng M, Chandraratna RA, Heyman RA, Allegretto EA (1997) Retinoic acid receptor α expression correlates with retinoid-induced growth inhibition of human breast cancer cells regardless of estrogen receptor status. *Cancer Res* 57: 2642–2650
121 Toma S, Isnardi L, Raffo P, Riccardi L, Dastoli G, Apfel C, LeMotte P, Bollag W (1998) RARα antagonist Ro 41-5253 inhibits proliferation and induces apoptosis in breast cancer cell lines. *Int J Cancer* 78: 86–94
122 Eckhardt K, Schmitt G (1994) A retinoic acid receptor α antagonist counteracts retinoid teratogenicity *in vitro* and reduced incidence and/or severity of malformations *in vivo*. *Toxicol Lett* 70: 299–308
123 Majewski S, Marczak M, Szmurlo A, Jablonska S, Bollag W (1995) Retinoids, interferon α, 1,25-dihydroxyvitamin D_3 and their combination inhibit angiogenesis induced by non-HPV- harboring tumor cell lines. RAR α mediates the antiangiogenic effect of retinoids. *Cancer Lett* 89: 112–124
124 Bernard BA, Shroot B (eds) (1993) *From Molecular Biology to Therapeutics. (Pharmacology and the Skin, Vol. 5)*. Karger, Basel

125 Lu X-P, Fanjul A, Picard N, Pfahl M, Rungta D, Nared-Hood K, Carter B, Piedrafita J, Tang S, Fabbrizio E et al (1997) Novel retinoid-related molecules as apoptosis inducers and effective inhibitors of human lung cancer cells *in vivo*. *Nat Med* 3: 686–690

126 Shroot B, Eustache J, Bernardon J-M (1990) Benzimidzole derivatives and their therapeutic and cosmetic use. U.S. Patent 44,920,140

127 Taneja R, Bouillet P, Boylan JF, Gaub M-P, Roy B, Gudas LJ, Chambon P (1995) Reexpression of retinoic acid receptor (RAR)$_\gamma$ or overexpression of RAR$_\alpha$ or RAR$_\beta$ in RAR$_\gamma$-null F9 cells reveals a partial functional redundancy between the three RAR types. *Proc Natl Acad Sci USA* 92: 7854–7858

128 Blanchet S, Favier B, Chevalier G, Kastner P, Michaille J-J, Chambon P, Dhouailly D (1998) Both retinoic acid receptors α (RARα) and γ (RARγ) are able to initiate mouse upper-lip skin glandular metaplasia. *J Invest Dermatol* 111: 206–212

129 Bollag W, Isnardi L, Jablonska S, Klaus M, Majewski S, Pirson W, Toma S (1997) Links between pharmacological properties of retinoids and nuclear retinoid receptors. *Int J Cancer* 70: 470–472

130 Hartman D, Bollag W (1993) Historical aspects of the oral use of retinoids in acne. *J Dermatol* 20: 674–678

131 Tallman MS (1998) Therapy of acute promyelocytic leukemia: all-*trans* retinoic acid and beyond. *Leukemia* 12 (Suppl. 1): S37–S40

132 Warrell RP Jr, de Thé H, Wang Z-Y, Degos L (1993) Acute promyelocytic leukemia. *New Engl J Med* 329: 177–189

133 Ledo A, Martin M, Geiger JM, Marrón JM (1988) Aciretin (Ro 10-1670) in the treatment of severe psoriasis: A randomized double-blind parallel study comparing aciretin and etretinate. *Int J Dermatol* 27: 656–33

134 Peck R, Bollag W (1991) Potentiation of retinoid-induced differentiation of HL-60 and U937 cell lines by cytokines. *Eur J Cancer* 27: 53–57

135 Muto Y, Moriwaki H, Shiratori Y (1998) Prevention of second primary tumors by an acyclic retinoid, polyprenoic acid, in patients with hepatocellular carcinoma. *Digestion* 59 (Suppl. 2): 89–91

136 Miller VA, Rigas JR, Benedetti FM, Verret AL, Tong WP, Kris MG, Gill GM, Crisp M, Loewen GR, Truglia JA et al (1996) Initial clinical trial of the retinoid receptor pan agonist 9-*cis* retinoic acid. *Clin Cancer Res* 2: 471–475

137 Kurie JM, Soo Lee J, Griffin T, Lippman SM, Drum P, Thomas MP, Weber C, Baser M, Massimini G, Hong WK (1996) Phase I trial of 9-*cis* retinoic acid in adults with solid tumors. *Clin Cancer Res* 2: 287–293

138 Rizvi NA, Marshall JL, Ness E, Yee J, Gill GM, Truglia JA, Loewen GR, Jaunakais B, Ulm EH, Hawkins MJ (1998) Phase I study of 9-*cis*-retinoic acid (ALRT1057 capsules) in adults with advanced cancer. *Clin Cancer Res* 4: 1437–1442

139 Miller VA, Benedetti FM, Rigas JR, Verret AL, Pfister DG, Straus D, Kris MG, Crisp M, Heyman R, Loewen GR et al (1997) Initial clinical trial of a selective retinoid X receptor ligand, LGD1069. *J Clin Oncol* 15: 790–795

140 Soignet SL, Benedetti F, Fleischauer A, Parker BA, Truglia JA, Crisp MR, Warrell RP Jr (1998) Clinical study of 9-*cis* retinoic acid (LGD1057) in acute promyelocytic leukemia. *Leukemia* 12: 1518–1521

141 Takeshita A, Shibata Y, Shinjo K, Yanagi M, Tobita T, Ohnishi K, Miyawaki S, Shudo K, Ohno R (1996) Successful treatment of relapse of acute promyelocytic leukemia with a new synthetic retinoid, Am80. *Ann Int Med* 124: 893–896

142 Tobita T, Takeshita A, Kitamura K, Ohnishi K, Yanagi M, Hiroka A, Karasuno T, Takeuchi M, Miyawaki S, Ueda R et al (1997) Treatment with a new synthetic retinoid, Am80, of acute promyelocytic leukemia relapsed from complete remission induced by all-*trans*-retinoic acid. *Blood* 90: 967–973

143 Yamakawa T, Kagechika H, Kawachi E, Hashimoto Y, Shudo K (1990) Retinobenzoic acids. 5. Retinoidal activities of compounds having a triethylsilyl or trimethylgermyl group(s) in human promyelocytic leukemic cells HL-60. *J Med Chem* 33: 1430–1437

144 Murakami K, Matsuura T, Sano M, Hashimoto A, Yonekura K, Sakukawa R, Yamada Y, Saiki I (1998) 4-[3,5-Bis(trimethylsilyl)benzamino]benzoic acid (TAC-101) inhibits the intrahepatic spread of hepatocellular carcinoma and prolongs the life-span of tumor-bearing animals. *Clin Exp Metastasis* 16: 633–643

145 Saurat JH (1997) Oral isotretinoin. Where now, where next? *Dermatology* 195 (Suppl. 1): 1–3: 38–40

146 Chandraratna RAS (1996) Tazarotene—first of a new generation of receptor-selective retinoids. *Brit J Dermatol* 135: 18–25

147 Weinstein GD, Krueger GG, Lowe NJ, Duvic M, Friedman DJ, Jegasothy BV, Jorizzo JL, Shmunes E, Tschen EH, Lew-Kaya DA et al (1997) Tazarotene gel, a new retinoid, for topical therapy of psoriasis. Vehicle-controlled study of safety, efficacy, and duration of therapeutic effect. *J Amer Acad Dermatol* 37: 85–92

148 Ishibashi Y (1995) Efficacy of Am-80 ointment on psoriasis—A bilateral-paired comparative study with betamethasone valerate ointment (phase III study). *Rinsho Iyaku* 11: 733–746

149 Ishibashi Y (1995) Clinical effect of Am-80 ointment on psoriasis and pustulosis plamaris et plantaris (phase III study). *Rinsho Iyaku* 11: 747–759

150 Barua AB, Olson JA (1989) Method of producing water-soluble glucuronic acid derivatives of vitamin A. U.S. Patent 4,885,463 (August 8, 1989)

151 Curley RW Jr, Abou-Issa H, Panigot JJ, Repa JJ, Clagett-Dame M, Alshafie G (1996) Chemopreventive activities of C-glucuronide/glycoside analogs of retinoid-O-glucuronides against breast cancer development and growth. *Anticancer Res* 16: 757–764

152 Abou-Issa H, Curley RW Jr, Panigot MJ, Tanagho SN, Sidhu BS, Alshafie GA (1997) Chemotherapeutic evaluation of N-(4-hydroxyphenyl) retinoid O-glucuronide in the rat mammary tumor model. *Anticancer Res* 17: 3335–3340
153 Balakrishnan V, Gilbert NE, Brueggemeier, Curley RW Jr (1997) N-linked glycoside/glucuronide conjugates of retinoids: Acitretin. *Bioorg Med Chem Lett* 7: 3033–3038

Vitamin A and retinoids: an update of biological aspects and clinical applications
M.A. Livrea (ed.)
© 2000 Birkhäuser Verlag Basel/Switzerland

Proposed role of gap junctional communication in retinoid-induced suppression of proliferation and inhibition of neoplastic transformation

J.S. Bertram

Cancer Research Center of Hawaii, University of Hawaii, 1236 Lauhala Street, Honolulu, HI 96813, USA

Introduction

With our increasing understanding of the molecular biology of cancer and of factors that predispose to a high risk of cancer, there seems little doubt that medical science will soon be able to achieve large decreases in the incidence of many forms of cancer through preventive means. The development of drugs with preventive potential, i.e. cancer chemopreventive agents, is therefore of high priority. The retinoids, the natural or synthetic analogs of retinoic acid, appear at present to offer the greatest potential for decreasing cancer incidence in such groups. Retinoic acid, as a locally produced hormone, produces many effects in target cells. In this review will be described effects of retinoids on gap junctional communication and the potential role of enhanced communication in decreasing proliferation of target cells and thus reducing their potential for neoplastic progression.

Chemoprevention studies in cultured cells

Because of the tight control of the experimental variables possible in studies conducted in cultured cells and because of the homogeneity of cell lines, many studies can be carried out which would be impossible in whole animal systems. This review will focus primarily on retinoid effects on the transformation of normal cells.

Inhibition of neoplastic transformation *in vitro: Studies in 10T1/2 cells*

The use of this system as a model for carcinogenesis has been extensively validated by many investigators [1]. Here, the application of a brief pulse of carcinogen applied to an actively growing cell population results, after a period of approximately five weeks, in the transformation of a small percentage of exposed cells to cells with a neoplastic phenotype. Such cells are not subject to normal growth control processes, and proliferate to form piled up transformed-foci of cells which can be readily distinguished from the normal background population. These cells are tumorigenic in immuno-suppressed mice, thus satisfying the ultimate criteria for malignant transformation. Earlier work performed in the author's laboratory, demonstrated that a wide range of retinoids was capable of completely suppressing this induction of neoplastic transformation [2]. This result was obtained when retinoids were applied long after exposure to the carcinogen and thus were not capable of interfering with the interaction of the carcinogen with the cells or with the subsequent processing of chemical lesions produced in the cells. Moreover, this inhibition was reversible upon withdrawal of the retinoids, indicating that action was not a consequence of selective cytotoxicity to carcinogen-initiated cells [2]. It was subsequently shown in clinical studies that the suppression of second primary neoplasms was also reversible, thus indicating the predictive value of this mouse cell culture system [3].

Mechanism of action of retinoids as inhibitors of neoplastic transformation: Proposed role of gap junctional communication

Our first hint that cell/cell communication could influence the expression of the neoplastic phenotype, came in very early studies during the development of the 10T1/2 cell line as an assay system for neoplastic transformation *in vitro*. Here we noted that as we increased the number of cells plated in culture dishes and subsequently exposed them to carcinogens, instead of seeing a proportional increase in the number of transformed foci which subsequently developed there was a paradoxical decrease [4].

In work conducted with the co-discoverer of gap junctions, Dr. Loewenstein, we proposed the hypothesis that the growth inhibition resulting from these cell/cell interactions was directly correlated with induction of gap junctional communication [5]. This work laid the groundwork for our later observations that the inhibition of neoplastic transformation caused by retinoids [6] and by carotenoids [7] was directly correlated with induced junctional communication. These results fit well with our previous observations, especially those involving population density effects on expression of neoplastic transformation. These results were now explicable in terms of the enhanced opportunities for interactions of initiated cells with normal cells when those initiated cells were present as small colonies surrounded by normal cells. In contrast, large colonies would contain central cells which would be physically shielded from communication with the surrounding normal cells. However, in the presence of retinoids or carotenoids, which we now know to enhance junctional communication within both normal and initiated cell populations, this physical barrier would be less effective and central cells would no longer be shielded from the influence of surrounding normal cells. In this situation, as we have shown, transformation is inhibited [8]. This model is shown in Figure 1.

Gap junctions and their proposed physiological role

Gap junctions are water-filled pores called connexons that connect adjacent cells in most organs of the body. These pores allow direct cytoplasmic to cytoplasmic communication of water-soluble molecules and ions. Because of the limiting size of the pore, only molecules less than about 1000 daltons can pass, thus excluding molecules such as mRNA and protein and thus

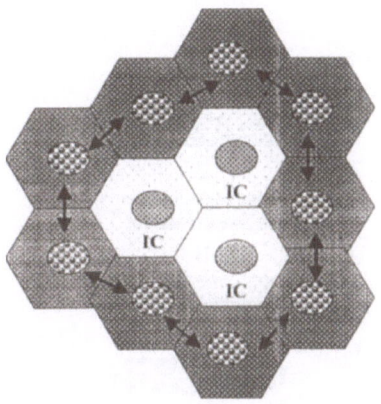

GJC deficient initiated cell [IC]:
Proliferative advantage

Carotenoid induced GJC$\updownarrow$:
no proliferation

Figure 1. Proposed relationship between junctional communication (GJC), growth control and colony size of initiated cells (IC). When the colony size is small or junctional communication is enhanced by retinoids or carotenoids, all cells are in effective communication with surrounding normal cells and transformation is suppressed. In contrast, cells in the center of large colonies in control situations are effectively shielded from the effects of surrounding normal cells and transformation can occur. While this diagram depicts the essentially two-dimensional situation in cell monolayer cultures, it seems equally applicable to the *in vivo* situation where initiated cells may grow as spheres, again surrounded by normal cells.

maintaining genetic identity of the cells. The existence of this communication network creates a syncytium through which cells can exchange nutrients, waste products and signalling molecules such as cyclic adenosine monophosphate (cAMP), Ca^{2+}-ions etc. [9]. Of relevance to the present discussion is our hypothesis, first formulated before this activity of retinoids and carotenoids was known [5], that gap junctions serve as conduits for antiproliferative signals generated by growth-inhibited normal cells. These signals would act to suppress the proliferation of carcinogen-initiated cells thereby preventing their transformation [8, 10].

The structural element of a gap junction is a trans-membrane protein called a connexin; six of these connexin molecules are known to assemble radially to enclose the central pore. This structure can then dock with a similar structure on a contacting adjacent cell to form a complete connexon. Thus, the structural unit of the gap junction is composed of 12 connexin molecules contributed equally by each of the communicating partners [11]. This arrangement is shown diagrammatically in Figure 2. Passage of molecules or ions through the central pore appears to be via passive diffusion down concentration gradients. Connexin 43, the connexin discussed below because of its inducibility by retinoids, is one of a family of about 14 genes and appears to be the most widely expressed connexin. Studies of so-called knock-out mice, in which individual connexin genes have been destroyed by homologous recombination, is providing compelling evidence for the actions of connexons in many processes (reviewed in [12]).

The consistent finding in studies of human or animal tumor cell lines and in studies of neoplastic transformation *in vitro*, is that tumor cells communicate poorly, if at all, with their nor-

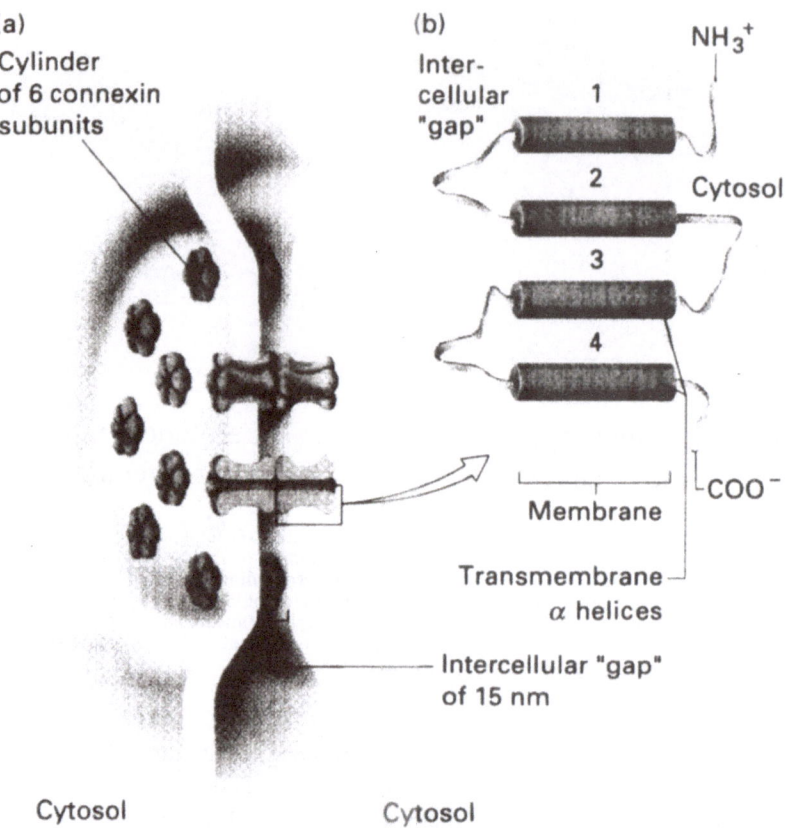

Figure 2. Proposed structural organization of connexin proteins into the plasma membrane to form a connexon unit. Panel A: Each hemi-connexon is composed of six identical protein units arranged to circle the water filled pore, hemi-connexons dock with equivalent structures in adjacent cells to form an intact gap junction. Panel B: topography of the connexin 43 protein in the plasma membrane. There are believed to be four trans-membrane spanning domains, the third of which is believed to line the water filled pore. The two extra-cellular loops are believed to dock with equivalent loops donated by connexins from the adjacent cell. (From [27] with permission).

mal counterparts. In some cases this is due to lack of expression of the connexin gene for reasons not yet known, in other situations in spite of expression of connexins, tumor cells are unable to undergo functional junctional communication with normal cells (reviewed in [13]). These findings led to the original hypothesis of growth control through junctional communication so eloquently proposed by Loewenstein [14]. In this hypothesis, signalling centers are created in a population of normal cells in direct proportion to the number of normal cells. Within the area dominated by the signalling center, proliferation would be inhibited. In this model, any cell deficient in the ability to receive or respond to such signals would be at a proliferative advantage, as is observed with tumor cells. Conversely, agents or actions which increase GJC would be expected to decrease proliferation of such cells when in contact with growth-inhibited normal cells. The retinoids and carotenoids have been shown to increase GJC between pre-malignant, initiated cells and their normal counterparts, and in support of the growth control hypothesis, this action is correlated with decreases in proliferation of communicating cells. Interestingly, as will be discussed in more detail below, in experiments conducted with retinoids in which communication was denied by plating at low cell densities, no decreases in prolifer-ation or morphological changes were induced in spite of upregulated connexin 43 expression [15]. Thus the requirement for cell/cell contact for the inhibitory action of retinoids on prolif-eration and the increases in cell spreading adds further weight to the notion that this aspect of retinoid action in 10T1/2 cells is mediated through gap junctional intercellular communication.

It is important to state here that these compounds have not been found to increase commu-nication between established tumor cells and normal cells [5]; this would be consistent with their inability in the 10T1/2 system to inhibit expression of the transformed phenotype i.e. growth of tumor cells in a background of normal cells, and with the experimental animal data showing that these compounds are active in the post-initiation phase of carcinogenesis prior to the establishment of tumors. These observations would also be consistent with the lack of abil-ity of retinoic acid to inhibit solid tumor growth in clinical trials in head and neck cancer [16]. Thus, in general, the actions of carotenoids and retinoids are considered to be preventive and not therapeutic. In a recent study of dysplastic regions of the oral cavity in patients with a prior history of oral carcinoma, we discovered that even in these pre-cancerous lesions major reduc-tions in connexin 43 expression had occurred [17]. Studies are underway to determine if retinoids can counter this decrease.

Evidence that retinoids upregulate connexin 43 expression and GJC in 10T1/2 cells

Correlation with retinoid-induced suppression of proliferation and inhibition of neoplastic transformation

Studies have been conducted with normal 10T1/2 cells, with carcinogen-initiated 10T1/2 cells and with fully transformed cells isolated from *in vitro* transformation assays. Studies with ini-tiated cells were made possible because of the ability of retinoids to reversibly inhibit trans-formation in the post-initiation phase of carcinogenesis. These cells had a normal phenotype in the presence of retinoids, but in the absence of retinoids after a three-four week latency period, cells underwent neoplastic transformation [18]. The addition of retinoids to confluent cultures of normal or initiated cells caused a dose-responsive increase in junctional communication, and this increase was tightly correlated with the ability of these retinoids to suppress neoplastic transformation in the same cells (Fig. 3). Increases in communication were observed within 18 h of treatment and were maintained for the four week duration of the experiment, i.e. the duration of the standard transformation assays. This enhanced communication was inhibited by the tumor promoter 12-0-tetradecanoylphorbol acetate (TPA), an effect consistent with the mutual antagonism of retinoids and TPA on carcinogenesis. Interestingly, initiated cells though highly sensitive to the effects of retinoids, were found to communicate poorly with the parental 10T1/2 cells regardless of retinoid treatment [6]. Thus, even in these early stages of carcino-

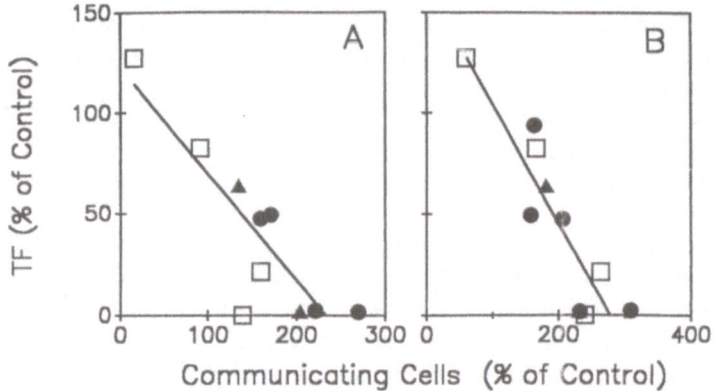

Figure 3. Correlation between transformation and junctional communication. Panel A. 10 T1/2 cells; panel B. initiated cells. Cells were exposed to various concentrations of retinol (●), retinoic acid (□), or the synthetic retinoid TTNPB (▲). A strong correlation was noted between suppression of transformation frequency (TF) and induction of junctional communication measured by dye transfer in confluent cell cultures. The Pearson correlation coefficient was –0.86 (P = 0.001) and –0.894 (P = 0.007) for 10 T1/2 cells and initiated cells respectively. Data from reference [6] with permission.

genesis, initiated cells find themselves junctionally isolated from normal cells. As previously observed in assays of transformation, all-*trans* retinoic acid itself had only a low activity in this system. This was later determined to be due to its rapid catabolism by P450 enzymes; we later demonstrated that simultaneous treatment with liarazole, an inhibitor of this enzyme, enhanced the activity of retinoic acid to upregulate connexin 43 expression by over 1000-fold [19]. The increased communication caused by retinoids was shown to be due to increased expression of connexin 43 at the message and protein level [20].

The enhanced communication induced by retinoids was found also to be correlated with enhanced growth control of normal 10T1/2 cells and their transformed counterparts. In these experiments, enhanced growth control was assessed as a reduction in the stable saturation density reached by cells after prolonged culture. As discussed earlier, a major phenotypic difference between normal cells and neoplasticly transformed cells is the ability of transformed cells to grow to high densities and their failure to restrict proliferation even at these high densities. Retinoids were shown to be capable of increasing homologous communication between both cell types and of proportionately reducing their saturation densities. These two effects were statistically highly correlated (Fig. 4). Microscopic examination of treated cultures revealed that cells still maintained a confluent monolayer and that this reduction in saturation density i.e. the number of cells per unit area of culture dish, was achieved by an increase in area occupied by each cell.

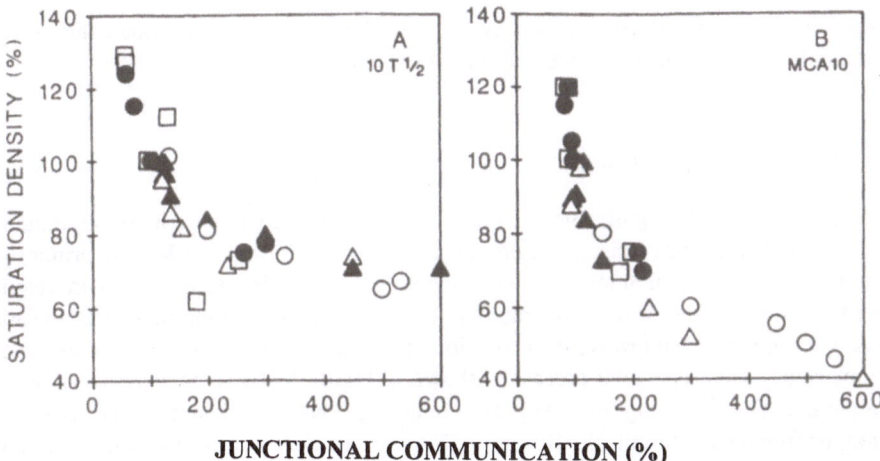

JUNCTIONAL COMMUNICATION (%)

Figure 4. Correlation between retinoid-induced decreases in saturation density and induction of gap junctional communication. Confluent cultures of 10 T1/2 cells (panel A) or methylcholanthrene transformed 10 T1/2 cells (panel B) were treated with various concentrations of retinoids for five days. At this time cells were probed by microinjection of Lucifer yellow and transfer of dye to adjacent cells was quantitated. Total cell counts were also performed at this time. Results are expressed as a percent of the untreated controls. Symbols: retinol (△); retinyl acetate (▲): all-*trans* retinoic acid (◆): 13-*cis* retinoic acid (□); TTNPB (○). The correlations were statistically highly significant, for 10 T1/2 cells and for transformed cells, P < 0.001. Data from reference [28] with permission.

Decreased saturation densities are associated with decreased rates of proliferation and require cell/cell contact

The preceding studies had examined the effects of retinoids in confluent cell cultures since it is in these cultures that transformation occurs. To determine if retinoids would have comparable effects on cell spreading in the absence of cell contact, we examined the effects of retinoids in 10T1/2 cells sparsely seeded in culture dishes and thus denied the ability to undergo junctional communication. Here, in spite of extensive expression of connexin 43, retinoid treatment did not cause increases in cell area nor did it decrease the proliferation rate of these logarithmically growing cells. Thus the radio-thymidine labeling index of control cultures was 19.3% while in retinoid treated cultures it was 18%. In contrast, in confluent cultures the control labeling index was 4.2% and this decreased to less than 0.1% in retinoid treated cultures. Similarly in logarithmically growing cultures retinoids did not alter the cell area occupied by the cell, whereas in confluent cultures the cell area was increased by 142% [15].

These studies have thus demonstrated that the actions of retinoids as inhibitors of neoplastic transformation are highly correlated with their abilities to increase connexin 43 expression, increase GJC, decrease saturation density and suppress proliferation in confluent cultures. However, it must be noted that these are only correlations and cannot prove a cause and effect relationship. More recent studies have utilized molecular methodology to over-express connexin 43 in a variety of cell types without the requirement for retinoid treatment. As will be discussed later in this chapter, these studies are also highly supportive of the role of junctional communication in growth control. However, before these studies are described, the role of proliferation in the carcinogenic process will be discussed.

Role of proliferation in carcinogenesis

To fully understand the significance of a reduction in proliferation rate caused by retinoids it is necessary to discuss the carcinogenic process in greater detail. It is now known, primarily from the work of Vogelstein in human colon carcinoma, that multiple genetic events are required in order for complete malignant transformation to occur [21]. These multiple genetic events must occur in a single cell and involve the activation of oncogenes and the inactivation of tumor suppressor genes. This sequential process is shown in Figure 5. These genetic events occur as a result of mutations resulting from exogenous or endogenous DNA damage, from errors in replication, or from errors in mitosis. Because DNA damage does not itself constitute a mutation, but can be converted to mutations during DNA replication, it is evident that the induction of mutation from all sources requires DNA replication. Thus, a reduction in the proliferation rate will lead to a reduction in mutation frequency for any given level of DNA damage. The requirement for proliferation in allowing sequential acquisition of mutations does not stop there. The

Multistep Carcinogenesis: requirement for DNA damage and proliferation

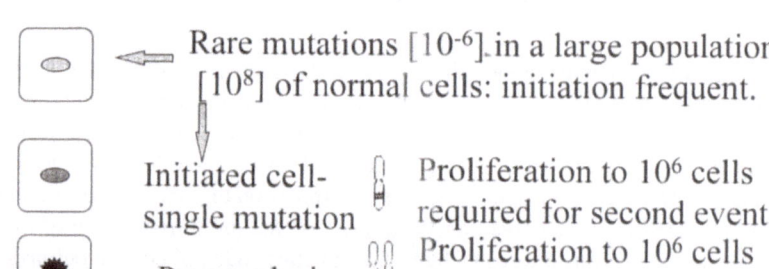

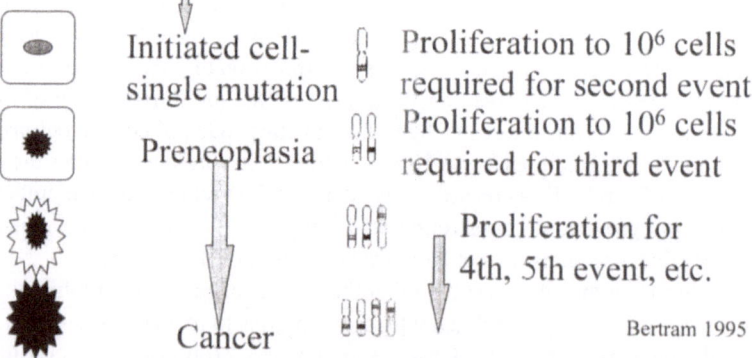

Rare mutations [10^{-6}] in a large population [10^8] of normal cells: initiation frequent.

Initiated cell-single mutation

Proliferation to 10^6 cells required for second event

Preneoplasia

Proliferation to 10^6 cells required for third event

Proliferation for 4th, 5th event, etc.

Cancer

Bertram 1995

Figure 5. The process of multistep carcinogenesis. This is drawn to emphasize the requirement for proliferation both for the production of mutations and the expansion of initiated cells to a population size sufficient for second mutations to occur in a single cell. The ability of carotenoids and retinoids to suppress proliferation by modifying junctional communication is proposed as one mechanism by which these agents are capable of acting as cancer preventives.

frequency of mutations at any given allele can range from 10^{-5} to 10^{-7} dependent upon the cell type and rate of DNA damage. Even the so-called mutator phenotypes do not exceed this rate by much. Thus, in a given organ, containing perhaps 10^7-10^8 stem cells, the induction of the initial mutation is a relatively common event. However, induction of a secondary and tertiary event in that single mutated cell has a vanishingly small probability. What is required is that this cell has a proliferative advantage over its neighbors and can replicate inappropriately to form a population of approximately 10^6 cells. In this population there will now exist a finite probability of a secondary mutation at another genetic locus—this cell bearing a double mutation must then itself expand its population to again achieve a mass in which a third mutation has a finite probability of occurring. The role of proliferation in carcinogenesis has been elegantly discussed by Ames [22].

In view of the above discussion, it can be appreciated that the suppression of aberrant proliferation, in our model caused by enhanced GJC between these minimally mutated cells and surrounding normal cells, could effectively inhibit the development of cells bearing multiple mutations required for full carcinogenic potential. In a clinical context, populations of cells which have not yet acquired the full neoplastic potential can be recognized in the colon as polyps, and in many epithelia as dysplasias.

This model cannot distinguish between the communication of growth inhibitory signals from normal to initiated or neoplastic cells, from diffusion of growth stimulating signals from these aberrant cells to the large mass of normal cells. Clearly, the production and transfer of growth inhibitory signals from normal cells is the most attractive of these two alternatives; it would allow the active participation of gap junctions in maintaining tissue homeostasis thereby providing one more reason for the strong evolutionary conservation of gap junctions. Moreover, these signals once characterized could be developed into a novel new therapy for the control of aberrant proliferation. Current work in our group is attempting to resolve these two issues and hopefully to identify this signal.

Studies of connexin function utilizing molecular methodology

The studies discussed above, relating connexin-mediated functional GJC with growth control, relied on correlations to prove the association. However, these correlations to not prove a cause and effect relationship. For example, the actions of retinoids on connexin 43 gene expression may go hand in hand with actions of retinoids on growth control but to be unrelated to these actions. To more firmly establish the role of GJC as important to the maintenance of the normal phenotype, a number of investigators have utilized the technique of genetic engineering to re-introduce or upregulate expression of connexins in tumor or normal cells. Published results have consistently reported that upregulated GJC is associated with either enhanced control of growth *in vitro*, or decreased tumorigenicity of cells when injected into immuno-suppressed mice (reviewed in [13]). Thus, increased GJC in the absence of retinoids decreases the neoplastic potential of these cells, and suggests that many of the actions of these agents may be due to their effects on GJC. It should be noted that the studies utilized established transformed cell lines, in which many genetic alterations may have occurred which would limit the full restoration of normal behavior. All these studies were performed utilizing constitutive promoters to drive the connexin 43 gene. One problem in interpreting these results is that the process of transfection involves the clonal selection of those few cells which stably integrate the gene and express it at useful levels. If, as we have reason to believe, individual cells within a population differ in their phenotypic properties, effects of the transfected gene are difficult to distinguish from pre-existing differences within cells of the original population. To circumvent this interpretational problem, we have recently constructed transformed cells in which the connexin 43 gene has been integrated under the control of a tetracycline- inducible promoter. In this system the addition of tetracycline causes rapid increases in transcription of connexin 43 which becomes functionally integrated into the plasma membrane. In this inducible system, cells in the non-induced state can serve as controls for induced cells, thus avoiding the problems of clonal variation discussed above. In research currently submitted for publication, we have shown that induction of connexin 43 in these human cervical carcinoma cells, causes reductions in indices of two markers of the neoplastic phenotype: a reduced ability to grow in an anchorage-independent manner, and a dramatically decreased growth rate as xenografts in immuno-compromised mice. However, in these established tumor cells we were unable to demonstrate that connexin 43 induction reduced the proliferation rate or saturation density of cells in monolayer culture (King et al., submitted for publication). This discrepancy between the results discussed above, utilizing normal and transformed 10T1/2 cells in which connexin 43 induction was induced by retinoids, and these studies involving established human tumor cells may simply reflect the fact that the human cells, because of their long history in cell culture, had lost the ability to respond to signals generated in monolayer culture. Unfortunately, for technical reasons, we have so far been unable to produce 10T1/2 cells or transformed derivatives of these cells containing an inducible connexin 43 gene.

Mechanisms by which retinoids increase connexin 43 expression

This issue is currently the focus of much research, unfortunately the situation is unclear as no known retinoic acid responsive elements (RAREs) are found in the 5' region of this gene. In some situations retinoic acid causes down-regulation of gene expression, for example expression of keratin 1 and loricrin is decreased in differentiating human and mouse keratinocytes; this action appears to be mediated through interactions with AP1 sites in the responsive genes [23, 24]. The rat and human 5' up-stream regions of the connexin 43 gene have been sequenced [25, 26] and both contain AP1 sites. We are currently examining the role of these sites in the regulation of connexin 43 transcription.

Conclusions

The studies discussed above have demonstrated that the induction of connexin 43 by retinoids and the consequent increase in homologous GJC between normal 10T1/2 cells, strongly correlates with the ability of these retinoids to suppress neoplastic transformation and increase growth control in confluent cultures. This increased growth control, which is expressed as a decrease in saturation density and in proliferation in confluent cells, is proposed to be directly implicated in the ability of retinoids to suppress transformation in the 10T1/2 system. Retinoids have been shown to increase connexin 43 expression in human skin and cultured keratinocytes. In contrast, decreased expression of connexin 43 has been seen in premalignant lesions of the oral mucosa. Thus, one aspect of the chemopreventive potential of retinoids at this site may be due to its ability to reverse these changes in connexin expression. Studies using molecular techniques in which connexin 43 is over-expressed in tumor cells have also demonstrated the ability of connexin 43 expression to decrease several aspects of the neoplastic phenotype. All these studies support the hypothesis of local growth control through gap junctional communication. The chemical nature of the transmitted signal and the physiological effects induced by connexin 43 expression have yet to be determined.

References

1   Bertram JS (1985) Neoplastic transformation in cell cultures: *in vitro/in vivo* correlations. *IARC Sci Pub* 67: 77–91
2   Merriman R, Bertram JS (1979) Reversible inhibition by retinoids of 3-methylcholanthrene-induced neoplastic transformation in C3H10T1/2 cells. *Cancer Res* 39: 1661–1666
3   Papadimitrakopoulou VA, Hong WK, Lee JS, Martin JW, Lee JJ, Batsakis JG, Lippman SM (1997) Low-dose isotretinoin versus β-carotene to prevent oral carcinogenesis: Long-term follow-up. *J Natl Cancer Inst* 89: 257–258
4   Ono Y, Ogita K, Kikkawa U, Igarashi K, Nishizuka Y (1988) The structure, expression, and properties of additional members of the protein kinase C family. *J Biol Chem* 263: 6927–6932
5   Mehta PP, Bertram JS, Loewenstein WR (1986) Growth inhibition of transformed cells correlates with their junctional communication with normal cells. *Cell* 44: 187–196
6   Hossain MZ, Wilkens LR, Mehta PP, Loewenstein WR, Bertram JS (1989) Enhancement of gap junctional communication by retinoids correlates with their ability to inhibit neoplastic transformation. *Carcinogenesis* 10: 1743–1748
7   Zhang L-X, Cooney RV, Bertram JS (1991) Carotenoids enhance gap junctional communication and inhibit lipid peroxidation in C3H/10T1/2 cells: Relationship to their cancer chemopreventive action. *Carcinogenesis* 12: 2109–2114
8   Mordan LJ, Martner JE, Bertram JS (1983) Quantitative Neoplastic Transformation of C3H/10T1/2 Fibroblasts: Dependence upon the Size of the Initiated Cell Colony at Confluence. *Cancer Res* 43: 4062–4067
9   Loewenstein WR, Rose B (1992) The cell-cell channel in the control of growth. *Semin Cell Biol* 3: 59–79
10 Mordan LJ, Bertram JS (1983) Retinoid effects on cell-cell interactions and growth characteristics of normal and carcinogen-treated C3H/10T1/2 cells. *Cancer Res* 43: 567–571
11 Beyer EC, Paul DL, Goodenough DA (1990) Connexin family of gap junction proteins. *J Membrane Biol* 116: 187–194
12 Nicholson SM, Bruzzone R (1997) Gap junctions: getting the message through. *Curr Biol* 7:R340–344
13 Neveu M, Bertram JS (1999) Gap junctions, neoplasia. *In*: EL Hertzberg (ed.): *Gap Junctions: In, Advances in cellular, molecular biology.* JAIPress, Greenwich ; *in press*
14 Loewenstein WR (1979) Junctional intercellular communication and the control of growth. *Biochim Biochim Biophys Acta* 560: 1–65
15 Hossain MZ, Bertram JS (1994) Retinoids suppress proliferation, induce cell spreading, and up-regulate connexin43 expression only in postconfluent 10T1/2 cells: Implications for the role of gap junctional communication. *Cell Growth Differ* 5: 1253–1261
16 Hong WK, Lippman SM, Itri LM, Karp DD, Lee JS, Byers RM, Schantz SP, Kramer A, Lotan R, Peters LJ et al (1990) Prevention of second primary tumors with isotretinoin in squamous-cell carcinoma of the head and neck. *New Engl J Med* 323: 795–800
17 Sakr W, Tabaska P, Kucuk O, Bertram JS (1996) Differential expression of connexin 43 in normal, preneoplastic, neoplastic squamous epithelium of the upper aerodigestive tract. *Proc AACR* 37: 269
18 Mordan LJ, Bergin LM, Budnick JL, Meegan R, Bertram JS (1982) Isolation of Methylcholanthrene-Initiated C3H/10T1/2 Cells by Inhibiting Neoplastic Progression with Retinyl Acetate. *Carcinogenesis* 3(3): 279–285
19 Acevedo P, Bertram JS (1995) Liarozole potentiates the cancer chemopreventive activity of and the up-regu-

lation of gap junctional communication and connexin43 expression by retinoic acid and β-carotene in 10T1/2 cells. *Carcinogenesis* 16: 2215–2222

20 Rogers M, Berestecky JM, Hossain MZ, Guo HM, Kadle R, Nicholson BJ, Bertram JS (1990) Retinoid-enhanced gap junctional communication is achieved by increased levels of connexin 43 mRNA and protein. *Mol Carcinogen* 3: 335–343

21 Kinzler KW, Vogelstein B (1996) Lessons from hereditary colorectal cancer. *Cell* 87: 159–170

22 Ames BN, Shigenaga MK, Gold LS (1993) DNA lesions, inducible DNA repair, and cell division: Three key factors in mutagenesis and carcinogenesis. *Environ Health Perspect* 101 Suppl. 5: 35–44

23 Lu B, Rothnagel JA, Longley MA, Tsai SY, Roop DR (1994) Differentiation-specific expression of human keratin 1 is mediated by a composite AP-1/steroid hormone element. *J Biol Chem* 269: 7443–7449

24 DiSepio D, Jones A, Longley MA, Bundman D, Rothnagel JA, Roop DR (1995) The proximal promoter of the mouse loricrin gene contains a functional AP-1 element and directs keratinocyte-specific but not differentiation-specific expression. *J Biol Chem* 18: 10 792–10 799

25 Lefebvre DL, Piersanti M, Bai XH, Chen ZQ, Lye SJ (1995) Myometrial transcriptional regulation of the gap junction gene, Connexin-43. *Reprod Fert Develop* 7: 603–611

26 Geimonen E, Jiang W, Ali M, Fishman GI, Garfield RE, Andersen J (1996) Activation of protein kinase C in human uterine smooth muscle induces *connexin*-43 gene transcription through an AP-1 site in the promoter sequence. *J Biol Chem* 271: 23 667–23 674

27 Milks LC, Kumar NM, Houghten R, Unwin N, Gilula NB (1988) Topology of the 32-kd liver gap junction protein determined by site-directed antibody localizations. *EMBO J* 7: 2967–2975

28 Mehta PP, Bertram JS, Loewenstein WR (1989) The actions of retinoids on cellular growth correlate with their actions on gap junctional communication. *J Cell Biol* 108: 1053–1065

Effects of all-*trans* retinoic acid and 13-*cis* retinoic acid on breast cancer cell lines

S. Toma[1,2,3], P. Raffo[2] and L. Isnardi[3]

[1] Department of Oncology, Biology and Genetics, National Institute for Cancer Research, Largo Rosanna Benzi, 10 I-16132 Genoa, Italy
[2] National Institute for Cancer Research, Largo Rosanna Benzi, 10, I-16132 Genoa, Italy
[3] Laboratory of Pre-Clinical Oncology, Advanced Biotechnology Center, Largo Rosanna Benzi 10, I-16132 Genoa, Italy

Introduction

For 20 years, it has been known that the retinoic acids inhibit mammary cell carcinoma growth *in vitro* with a variable activity; some cell lines being sensitive and others almost unresponsive to these compounds. In particular, the sensitivity to the retinoic acid growth-inhibitory effect is typically related to the presence of estrogen receptors (ERs) in the evaluated cell lines; the effect of retinoic acid being less, or not at all, evident in estrogen receptor negative cell lines ([1–14]; see also Tab. 1).

Estrogen and progesterone receptors

Based on these observations, a series of studies have evaluated the relationship between the estrogen and/or progesterone receptor-status and the retinoid activities. The first observations indicated the ability of the retinoic acid to influence the levels of progesterone receptor (PgR) [15], in terms of a decrease in PgR mRNA and protein in ER+ breast cancer cell lines, probably through an inhibition of PR gene transcription, concomitant of an impaired progestin responsiveness [16].

Further studies have since then evaluated the effects of retinoic acid (RA) on estrogen receptor (ER) expression, obtaining contrasting results: several authors have evidenced the ability of retinoids to downregulate, even if only for a short time-course, ER mRNA and protein in a ER+ breast cancer cell line [17]; other studies, probably due tue some differences in biological properties and karyotype of the cell lines used in the different laboratories, have failed to evidence this regulation of ER after a longer treatment [18, 19]. These studies confirmed the ability of retinoids to inhibit estrogen-stimulated gene expression, probably through a direct or indirect interference with the ER at the estrogen response element (ERE) level [19–21]. At the same time, the importance of ER basal levels in determining retinoic acid responsiveness was confirmed by the fact that the transfection of ER in ER− breast cancer cell lines resulted in a significant growth inhibition by retinoids in the same cells [17]. The hypothesis that the selected

Table 1. Antiproliferative effect of retinoids on different breast cancer cell lines *in vitro*.

Retinoid	Cell line	Effect on proliferation	Author
all-trans retinoic acid (ATRA)	MCF-7	reduced	Ueda, 1980 [4]
Retinol	MDA.MB.231	low inhibition	Fraker, 1984 [5]
ATRA	Hs578T BT20 734B	low inhibition low inhibition low inhibition	Marth, 1984 [7]
HPR ATRA	ZR 75.1 734B	reduced reduced	Marth, 1985 [8]
ATRA 13cRA Etretinate Arotinoids	T47D	reduced reduced reduced reduced	Wetherall, 1986 [10]
ATRA	MCF-7 ZR 75.1 T47D MDA.MB.231 Hs578T	reduced[1] reduced[1] reduced[1] unaltered unaltered	Fontana, 1987 [11]
ATRA	MCF-7 MDA.MB.231	reduced unaltered	Fontana, 1987 [11]
ATRA	MCF-7 ZR 75.1 MDA.MB.231	reduced reduced unaltered	Fontana, 1990 [13]
ATRA	T47D	reduced[2]	Koga, 1991 [14]

[1] with Tamoxifen (TAM) added
[2] synergy with vitD3

ER[+] breast cancer cell lines may retain the ability to be modulated by retinoids, not only in terms of a reduced expression of estrogen-regulated genes, but also of a direct regulation of the ER itself, represents an important research field requiring further evaluation in order to better explain the biological interactions between the retinoids and the steroid-hormone patterns of activities.

Retinoid receptors

In successive studies, several authors have examined another aspect of the matter: the role of the retinoic acid receptors, the most important mediators of the retinoid activity, and the ability of the estrogens to influence the levels of these receptors. The first evidence observed in one of these studies was that the expression of the basal mRNA levels of retinoic acid receptors (RARs) was different in the various evaluated breast cancer cell lines: in particular, RARα was expressed in all the examined cell lines, but its expression was higher in ER$^+$ cells; RARγ was expressed in all the cell lines, without any difference between ER$^+$ and ER$^-$ cells; RAR β expression was, interestingly, different in the various cell lines, and more prominent in ER$^-$ cells [22]. The treatment with all-*trans*-retinoic acid in these studies failed to induce significant modulations of RARs, except for an increase in RARβ in T47D cells, the same cell line in which retinoids have been previously shown to be more active in terms of an influence on the estrogen-action pathways. Interestingly, progestins were able to decrease RARα expression, while 17-β-estradiol caused a slight increase of the same receptor, indicating the existence of a mutual interaction between the pathways of steroid hormones and the retinoids in breast cancer cells [22]. The importance, suggested by these authors, of RARα basal levels in explaining the retinoic acid responsiveness of ER$^+$ cell lines, is confirmed also by other studies [23] showing that RARα basal levels are higher in ER$^+$ breast tumors; in the same study, Roman et al. also confirm the ability of estradiol to increase, and of progestins to decrease, RARα gene expression, and suggest that the higher RARα levels in ER$^+$ breast cancer cell lines and tumors occur due to an estradiol augmentation of RARα gene expression, and that the induction caused by the estrogens normally present in serum is sufficient to explain the observed *in vitro* effects.

The central role of RARα in mediating the antiproliferative activity of retinoids in breast cancer cell lines is confirmed by transfection experiments [24] showing that ER$^-$ breast cancer cell lines, transfected with ER, show increased levels of RARα and an augmented sensitivity to the antiproliferative effects of retinoic acid. A further observation is that ER$^-$ breast cancer cells become sensitive to the retinoic acid growth inhibitory effect when stably transfected with high levels of RARα cDNA [25].

The difference in RARα basal expression levels between the different breast cancer cell lines correlated well with the differences in sensitivity to retinoids, whether or not the cells expressed ER [26, 27], and the molecules with specific affinities for RARα were the most effective antiproliferative agents [28].

With this in mind, the recent interest has been focused on RARβ, starting from the observed differences of RARβ basal levels in the different breast cancer cell lines and breast cancer specimens. In particular, Deng [29] has demonstrated the loss of heterozygosity on chromosome 3p24 in breast cancer specimens and the morphologically normal appearance of the adjacent tissue. Interestingly, this locus includes the region coding for RARβ. Successive studies [30] have confirmed that this receptor is almost completely absent in the human breast neoplastic tissue and also in the morphologically "normal appearing" tissue adjacent to the tumor, suggesting that RARβ is a recessive gene, which is relevant to malignancy and useful for defining a predisposed region from which the cancer arises. Other *in vivo* clinical studies [31] have also demonstrated that the level of expression of RARβ progressively decreases from ductal carcinoma *in situ* to invasive cancer, further indicating a role of the receptor in breast carcinogenesis. Interestingly, RARβ expression increases in senescing normal breast cells but not in established ER$^+$ tumor cell lines [32]. With regard to the biological mechanisms of this reduced expression, some loss of transcription factors is probably involved in the phenomenon, since the authors failed to observe any evident mutation in the β-retinoic acid response element (β-RARE) promoter. In established breast cancer cell lines the relationship between RARβ basal levels and sensitivity to retinoic acid is not simple; classical studies have demonstrated a loss of RARβ in a number of breast cancer cell lines [22], and these authors state that in the majority of cases the ER⁻ cell lines expressed the receptor, while most of the ER⁺ cells failed to do so. However, other studies failed to show significant levels of the same receptor in several ER⁻ cell lines [32]. A possible explanation may come from the existence of mutated RARβ forms in ER⁻ cells, that may or not be recognized, depending on the detection methods used. These observations need to be further evaluated, and confronted with other cellular systems, e.g. head and neck cancer cell lines, where an inverse relationship has already been demonstrated between RARβ levels and differentiation markers.

In some ER⁺ breast cancer cell lines, the treatment with retinoic acid caused an increase from no expression at all to detectable levels of RARβ [15, 16, 22], concomitant of a modulation of the differentiative status of the same cells. ER⁺ breast cancer cell lines seemed to be the only cells able to respond to retinoic acid with an induction of RARβ, and this seemed to be one of the reasons for the observed differences between ER⁺ and ER⁻ cells in terms of retinoid responsiveness [33]. The induction of RARβ by retinoids correlates in some cases with the activity of the compounds; moreover, the transfection of a functional RARβ in ER⁻ retinoid-resistant breast cancer cell lines rendered them sensitive to retinoids, while, in contrast, ER⁺ retinoid-responsive mammary carcinoma cells transfected with RARβ antisense RNA became retinoid-resistant. The observation, made also by other authors, of an increased sensitivity to RA of both ER⁺ and ER⁻ breast cancer cell lines when transfected with a RARβ expressing vector [35], confirms the central role of this receptor in mediating the RA effects in this system. Correlating with the previously presented observations, the introduction of RARα in ER⁻ cells also restored RA sensitivity, but through the induction of endogenous RARβ [34]. We can therefore conclude that the induction of RARβ by RA in some cases correlates with an activity of retinoids, and that the autoregulation of the RARβ gene presumably plays a critical role in amplifying the RA response.

To further complicate this biological-activity scheme, as well as the central importance observed for RARα and RARβ in mediating RA activity in breast cancer, recent studies have demonstrated the relevance of the third member of the RAR family, RARγ. This receptor also seems to play a role in controlling the growth of breast cancer cells [36], being elevated by RA treatment in several ER⁺ and ER⁻ cells, concomitant of retinoid responsiveness [37]. These data are also confirmed by the retained biological activity of RARγ-selective retinoids in comparison with RA, as observed by successive studies [38].

The demonstrated auxiliary role exerted by retinoid x receptors (RXRs) [39] in mediating the retinoid effect on breast cancer cell lines is not of great relevance when we analyze RA activity because of the low binding and transactivating activity of this retinoid for this class of receptors. These observations are important only when considering reversal of RA resistance through the use of RXR-selective compounds [40].

In parallel with the studies aimed at the identification of the primary biological response genes to RA and to the evaluation of the role of retinoic acid receptors in this context, other researchers have evaluated the different final macroscopic responses to RA of breast cancer cell lines. These latter studies have evaluated the three possible terminal phenomena that can occur, separately or in association, in RA-sensitive cells:

– inhibition of cell growth;
– induction of programmed cell death (apoptosis) [41];
– modulation of cell differentiation.

Which of these three mechanisms takes place because of the retinoid activity depends fundamentally on the basic biological characteristics of the treated cells.

In normal human mammary epithelial cells, which characteristically express RARβ and increase its levels when senescing [32] and after RA treatment [42], RA exerts essentially an antiproliferative effect through G1 arrest, indicating in this simple growth inhibition a mechanism sufficient to prevent the malignant transformation of these cells [42].

Malignant ER$^+$ breast cancer cells, which generally fail to express RARβ but retain high levels of RARα, are probably unable to respond to retinoids like the normal mammary epithelial cells only in terms of a proliferation inhibition. Therefore other mechanisms have to take place.

More differentiated and slowly proliferating cells, like T47D and ZR 75.1, concomitant of a RARβ induction, undergo a growth inhibition [43], generally coupled with a modulation of the cellular differentiation in terms of a reduction of the overexpressed ER and/or PgR [15, 16, 44]. The evaluation of an apoptotic effect in these cells leads to contrasting results [34, 44] (see also Fig. 1).

Other ER$^+$ breast cancer cell lines, rapidly proliferating even though retaining a differentiated phenotype, like MCF-7 cells, respond to all-*trans*- and 13-*cis*-retinoic acid (and, to a lesser extent, also to other all-*trans* retinoic acid [ATRA] metabolites) not only in terms of proliferation inhibition [45] with G1 arrest [46, 47] but also in terms of apoptosis induction [44, 47–49] (see also Fig. 1 and 2). In order to explain the mechanism of apoptosis induction by retinoids in MCF-7 a role for RARβ was suggested, but contrasting data are presented on the inducibility of the receptor after treatment of these cells [34, 35]. With regard to the molecular biology of apoptosis induction, however, we already know from our cellular model that other molecules are involved in this active mechanism: breast cancer apoptosis induced by retinoids is clearly p53-independent and passes through a down-regulation of bcl-2 protein [44]. In contrast, only limited and unconfirmed observations were presented on the ability of retinoids to modulate differentiation (e.g. to reduce ER levels) in these cells [17].

Finally, when we evaluate ER$^-$ cells, RA and 13-*cis* retinoic acid (13cRA) seem to be generally less effective, except when used at very high concentrations, than in ER$^+$ cells, both in terms of proliferation inhibition and apoptosis induction, and modulation of cellular differentiation fails to occur under any experimental condition (see also Figs. 1 and 2). Probably one or more alterations in the complex pathway of retinoid action occurs in these cells, and these anomalies make it difficult for the retinoids to exert their effects (e.g. low levels of RARα, low levels or altered forms of RARβ).

Combined treatment with RA and other drugs

Another interesting and current research field on the *in vitro* activity of retinoids on breast cancer cell lines focuses on the evaluation of their association with hormones, biological response modifiers or chemotherapeutic drugs, in order to increase their antineoplastic activity.

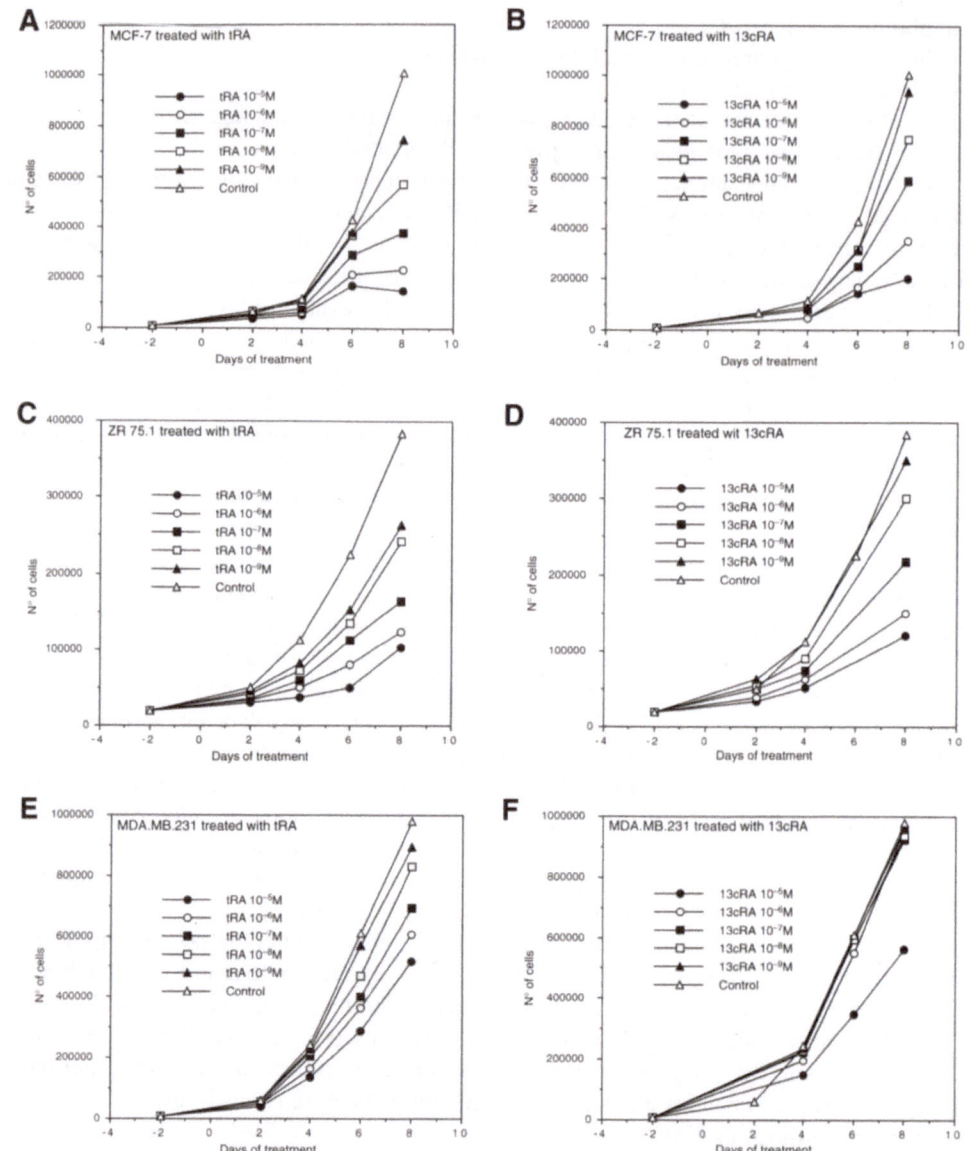

Figure 1. Growth curves of ER$^+$ MCF-7 (A, B) and ZR 75.1 (C,D), and ER$^-$ MDA.MB.231 (E,F) cells during continuous exposure to 10^{-5} M to 10^{-9} M of ATRA or 13cRA, for a period of eight days. Results are presented as the mean of three separate experiments.

Among the first class of compounds, classical studies have demonstrated the potentiating effect of a combined treatment with Tamoxifen and retinoids in breast cancer cells [10, 11, 14],

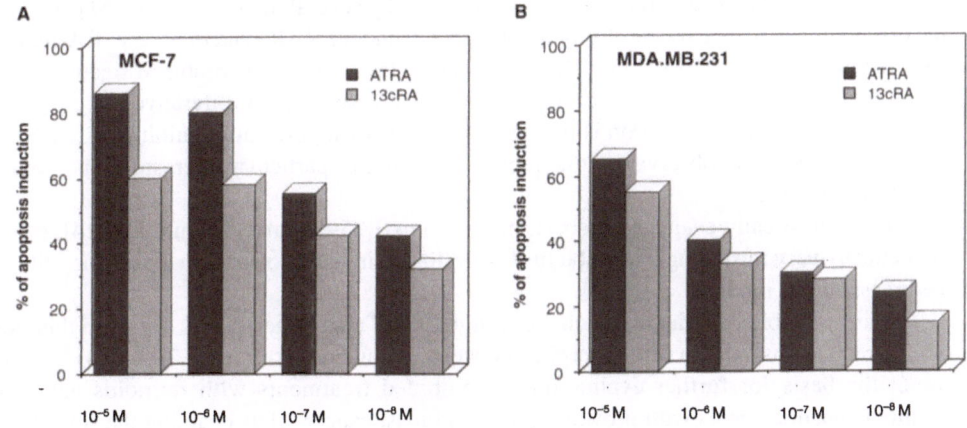

Figure 2. Dose-dependent induction of apoptosis in MCF-7 (A) and MDA.MB.231 (B) cells after six days of treatment with ATRA or 13cRA. Results represent the mean of three independent experiments.

and our group has also confirmed an additive effect of this combination in ER[+] cells [50] (Fig. 3A).

With regard to the biological response modifiers, a synergism exists between retinoids and interferons, already demonstrated *in vitro* for several neoplasms, comprising ER[+] breast cancer

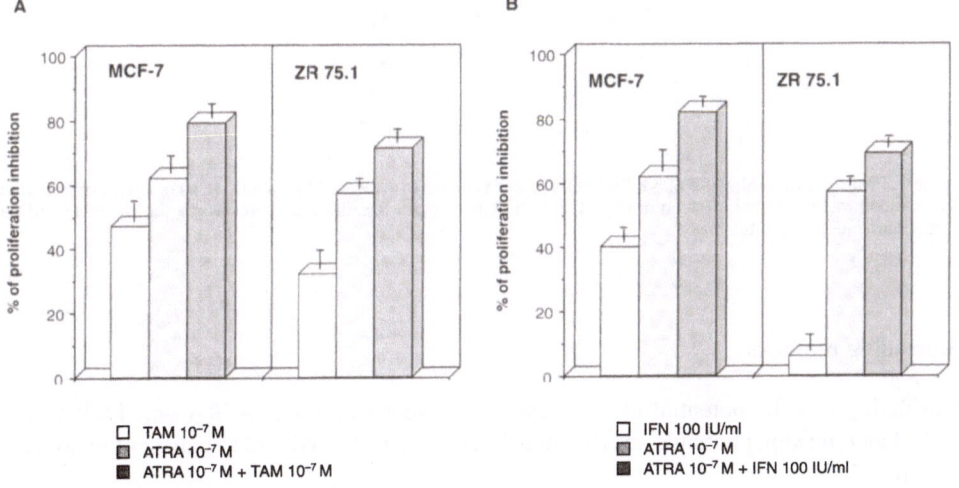

Figure 3. A. Proliferation inhibition of MCF-7 and ZR 75.1 cell lines after exposure to ATRA, TAM or their combination for eight days. B. Proliferation inhibition of the same cells after exposure to ATRA, IFN or their combination for eight days. Results are presented as the mean +/– SD of three distinct experiments.

cells. This was confirmed in this latter cellular model by several authors [38, 50, 51] (Fig. 3B) and essentially explained by the ability of interferon to induce RARγ and to prevent RA-induced downregulation of CRABPII [36]. The addition of interferon to the combined treatment with RA + Tamoxifen is also able to produce a further increase in antiproliferative effects [50] in ER+ cells. Noteworthy, interferon is the only agent able to significantly inhibit proliferation of ER− cell lines [51], an observation that renders this drug of particular interest in this group of breast cancer cells.

Another classically evaluated association, RA plus 1,25-dhydroxyvitamin D3 [14], even if theoretically very intriguing, has to be further evaluated in order to confirm a possible utility in the breast cancer model.

The last possible combination is that of retinoids and chemotherapeutic drugs: in this field, our interesting observation of a synergism between RA and doxorubicin [52] (Fig. 4) may represent the basis for further evaluation of combined treatments with retinoids and other chemotherapeutic agents with proven activity on breast cancer. Thus enabling the selection of a better association for use in a specific cellular model of mammary carcinoma.

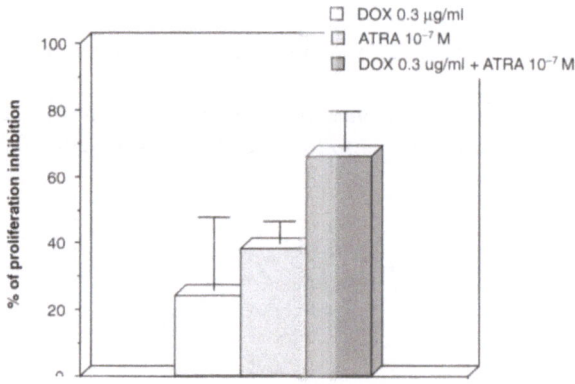

Figure 4. Proliferation inhibition of MCF-7 cells after exposure to ATRA (72 h), DOX (1 h) or their combination. Cell number was evaluated after eight days from DOX treatment. Results are expressed as the mean +/− SD of three separate experiments.

Concluding remarks

Considering that the potential of the classic pan-agonist retinoids ATRA and 13cRA have reached their maximal possible activity, future perspectives for retinoid treatment of breast cancer will include:

– a better evaluation of biological parameters potentially predictive of retinoid responsiveness in this specific cellular model, to be used to select cellular subpopulations potentially more responsive to the treatment;

- the study of the *in vitro* biological activity of new synthetic retinoids, with selective binding and transactivation activity for a specific RAR and/or RXR (selective retinoids), to increase the antiproliferative, apoptotic and/or differentiating activities in specific breast cancer cell lines [53], and to reduce the *in vivo* toxicity. In this field, the study of compounds able to bind specific receptors but unable to induce transcriptional activation (antagonist retinoids) will be particularly interesting. These molecules, in fact, retain antineoplastic activity in breast cancer cell lines through interference with other transcriptional factors like AP-1, but are gravated by fewer side-effects [54].
- the more extensive evaluation of combined treatments with retinoids and other compounds with different molecular mechanisms of action, especially those that display a significative activity in retinoid-resistant breast cancer cell lines, in order to retain an antineoplastic effect over the broad range of mammary carcinoma subpopulations (ER⁻, ER⁺, more or less differentiated). In this field, interest should be focused on interferons and classic chemotherapeutic agents, e.g. Doxorubicin, that have already demonstrated an activity as single agents and, in combination with the classic retinoids, in breast cancer cells.

References

1   Lotan R, Nicolson GL (1977) Inhibitory effects of retinoic acid or retinyl acetate on the growth of untransformed, transformed and tumor cells *in vitro*. *J Nat Cancer Inst* 59: 1717–1722
2   Lotan R (1979) Different susceptibilities of human melanoma and breast carcinoma cell lines to retinoic acid-induced growth inhibition. *Cancer Res* 39: 1014–1019
3   Lacroix A, Lippman ME (1979) Binding of retinoids to human breast cancer cell lines and their effects on cell growth. *J Clin Invest* 65: 586–591
4   Ueda H, Takenawa T, Millian JC, Gesell MS, Brandes D (1980) The effects of retinoids on proliferation capacities and macromolecular synthesis in human breast cancer MCF-7 cells. *Cancer Res* 46: 2203–2209
5   Fraker LD, Halter SA, Forbes JT (1984) Growth inhibition by retinol of a human breast cancer cell line *in vitro* and in athymic mice. *Cancer Res* 44: 5757–5763
6   Taylor CM, Blanchard B, Wetherall NT (1984) The antiestrogen effects of synthetic retinoids and arotinoids on ovarian and breast cancer cell lines. *Proc Amer Assoc Cancer Res* 25: 360
7   Marth C, Bock G, Daxenbichler G (1984) Effect of retinoic acid and 4-hydroxy-tamoxifen on human breast cancer cell lines. *Biochem Pharmacol* 33: 2217–2221
8   Marth C, Bock G, Daxenbichler G (1985) Effects of 4-hydroxy-phenylretinamide and retinoic acid on proliferation and cell cycle of cultured human breast cancer cells. *J Nat Cancer Inst* 75: 871–875
9   Marth C, Daxenbichler G, Dapunt O (1986) Synergistic antiproliferative effect of human recombinant interferons and retinoic acid in cultured breast cancer cells. *J Nat Cancer Inst* 77: 1197–1202
10 Wetherall NT, Taylor CM (1986) The effects of retinoid treatment and antiestrogens on the growth of T47D human breast cancer cells. *Eur J Cancer Clin Oncol* 22: 53–59
11 Fontana JA (1987) Interaction of retinoids and tamoxifen on the inhibition of human mammary carcinoma cell proliferation. *Exp Cell Biol* 55: 136–144
12 Fontana JA, Miksis G, Miranda DM, Durham JP (1987) Inhibition of human mammary carcinoma cell proliferation by retinoids and intracellular cAMP-elevating compounds. *J Nat Cancer Inst* 78: 1107–1112
13 Fontana JA, Mezu AB, Cooper BN, Miranda D (1990) Retinoid modulation of estradiol-stimulated growth and of protein synthesis and secretion in human breast carcinoma cells. *Cancer Res* 50: 1997–2002
14 Koga M, Sutherland RL (1991) retinoic acid acts synergistically with 1,25-dihydroxyvitamin D3 or antiestrogens to inhibit T47D human breast cancer proliferation. *J Steroid Biochem Mol Biol* 39: 455–460
15 Clarke CL, Roman SD, Graham J, Koga M, Sutherland RL (1990) Progesterone receptor regulation by retinoic acid in the human breast cancer cell line T47D. *J Biol Chem* 265(21): 12 694–12 700
16 Clarke CL, Graham J, Roman SD, Sutherland RL (1991) Direct transcriptional regulation of the progesterone receptor by retinoic acid diminishes progestin responsiveness in the breast cancer cell line T47D. *J Biol Chem* 266 (28): 18 969–18 979
17 Rubin M, Fenig E, Rosenauer A, Menendez-Botet C, Achkar C, Bentel JM, Yahalom J, Mendelsohn J, Miller WHJr (1994) 9-*cis*-retinoic acid inhibits growth of breast cancer cells and own-regulates estrogen receptor RNA and protein. *Cancer Res* 54: 6549–6556

18 Guilbaud NF, Gas N, Dupont MA, Valette A (1990) Effects of differentiation-inducing agents on maturation of human MCF-7 breast cancer cells. *J Cell Physiol* 145: 162–172

19 Fontana JA, Nervi C, Shao Z-M, Jetten AM (1992) Retinoid antagonism of estrogen-responsive transforming growth factor a and pS2 gene expression in breast carcinoma cells. *Cancer Res* 52: 3938–3945

20 Demirpence E, Pons M, Balaguer P, Gagne D (1992) Study of an antiestrogenic effect of retinoic acid in MCF-7 cells. *Biochem Biophys Res Commun* 183: 100–106

21 Demirpence E, Balaguer P, Trousse F, Nicolas J-C, Pons M, Gagne D (1994) Antiestrogenic effects of all-*trans*-retinoic acid and 1,25-dihydroxyvitamin D3 in breast cancer cells occur at the estrogen response element level but through different molecular mechanisms. *Cancer Res* 54: 1458–1464

22 Roman SD, Clarke CL, Hall RE, Alexander IE, Sutherland RL (1992) Expression and regulation of retinoic acid receptors in human breast cancer cells. *Cancer Res* 52: 2236–2242

23 Roman SD, Ormandy CJ, Manning DL, Blamey RW, Nicholson RI, Sutherland RL, Clarke CL (1993) Estradiol induction of retinoic acid receptors in human breast cancer cells. *Cancer Res* 55: 5940–5945

24 Sheikh MS, Shao Z-M, Chen J-C, Hussein A, Yetten AM, Fontana JA (1993) Estrogen receptor-negative breast cancer cells transfected with estrogen receptor exhibit increased RAR a gene expression and sensitivity to growth inhibition by retinoic acid. *J Cell Biochem* 53: 394–404

25 Sheikh MS, Shao Z-M, Li X-S, Dawson M, Yetten AM, Wu S, Conley BA, Garcia M, Rochefort H, Fontana JA (1994) Retinoid-resistant estrogen receptor-negative human breast carcinoma cells transfected with retinoic acid receptor a acquire sensitivity to growth inhibition by retinoids. *J Biol Chem* 269(34): 21 440–21 447

26 Rishi AK, Gerald TM, Shao Z-M, Li X-S, Baumann RG, Dawson MI, Fontana JA (1996) Regulation of the human retinoic acid receptor a gene in the estrogen receptor negative human breast carcinoma cell lines SKBR-3 and MDA.MB.231. *Cancer Res* 56: 5246–5252

27 Fitzgerald P, Teng M, Chandraratna RAS, Heyman RA, Allegretto EA (1997) retinoic acid receptor a expression correlates with retinoid-induced growth inhibition of human breast cancer cells regardless of estrogen receptor status. *Cancer Res* 57: 2642–2650

28 Dawson MI, Chao W-R, Pine P, Jong L, Hobbs PD, Ridd CK, Quick TC, Niles RM, Zhang X-K, Lombardo A et al (1995) Correlation of retinoid binding affinity to retinoic acid receptor a with retinoid inhibition of growth of estrogen receptor-positive MCF-7 mammary carcinoma cells. *Cancer Res* 55: 4446–4451

29 Deng G, Lu Y, Zlotnikov G, Thor AD, Smith HS (1996) Loss of heterozygosity in normal tissue adjacent to breast carcinomas. *Science* (Washington DC) 274: 2057–2059

30 Widschwendter M, Berger J, Daxenbichler G, Müller-Holzner E, Widschwendter A, Mayr A, Marth C, Zeimet AG (1997) Loss of retinoic acid receptor b expression in breast cancer and morphologically normal adjacent tissue but not in the normal breast tissue distant from the cancer. *Cancer Res* 57: 4158–4161

31 Xu XC, Sneige N, Liu X, Nandagiri R, Lee JJ, Lukmanji F, Hortobagyi G, Lippman SM, Dhingra K, Lotan R (1997) Progressive decrease in nuclear retinoic acid receptor b messenger RNA level during breast carcinogenesis. *Cancer Res* 57: 4992–4996

32 Swisshelm K, Ryan K, Lee X, Tsou HC, Beacocke M, Sager R (1994) Down-regulation of retinoic acid receptor b in mammary carcinoma cell lines and its up-regulation in senescing normal mammary epithelial cells. *Cell Growth Differ* 5: 133–141

33 Zhang KK, Liu Y, Lee MO (1996) Retinoid receptors in human lung cancer and breast cancer. *Mutat Res* 350: 267–277

34 Liu YI, Lee MO, Wang HG, Li Y, Hashimoto Y, Klaus M, Reed JC, Zhang XK (1996) Retinoic acid receptor b mediates the growth inhibitory effect of retinoic acid by promoting apoptosis in human breast cancer cells. *Mol Cell Biol* 16(3): 1138–1149

35 Seewaldt VL, Johnson BS, Parker MB, Collins SJ, Swisshelm K (1995) Expression of retinoic acid receptor b mediats retinoic acid-induced growth arrest and apoptosis in breast cancer cells. *Cell Growth Differ* 6: 1077–1088

36 Widschwendter M, daxenbichler G, Dapunt O, Marth C (1995) Effects of retinoic acid and g-interferon on expression of retinoic acid receptor and cellular retinoic acid-binding protein in breast cancer cells. *Cancer Res* 55: 2135–2139

37 Fanjul AN, Bouterfa H, Dawson M, Pfahl M (1996) Potential role for retinoic acid receptor g in the inhibition of breast cancer cells by selective retinoids and interferons. *Cancer Res* 56: 1571–1577

38 Widschwendter M, Daxenbichler G, Culig Z, Michel S, Zeimat AG, Mortl MG, Widschwendter A, Marth C (1997) Activity of retinoic acid receptor-γ selectively binding retinoids alone and in combination with interferon-γ in breast cancer cell lines. *Int J Cancer* 71(3): 497–504

39 Shutt A, Rosenberger M, Covey A, Apfel C, Hsu MC (1996) Receptor specificity in the antiproliferative response of breast carcinoma cell lines to retinoids. *Proc Am Soc Clin Oncol* 37: A1577

40 Wu Q, Dawson MI, Zhang X, Hobbs PD, Agadir A, Jong L, Li Y, Liu R, Lin B, Zhang XK (1997) Inhibition of *trans*-retinoic acid-resistant human breast cancer cell growth by retinoid X receptor-selective retinoids. *Mol Cell Biol* 17(11): 6598–6608

41 Wyllie AH (1993) Apoptosis. *Brit J Cancer* 67: 205–208

42 Seewald VL, Kim JH, Caldwell LE, Johnson BS, Swisshelm K, Collins SJ (1997) All-*trans*-retinoic acid mediates G1 arrest but not apoptosis in normal human mammary epithelial cells. *Cell Growth Differ* 8: 631–641

43 Wilcken NRC, Scarcenc B, Musgrove EA, Sutherland RL (1996) Differential effects of retinoids and anti-

estrogens on cell cycle progression and cell cycle regulatory genes in human breast cancer cells. *Cell Growth Differ* 7: 65–74

44 Toma S, Isnardi L, Raffo P, Dastoli G, De Francisci E, Riccardi L, Palumbo R, Bollag W (1997) Effects of all-*trans*-retinoic acid and 13-*cis*-retinoic acid on breast cancer cell lines: growth inhibition and apoptosis induction. *Int J Cancer* 70(5): 619–627

45 Van Heusden J, Wouters W, Ramaekers FCS, Krekels MDWG, Dillen L, Borgers M, Smets G (1998) All-*trans*-retinoic acid metabolites inhibit the proliferation of MCF-7 human breast cancer cells *in vitro*. *Brit J Cancer* 77(1): 26–32

46 Zhu WY, Jones CS, Kiss A, Matsukuma K, Amin S, De Luca LM (1997) retinoic acid inhibition of cell cycle progression in MCF-7 human breast cancer cells. *Exp Cell Res* 234(2): 293–299

47 Mangiarotti R, Danova M, Alberici R, Pellicciari C (1998) All-*trans* retinoic acid (ATRA)-induced apoptosis is preceded by G1 arrest in human MCF-7 breast cancer cells. *Brit J Cancer* 77(2): 186–191

48 Zhi-Ming Shao Dawson MI, Xiao Su Li Rishi AK, Sheikh MS, Qi-Xia Han Ordonez JV, Shroot B, Fontana JA (1995) p53 independent G0/G1 arrest and apoptosis induced by a novel retinoid in human breast cancer cells. *Oncogene* 11: 493–504

49 Bollag W, Isnardi L, Jablonska S, Klaus M, Majewski S, Pirson W, Toma S (1997) Links between pharmacological properties of retinoids and nuclear retinoid receptors. *Int J Cancer* 70: 470–472

50 Toma S, Raffo P, Isnardi L, Riccardi L, De Francisci E, Maselli G, Dastoli G, Palumbo R (1997) Associations of retinoids, tamoxifen and a-interferon 2a in human breast cancer: *in vitro* and *in vivo* studies. *Int J Oncol* 10: 597–607

51 Toma S, Monteghirfo S, Tasso P, Nicolò G, Spadini N, Palumbo R, Molina F (1994) Antiproliferative and synergistic effect of interferon α-2a, retinoids and their association in established human cancer cell lines. *Cancer Lett* 82: 209–216

52 Toma S, Maselli G, Dastoli G, De Francisci E, Raffo P (1997) Synergistic effect between doxorubicin and a low dose of all-*trans*-retinoic acid in MCF-7 breast cancer cell line. *Cancer Lett* 116: 103–110

53 Toma S, Isnardi L, Raffo P, Riccardi L, Bollag W (1998) Induction of apoptosis in MCF-7 carcinoma cell line by RAR and RXR selective retinoids. *Anticancer Res* 18: 935–942

54 Toma S, Isnardi L, Raffo P, Riccardi L, Dastoli G, Apfel C, LeMotte P, Bollag W (1998) RARα antagonist RO 41-5253 inhibits proliferation and induces apoptosis in breast cancer cell lines. *Int J Cancer* 78: 86–94

Vitamin A and retinoids: an update of biological aspects and clinical applications
M.A. Livrea (ed.)
© 2000 Birkhäuser Verlag Basel/Switzerland

Retinoids and interferons: combination studies in human cancer

R. Lotan[1], J.L. Clifford[2] and S.M. Lippman[2]

[1] *Department of Thoracic/Head and Neck Medical Oncology, University of Texas, M.D. Anderson Cancer Center, Houston, TX 77030, USA*
[2] *Department of Clinical Cancer Prevention, University of Texas, M.D. Anderson Cancer Center, Houston, TX 77030, USA*

Introduction

The clinical use of biological response modifiers (BRMs) that augment host antitumor immune responses and of cytostatic agents that stimulate tumor cell differentiation has gained interest in recent years as an alternative or complementary approach to the therapy of certain types of cancer. Interferons (IFNs) are the most intensively studied BRMs and retinoids (vitamin A analogs) are the most recognized differentiation-inducing agents. IFNs and retinoids share similar antitumor activities *in vitro* and *in vivo*. Their use as single agents in clinical trials resulted in limited success in some types of cancer. *In vitro* studies have shown additive or synergistic growth inhibitory and differentiation-inducing effects of the combination of interferons and retinoids. Recent clinical trials have highlighted the efficacy of the combination of IFNs and retinoic acid (RA) in the treatment of several types of cancer. This chapter summarizes reports on the clinical antitumor activity of the combination of IFNs and retinoids, outlines possible mechanisms of interaction between these two agents based on their putative mechanisms of action, and highlights the potential application of this combination in cancer therapy.

Retinoids, interferons and cancer

Retinoids are a group of compounds, which include natural metabolites of vitamin A (retinol) as well as a variety of synthetic derivatives and analogs. Retinoids have been shown to modulate various fundamental cellular processes including proliferation, differentiation, apoptosis, and malignant transformation [1, 2]. Retinoids inhibit the proliferation and colony formation of various transformed cell lines and induce the differentiation of malignant cells including promyelocytic and premonocytic leukemias, embryonal carcinoma, neuroblastoma, melanoma, rhabdomyosarcoma, and breast carcinoma. Because they can restore regulation of differentiation and growth in certain premalignant and malignant cells *in vitro* and *in vivo*, certain retinoids are under investigation as therapeutic and chemopreventive agents for a variety of cancers [3, 4].

Retinoids are effective in the suppression of premalignant lesions including keratoacanthoma, oral leukoplakia, laryngeal dysplasia, bronchial metaplasia and dysplasia, and cervical dysplasia [4]. In addition, retinoids are effective in reducing the incidence of second primary tumors in the aerodigestive tract [4].

Laboratory and clinical studies also indicate that retinoids have a potential as therapeutic agents. Significant therapeutic activity has been observed with all-*trans*-retinoic acid in acute promyelocytic leukemia [5]. Retinoic acid was effective in the treatment of basal and squamous cell carcinoma (SCC) of the skin in several studies, including a study with *Xeroderma pigmentosum* patients, and in cutaneous T cell lymphoma (CTCL) [3, 6]. A number of phase II trials of retinoids are currently under way in a range of solid tumors including cancers of the skin, head and neck, lung, cervix, breast, bladder and prostate as well as neuroblastoma. Unfortunately, few responses were observed among patients with established solid tumors after treatment with certain retinoids as single agents [6].

IFNs were originally discovered as mediators of an antiviral state in virus-infected cells. They are cytokines that act as potent BRMs. In humans there are at least 26 IFN genes, which are classified according to the cells that produce them. Thus, IFN-α is leukocyte IFN, IFN-β is fibroblast IFN, and IFN-γ is immune IFN [7, 8]. It is now well established that besides their effects on the immune response, IFNs also regulate proliferation, differentiation and apoptosis [8, 9]. IFNs inhibit *in vitro* proliferation and colony formation in agar of malignant cells including myeloma, Burkitt's lymphoma, erythroleukemia, breast carcinoma, neuroblastoma, and lung cancer. In a few cases, IFNs also induced cell differentiation (e.g. neuroblastoma). Clinical trials demonstrated the efficacy of IFNs in patients with hairy cell leukemia, chronic myelogenous leukemia, Kaposi sarcoma, CTCL, multiple myeloma, melanoma, skin carcinoma, and renal carcinoma.

Effects of retinoids in combination with IFNs

Retinoids combined with IFNs exhibited encouraging clinical activity in untreated locally advanced cervical squamous cell carcinoma (SCC) [9–12], cutaneous SCC [13, 14], and renal cell carcinoma [15]. However, negative results were observed in patients with advanced malignant melanoma, head and neck and lung squamous cell carcinoma [6].

Skin cancer

In the first phase II trial of the combination of agents in advanced SCC of the skin, 28 evaluable patients received 13-*cis*-RA (1 mg/kg/d) plus IFN-α2a (3 million units/d) and their overall major response rate was 68% with a median response duration of approximately five months [13]. This trial also demonstrated a significant inverse correlation between response and disease extent; a significantly higher response rate in chemotherapy-naive patients; and considerable toxicity (grade 3 or greater fatigue occurring in 12 patients, possibly due to study population age—median 67 years). A subsequent phase II trial of similar doses of 13-*cis*-RA plus IFN targeted unresectable, recurrent and/or metastatic SCCs from several sites [14]. A 50% major response rate was observed among 32 evaluable patients in the two largest study subsets (head and neck and skin cancers). Major responses at other sites including vulvar, esophageal and penile cancers were also documented.

Cervical cancer

Treatment of 55 patients with previously untreated, locally advanced primary cervical SCC with the combination of 13-*cis*-RA and IFN-α has achieved an overall major response rate of 45% [10, 11]. Promising pilot data on response to the above combination integrated with radiotherapy (RT) [12], have led to a phase III study, which compared the activity of 13-*cis*-RA/IFN-α/RT versus RT-alone. This trial showed significantly improved survival for patients with locally advanced cervical cancer who received the combined regimen with RT [16]. A similar triple combination of 13-*cis*-RA plus IFN-α plus RT in 15 patients with locally advanced carcinoma of the cervix resulted in a 47% response rate (33% complete remissions), while patients treated with concurrent chemoradiotherapy had a 42% response rate (17% complete remissions). This regimen was tolerable despite severe toxicity [17]. *In vitro* studies with human cervical carcinoma cell lines have demonstrated that 13-*cis*-RA plus IFN-α acted as radiosensitizers in some of the cell lines [18–20]. A randomized phase II trial, which evaluated all-*trans*-retinoic acid (ATRA)/IFN-α and 13-*cis*-RA/IFN-α in 60 patients with recurrent chemotherapy-naive cervical SCC, showed low major response rates in both groups (5% and 8%, respectively) and even the confirmed responses were partial and not complete [21]. Both regimens were generally well-tolerated, producing expected toxicities consistent with each agent's toxic effects. Other groups reported no responses to RA/IFN combinations in a study with 13 patients who had been pretreated heavily with RT-plus-chemotherapy for recurrent SCC of the cervix [22] and in another study with 26 patients with recurrent chemotherapy-naive cervical cancer to ATRA/IFN-α [23].

Renal cell carcinoma (RCC)

A phase II trial conducted at Memorial Sloan Kettering Cancer Center involving 44 evaluable RCC patients who received 13-*cis*-RA (1 mg/kg/d) and IFN-α (3–9 million units/d) resulted in a major response rate of 30% (13 patients), including three complete responses [15]. Responses were durable (>10 months in seven patients) and were observed in sites typically resistant to IFN alone. This response is remarkable considering that the overall response rate of RCC patients treated with IFN-α as a single agent was 10% in 149 patients treated in the same institute and 12% in 628 cases treated in various institutes. The overall response rate to IFN-α combined with vinblastin or interleukin-2 was 10% in 149 patients. The responses to the above treatments were better than the poor overall response rate (<2%) to cytotoxic and hormonal agents reported in 14 clinical trials with a total of 237 patients [15]. A study of 13-*cis*-RA (1 mg/kg/d) and IFN-α (3 million units/d) in eleven patients who were pretreated with chemotherapy and immunotherapy resulted after a median of eight months in two partial responses, five disease stabilization and four progressive disease [24]. A similar combination of agents was found to have no efficacy in RCC patients who were ineligible for another trial because of poor prognosis [25].

Recently, 13-*cis*-RA was added to an outpatient regimen of interleukine 2 (IL2) and IFN-α in previously untreated patients with metastatic RCC. Among 47 patients, eight (17%) showed response (one complete response, seven partial responses) and four additional patients experienced a minor response in lung or soft tissue metastases. The median time to response was 42 weeks and median survival was 74 weeks [26]. The authors concluded that outpatient study with this combination is feasible and that the modest efficacy with prolonged survival justify randomized trials to determine whether the combination is more effective than single agent therapy.

The combination of RA plus IFN plus subcutaneous interleukin-2 plus 5-fluorouracil was given to 44 patients with progressive metastatic RCC and resulted in a major response rate of 44%, including six complete responses. Responding cancers included lung, liver and adrenal cancers. Responses were durable in 19 of the 20 responding cases and prolonged stabilization of disease occurred in 21 patients (47%). Another study of RA plus IFN in 19 patients with metastatic disease that was refractory to IFN, interleukin-2 and chemotherapy achieved a major response rate of 21% (4/19) [27].

Pancreatic cancer

A phase II pilot study evaluated the activity of a combination of 13-*cis*-RA and IFN-β with several chemotherapeutic agents in 23 chemotherapy-naïve pancreatic cancer patients and found that eight patients responded (35%) and eight had stable disease. Median time to progression and survival for all patients were, respectively, 6.1 and 11 months [28]. Although the clincial activity was significant, toxicity was severe.

Chronic myelogenous leukemia (CML)

Administration of 13-*cis*-RA in combination with IFN-α to 43 patients with chronic myelogenous leukemia was associated with a significant fall in the white blood cell count of patients with chronic-phase disease and with a fall in the percentage S-phase cells in CML patients regardless of the stage of their leukemia. In two thirds of the patients, there was an increase in marrow apoptosis. Bcl-2 and myc expression was suppressed in a minority of patients and such expression appears to be associated with response to a treatment regimen which includes RA/IFN. These effects, if seen in other malignant diseases, could account for the therapeutic benefit which has been associated with the administration of this combination of biological agents to patients with malignant disease [29].

Laryngeal dysplasia

Recently, 13-*cis*-RA plus IFN-α achieved striking clinical activity in very high-grade premalignant lesions of the larynx [30]. These data indicate that this combination may be effective in some chemoprevention regimens in the head and neck.

Other sites

ATRA in combination with IFN-α induced an objective response of six months in a child with recurrent neuroblastoma and another child with recurrent Wilm's tumor had histological maturation of multiple tumors [31].

Mechanisms of the combination of retinoids and interferons

A variety of *in vitro* studies and a few *in vivo* studies of human cancer xenografts have demonstrated additive or synergistic effects of retinoids and interferons on the growth inhibition, apoptosis and, in a few cell types, differentiation [32–40]. These findings suggest that these agents can exert direct effects on the tumor cells. In addition, several lines of evidence from *in vitro* and animal models indicate that at least part of the antitumor effect of retinoids is due to inhibition of angiogenesis and tumor vascularization [41–43]). Analogous studies with IFNs have revealed their ability to inhibit tumor neovascularization [44].

A synergistic inhibition of blood vessel formation at the cutaneous site of injection of transformed keratinocytes or HeLa carcinoma cells was observed in mice receiving one of several retinoids in combination with IFN-α [45]. More recently, it was found that HIV Tat protein-induced angiogenesis *in vitro* and *in vivo* was inhibited synergistically by 13-*cis*-RA and IFN-β and it was suggested that this combination could be useful for treatment of Kaposi sarcoma patients [46]. Furthermore, ATRA and IFN-α were found to act synergistically to suppress angiogenic activity of head and neck squamous carcinoma cells *in vitro* as evidenced by suppression of IL-8 secretion. In addition, the combination of agents reduced synergistically the response of endothelial cells to both IL-8 and tumor-conditioned medium. Finally, ATRA and IFN-α exhibited antitumor activity and anti-angiogenic activity in a nude mouse model under conditions in which each agent alone was inactive [47]. These findings suggest that inhibition of angiogenesis should be considered as one of the mechanisms by which the RA/IFN combination exerts its anti-tumor effects *in vivo*.

Retinoids and interferon signalling pathways and their cross-talk

Numerous studies indicate that IFNs and retinoids exert similar effects on malignant cells *in vitro*. Each agent can arrest cells in the G1 phase of the cell cycle, inhibit ornithine decarboxylase induction, induce 2-5-oligoadenylate synthase (2-5-OASE), suppress the expression of certain oncogenes, alter the phosphorylation of the RB tumor suppressor gene, suppress human papilloma virus E6/E7 gene expression, and induce apoptosis. Many of the effects of IFNs and retinoids on tumor cells are most likely mediated by changes in gene expression.

The combined effects of IFNs and retinoids may result from activation of the same genes or distinct genes depending on the cell type. For example, both agents induce 2-5-OASE and p21 gene transcription in some cells. In other cells, one agent can potentiate the effect of the other by different mechanisms. For example, ATRA increased the binding of ^{125}I-labeled IFN-α to various tumor cell lines by increasing the number of receptors. ATRA also stabilized 2-5-OASE mRNA induced by IFN-γ [48]. The agents may work through different mechanisms as well. For example, human neuroblastoma cells are induced to differentiate by either agent alone, but the combination of agents leads to synergistic effects. The synergy is probably effected via distinct mechanisms, as indicated by the finding that ATRA increases inositol 1,4,5-trisphosphate and 1,2-diacylglycerol (PKC activator), whereas IFN-γ has no effect on the level of these second messengers.

Retinoid signalling

Retinoids are thought to exert most of their effects primarily through nuclear receptors. The nuclear retinoid receptors are DNA-binding, ligand-dependent transcriptional transactivators, which belong to the superfamily of steroid hormone receptors [48–50]. Two types of retinoid receptors have been identified. These are designated retinoic acid receptors (RARs) and retinoid X receptors (RXRs). RARs bind both all-*trans*-retinoic acid and 9-*cis*-retinoic acid, whereas RXRs bind only 9-*cis*-retinoic acid [49–51]. Each type includes three subtypes (α, β, and γ). RARs and RXRs form either homodimers or heterodimers that bind to a specific DNA sequence, the retinoic acid response element (RARE), in the promoter regions of genes to activate or suppress gene transcription depending on the cell type [49–51]. In humans, the genes encoding these receptors are located on different chromosomes. Each receptor subtype is thought to regulate the expression of distinct genes because each exhibits specific patterns of expression during embryonal development and different distributions in adult tissues [49–51].

Interferon signalling

IFN-α and β family members bind to the interferon-α receptor-2 (IFNAR-2) subunit and trigger its dimerization with IFNAR-1 and phosphorylation of Janus kinase 1 (JAK1) and another JAK family member, tyrosine kinase 2 (Tyk2), which are associated with the receptor [9, 52–54]. JAK1 and Tyk2 subsequently phosphorylate and activate cytoplasmic proteins including the signal transducing activator of transcription 2 (STAT2) and then STAT1 [52–54]. After translocation to the nucleus, these proteins form a complex with a third protein, p48, which is called IFN-stimulated gene factor 3 (ISGF-3) [56, 57]. ISGF-3 binds to consensus DNA sequences designated IFN-stimulated response elements (ISREs) found in the promoters of most IFN α/β responsive genes and activates their transcription [56, 57].

IFN-γ binds as a dimer to two IFN-γ receptor 1 (IFNGR-1) molecules, which each dimerize with an IFNGR-2 subunit [58]. JAK1 and JAK2, which are bound to IFNGR-1 and IFNGR-2, respectively, then phosphorylate STAT1. The phosphorylated STAT1 enters the nucleus, forms homodimers by binding to IFN-γ activated sites (GAS), which are distinct from ISREs and are found in the promoters of a large number of genes, and activates transcription [9, 58].

Cross-talk between retinoid and interferon signalling pathways

Evidence that retinoids and IFN-α interact favorably at the molecular level is provided by *in vitro* and *in vivo* preclinical data [36–38, 44, 45, 59–63]. Recent advances in the understanding of the mechanisms of action of retinoids and IFNs at the molecular level enabled the investigation of cross-talk between their signalling pathways that might provide a mechanistic rationale for the use of retinoid and IFN combinations in cancer therapy.

RA can augmant IFN signalling by upregulating the expression of IFN-γ [62] as well as proteins components of the IFN signalling pathway, including STAT1, STAT2 and p48 [59, 61–66, 68–70]. For example, in embryonal carcinoma cells, pretreatment of the cells with ATRA rendered them responsive to induction of gene expression by IFN by increasing the level of STAT-1 [59]. Recently it has been shown that RA can directly induce transcription of the STAT1 gene through a RARE in its promoter (64). Retinoids can also directly induce the expression of ISGs [59, 61–69] including IFN regulatory factor-1 (IRF-1) and 2-5-OASE [68–70]. The cross-regulation of genes in this class suggests an overlap in the mechanism of proliferation control by retinoids and IFNs [70, 71].

There are examples of the upregulation of nuclear retinoid receptors by IFN. In breast cancer cell lines IFN-γ could upregulate RAR-γ expression [72] and in the NB4 acute promyelocytic leukemia cell line, IFN-α,β and γ could upregulate RAR-α expression [73]. Thus, a mutual enhancement of signalling between retinoid and IFN pathways may explain some of their additive and synergistic effects and could be exploited therapeutically. Although few of the target genes that are regulated by either RA or IFN have been implicated in the interaction between the two agents, it appears that the p21 protein, which can interact with and inhibit cyclin-dependent kinases, causing cells to arrest in the G1 phase of the cell cycle [74], represents a potential intersecting point between the retinoid and IFN signalling pathways. RA and IFNs directly regulate the transcription of the p21 through their respective upstream promoter regulatory elements, suggesting a common mechanism of cell cycle regulation by these agents [75, 76].

Future directions

Overall, the clinical data profiles of 13-*cis*-RA/IFN-α are promising and encourage further study of these agents with concomitant radiation to establish the optimal application in locally advanced, untreated cervical cancer. Adding cisplatin to the RA/IFN combination is suggested by cisplatin's activity in advanced SCC of several sites and the ability of both RA and IFN-α to potentiate cisplatin antitumor activity in several *in vitro* and human tumor xenograft systems [77, 78]. Triple combination of a retinoid, IFN and vitamin E should also be explored further in view of the efficacy of this regimen in some premalignant lesions. Further investigation into the precise mechanisms of inhibition of tumor angiogenesis by both IFNs and retinoids will be necessary in order to better exploit their use in combination antiangiogenic therapy.

Acknowledgement
Work in the authors' laboratory was supported in part by United States Public Health Service grant PO1 CA68233 from the National Cancer Institute.

References

1   Gudas LJ, Sporn MB, Roberts AB (1994) Cellular biology and biochemistry of the retinoids. *In*: MB Sporn, AB Roberts (eds): *The Retinoids: Biology, Chemistry, and Medicine*. Raven Press, New York, 443–520
2   Lotan R (1995) Cellular biology of the retinoids. *In*: L Degos, DR Parkinson (eds): *Retinoids in Oncology*. Springer, Berlin, 27–42
3   Smith MA, Parkinson DR, Cheson BD, Friedman MA (1992) Retinoids in cancer therapy. *J Clin Oncol* 10: 839–864
4   Lotan R (1996) Retinoids in cancer chemoprevention. *FASEB J* 10: 1031–1039
5   Tallman MS, Andersen JW, Schiffer CA, Appelbaum FR, Feusner JH, Ogden A, Shepherd L, Willman C, Bloomfield CD, Rowe JM et al (1997) All-*trans*-retinoic acid in acute promyelocytic leukemia. *New Engl J Med* 337: 1021–1028
6   Pastorino U, Parkinson DR, Chiesa F (1995) Retinoids in the prevention and therapy of solid tumors. *In*: L Degos, DR Parkinson (eds): *Retinoids in Oncology*. Springer-Verlag, New York, 93–111
7   Diaz MO, Bohlander S, Allen G (1993) Nomenclature of human interferon genes. *J Interferon Res* 13: 234–243
8   Kalvakolanu DV, Borden EC (1996) An overview of the interferon system: signal transduction and mechanisms of action. *Cancer Invest* 14: 25–53
9   Lippman SM, Lotan R, Schleuniger U (1997) Retinoid-interferon therapy of solid tumors. *Int J Cancer* 70: 481–483
10 Lippman SM, Kavanagh JJ, Paredes-Espinoza M, Delgadillo-Madrueno F, Paredes-Casillas P, Hong WK, Holdener E, Krakoff IH (1992) 13-*cis* retinoic acid plus interferon-α2a: highly active systemic therapy for squamous cell carcinoma of the cervix. *J Nat Cancer Inst* 84: 214–245
11 Lippman SM, Kavanagh JJ, Paredes-Espinoza M, Delgadillo-Madrueno F, Paredes-Casillas P, Hong WK, Massimini G, Holdener EE, Krakoff IH (1993) 13-*cis* retinoic acid plus interferon α2a: in locally advanced squamous cell carcinoma of the cervix. *J Nat Cancer Inst* 85: 499–500
12 Kavanagh JJ, Lippman SM, Paredes-Espinoza M (1996) The combination of 13-*cis*-retinoic acid and interferon a2a with radiation therapy in squamous cell carcinoma of the cervix. *Int J Gynecol Cancer* 6: 439–444
13 Lippman SM, Parkinson DR, Itri LM, Weber RS, Schantz SP, Ota DM, Schusterman MA, Krakoff IH, Gutterman JU, Hong WK (1992) 13-*cis*-retinoic acid and interferon-2a. Effective combination therapy for advanced squamous cell carcinoma of the skin. *J Nat Cancer Inst* 84: 235–240
14 Toma S, Palumbo R, Vincenti M (1994) Efficacy of recombinant interferon-α and 13-*cis*-retinoic acid in the treatment of squamous cell carcinomas. *Ann Oncol* 5: 463–465
15 Motzer RJ, Schwartz L, Law TM, Murphy BA, Hoffman AD, Albino AP, Vlamis V, Nanus DM (1995) Interferon α-2a and 13-*cis* retinoic acid in renal cell carcinoma: antitumor activity in a phase II trial and interactions *in vitro*. *J Clin Oncol* 13: 1950–1957
16 Antonadou D, Cardamakis E, Iliopoulos P (1996) Comparative study between exclusive irradiation or combined with IFN-α-2a and isotretinoin in stage IIb and III cervical carcinoma. *Amer Soc Ther Rad Oncol* 36: (Abstr 122), 121
17 Park TK, Lee JP, Kim SN, Choi SM, Kudelka AP, Kavanagh JJ (1998) Interferon-α 2a, 13-*cis*-retinoic acid and radiotherapy for locally advanced carcinoma of the cervix: a pilot study. *Eur Gynaecol Oncol* 19: 35–38
18 Hoffman W, Bamberg M, Rodemann HP (1994) Antiproliferative effects of ionizing radiation, all-*trans*-retinoic acid and interferon-a on cultured human squamous cell carcinomas. *Radiat Oncol Invest* 2: 12–19
19 DeLaney TF, Afridi N, Taghian AG, Sanders DA, Fuleihan NS, Faller DV, Nogueira CP (1996) 13-*cis*-retinoic acid with α-2a-interferon enhances radiation cytotoxicity in head and neck squamous cell carcinoma *in vitro*. *Cancer Res* 5: 2277–2280
20 Ryu S, Kim OB, Kim SH, He SQ, Kim JH (1998) *In vitro* radiosensitization of human cervical carcinoma cells by combined use of 13-*cis*-retinoic acid and interferon-α2a. *Int J Radiat Oncol Biol Phys* 41: 869–873
21 Weiss GR, Liu PY, Alberts DS, Peng Y-M, Fisher E, Xu MJ, Scudder SA, Baker LH, Moore DF, Lippman SM (1999) 13-*cis*-retinoic acid or all-*trans*-retinoic acid plus interferon α 2a in recurrent cervical cancer: A Southwest Oncology Group Phase II Randomized trial. *Gynecol Oncol* 71: 386–390
22 Hallum AV 3rd Alberts DS, Lippman SM, Inclan L, Shamdas GJ, Childers JM, Surwit EA, Modiano M, Hatch KD (1995) Phase II study of 13-*cis*-retinoic acid plus interferon-a2a in heavily pretreated squamous carcinoma of the cervix. *Gynecol Oncol* 56: 382–386
23 Wadler S, Schwartz EL, Haynes H, Rameau R, Quish A, Mandeli J, Gallagher R, Hallam S, Fields A, Goldberg G et al (1997) All-*trans* retinoic acid and interferon-α-2a in patients with metastatic or recurrent carcinoma of the uterine cervix. *Cancer* 79: 1574–1580
24 Casali A, Sega FM, Casali M, Serrone L, Terzoli E (1998) 13-*cis* retinoic acid and interferon α-2a in the treatment of metastatic renal cell carcinoma. *J Exp Clin Cancer Res* 17: 227–229
25 Escudier B, Ravaud A, Berton D, Chevreau C, Douillard JY, Dietrich PY (1998) Phase II study of interferon-α and all-*trans* retinoic acid in metastatic renal cell carcinoma. *J Immunother* 21(1): 62–64
26 Stadler WM, Kuzel T, Dumas M, Vogelzang NJ (1998) Multicenter phase II trial of interleukin-2, interferon-α, and 13-*cis*-retinoic acid in patients with metastatic renal cell carcinoma. *J Clin Oncol* 16(5): 1820–1825
27 Atzpodion J, Buer J, Probat M, Duensing S, Kirchner H, Ganser A (1996) Clinical and preclinical role of 13-

cis-retinoic acid in renal cell carcinoma: Hannover experience. *Proc Am Soc Clin Oncol* 15: (Abst. 625), 247

28 Recchia F, Sica G, Casucci D, Rea S, Gulino A, Frati L (1998) Advanced carcinoma of the pancreas: phase II study of combined chemotherapy, β-interferon, and retinoids. *Amer J Clin Oncol* 21: 275–278

29 Handa H, Hegde UP, Kotelnikov VM, Mundle SD, Dong LM, Burke P, Rose S, Hsu WT, Gaskin F, Raza A et al (1997) The effects of 13-*cis* retinoic acid and interferon-α in chronic myelogenous leukemia cells *in vivo* in patients. *Leuk Res* 21: 1087–1096

30 Papadimitrakopoulou VA, Shin DM, layman G, El-Naggar A, Goepfert HJ, Hong WK (1997) Efficacy of biochemoprevention in reversal of advanced premalignant lesions in the upper aerodigestive tract. *Proc Am Soc Clin Oncol* 16: (abst 1366), 1557

31 Adamson PC, Reaman G, Finklestein JZ, Feusner J, Berg SL, Blaney SM, O'Brien M, Murphy RF, Balis FM (1997) Phase I trial and pharmacokinetic study of all-*trans*-retinoic acid administered on an intermittent schedule in combination with interferon-α2a in pediatric patients with refractory cancer. *J Clin Oncol* 15: 3330–3337

32 Bollag W (1991) Retinoids and interferon: a new promising combination? *Brit J Haematol* 79: 87–91

33 Cornaglia-Ferraris P, Mariottini GL, Ponzoni M (1992) Gamma-interferon and retinoic acid synergize in inhibiting the growth of human neuroblastoma cells in nude mice. *Cancer Lett* 31: 215–220

34 Lotan R, Dawson MI, Zou CC, Jong L, Lotan D, Zou CP (1995) Enhanced efficacy of combinations of retinoic acid and retinoid X receptor-selective retinoids and a-interferon inhibition of cervical carcioma cell proliferation. *Cancer Res* 55: 232–236

35 Giandomenico V, Lancillotti F, Fiorucci G, Percario ZA, Ribavene R, Malorni W, Affabris E, Romeo G (1997) Retinoic acid and IFN inhibition of cell proliferation is associated with apoptosis in squamous carcinoma cell lines: Role of IRF-1 and TGase II-dependent pathways. *Cell Growth Differ* 8: 91–100

36 Widschwendter M, Daxenbichler G, Bachmair F, Muller E, Zeimet AG, Windbichler G, Uhl-Steidl M, Lang T, Marth C (1996) Interaction of retinoic acid and interferon-α in breast cancer cell lines. *Anticancer Res* 16: 369–374

37 Agarwal C, Hembree JR, Rorke PA, Eckert RL (1994) Interferon and retinoic acid suppress the growth of human papillomavirus type-16 immortalized cervical epithelial cells, but only interferon suppresses the level of the human papillomavirus transforming oncogenes. *Cancer Res* 54: 2108–2112

38 Lancillotti F, Giandomenico V, Affabris E, Fiorucci G, Romeo G, Rossi G (1995) Interferon α-2b and retinoic acid combined treatment effects proliferation and gene expression of human cervical carcinoma cells. *Cancer Res* 55: 3158–3164

39 Jozan S, Courtade M, Mathieu-Boue A, Lochon I, Bugat R (1998) Cytotoxic effect of interferon α 2a in combination with all-*trans* retinoic acid or *cis*-platin in human ovarian carcinoma cell lines. *Anti-Cancer Drugs* 9: 229–238

40 Widschwendter M, Daxenbichler G, Culig Z, Michel S, Zeimet AG, Mortl MG, Widschwendter A, Marth C (1997) Activity of retinoic acid receptor-γ selectively binding retinoids alone and in combination with interferon-γ in breast cancer cell lines. *Int J Cancer* 71: 497–504

41 Majewski S, Szmurlo A, Marczak M, Jablonska S, Bollag W (1993) Inhibition of tumor cell-induced angiogenesis by retinoids, 1,25-dihydroxyvitamin D3 and their combination. *Cancer Lett* 75: 35–39

42 Liaudet-Coopman EDE, Berchem GJ, Wellstein A (1997) *In Vivo* inhibition of angiogenesis and induction of apoptosis by retinoic acid in squamous cell carcinoma. *Clin Cancer Res* 3: 179–184

43 Lingen MW, Polverini PJ, Bouck NP (1996) Inhibition of squamous cell carcinoma angiogenesis by direct interaction of retinoic acid with endothelial cells. *Lab Invest* 74: 476–483

44 Sidky YA, Borden EJC (1987) Inhibition of angiogenesis by interferons: effects on tumor- and lymphocyte-induced vascular responses. *Cancer Res* 47: 5155–5161

45 Majewski S, Szmurlo A, Marczak M, Jablonska S, Bollag W (1994) Synergistic effect of retinoids and interferon α on tumor-induced angiogenesis: anti-angiogenic effect on HPV-harboring tumor cell lines. *Int J Cancer* 57: 81–85

46 Iurlaro M, Benelli R, Masiello L, Rosso M, Santi L, Albini A (1998) β Interferon inhibits HIV-1 Tat-induced angiogenesis: synergism with 13-*cis* retinoic acid. *Eur J Cancer* 34: 570–576

47 Lingen MW, Polverini PJ, Bouck NP (1998) Retinoic acid and interferon a act synergistically as antiangiogenic and antitumor agents against human head and neck squamous carcinoma. *Cancer Res* 58: 5551–5558

48 Higuchi T, Hannigan GE, Malkin D, Yeger H, Williams BR (1991) Enhancement by retinoic acid and dibutyryl cyclic adenosine 3': 5'-monophosphate of the differentiation and gene expression of human neuroblastoma cells induced by interferon. *Cancer Res* 51: 3958–3964

49 Lotan R, Clifford JL (1991) Nuclear receptors for retinoids: mediators or retinoid effects on normal and malignant cells. *Biomed Pharmacotherapy* 45: 145–156

50 Chambon P (1996) A decade of molecular biology of retinoic acid receptors. *FASEB J* 10: 940–954

51 Mangelsdorf DJ, Evans RM (1995) The RXR heterodimers and orphan receptors. *Cell* 83: 841–850

52 Heldin CH (1995) Dimerization of cell surface receptors in signal transduction. *Cell* 80: 213–233

53 Darnell JEJr (1997) STATs and gene regulation. *Science* 277: 1630–1635

54 Silvennoinen O, Ihle JN, Schlessinger J, Levy DE (1993) Interferon-induced nuclear signalling by Jak protein tyrosine kinases. *Nature* 366: 583–585

55 Leung S, Qureshi SA, Kerr IM, Darnell JE, Stark GR (1995) Role of STAT2 in the α interferon signalling

pathway. *Mol Cell Biol* 15: 1312–1317
56 Schindler C, Fu XY, Improta T, Aebersold R, Darnell Jr, JE (1992) Proteins of transcription factor ISGF-3: one gene encodes the 91- and 84-kDa ISGF-3 proteins that are activated by interferon a. *Proc Natl Acad Sci USA* 89: 7836–7839
57 Qureshi SA, Salditt-Georgieff M, Darnell Jr, JE (1995) Tyrosine-phosphorylated Stat 1 and Stat 2 plus a 48-kDa protein all contact DNA in forming interferon-stimulated-gene factor 3. *Proc Natl Acad Sci USA* 92: 3829–3833
58 Haque JS, William RG (1998) Signal transduction in the interferon system. *Semin Oncol* 25 (suppl)1: 14–22
59 Kolla V, Lindner DJ, Weihua X, Borden EC, Kalvakolanu DV (1996) Modulation of interferon (IFN)-inducible gene expression by retinoic acid. *J Biol Chem* 271: 10 508–10 514
60 Lindner DJ, Borden EC, Kalvakolanu DV (1997) Synergistic antitumor effects of a combination of interferons and retinoic acid on human tumor cells *in vitro* and *in vivo*. *Clin Cancer Res* 3: 931–937
61 Gianni M Terao M, Fortino I, LiCalzi M, Viggiano V, Barbui T, Rambaldi A, Garattini E (1997) Stat 1 is induced and activated by all-*trans* retinoic acid in acute promyelocytic leukemia cells. *Blood* 89: 1001–1012
62 Cippitelli M, Ye J, Viggiano V, Sica A, Ghosh P, Gulino A, Santoni A, Young HA (1996) Retinoic acid-induced transcriptional modulation of the human interferon-γ promotor. *J Biol Chem* 271: 26 783–26 793
63 Lotan R, Dawson MI, Zou CC, Jong L, Lotan D, Zou CP (1995) Enhanced efficacy of combinations of retinoic acid- and retinoid X receptor-selective retinoids and a-interferon in inhibition of cervical carcinoma cell proliferation. *Cancer Res* 55: 232–236
64 Weihua X, Kolla V, Kalvakolanu DV (1997) Modulation of interferon action by retinoids. *J Biol Chem* 272: 9742–9748
65 Harada H, Willison K, Sakakibara J, Miyamoto M, Fujita T, Taniguchi T (1990) Absence of the type I IFN system in EC cells: transcriptional activator (IRF-1) and repressor (IRF-2) genes are developmentally regulated. *Cell* 63: 303–312
66 Matikainen S, Ronni T, Lehtonen A, Sareneva T, Melen K, Nordling S, Levy DE, Julkunen I (1997) Retinoic acid induces signal transducer and activator of transcription (STAT) 1, STAT2, and p48 expression in myeloid leukemia cells and enhances their responsiveness to interferons. *Cell Growth Differ* 8: 687–698
67 Lippman S, Glisson BS, Kavanagh JJ, Lotan R, Hong WK, Paredes-Espinoza M, Hittelman WN, Holdener EE, Krakoff IH (1993) Retinoic acid and interferon combination studies in human cancer. *Eur J Cancer* 29A (suppl) 5: 9–13
68 Pelicano L, Li F, Schindler C, Chelbi-Alix MK (1997) Retinoic acid enhances the expression of interferon-induced proteins: evidence for multiple mechanisms of action. *Oncogene* 6: 2349–2359
69 Yu M, Tong JH, Mao M, Kan LX, Liu MM, Sun YW, Fu G, Jing YK, Yu L et al (1997) Cloning of a gene (RIG-G) associated with retinoic acid-induced differentiation of acute promyelocytic leukemia cells and representing a new member of a family of interferon-stimulated genes. *Proc Natl Acad Sci USA* 94: 7406–7411
70 Matikainen S, Lehtonen A, Sareneva T, Julkunen I (1998) Regulation of IRF and STAT gene expression by retinoic acid. *Leuk Lymphoma* 30: 63–71
71 Gaboli M, Gandini Delva L, Wang ZG, Pandolfi PP (1998) Acute promyelocytic leukemia as a model for cross-talk between interferon and retinoic acid pathways: from molecular biology to clinical applications. *Leuk Lymphoma* 30: 11–22
72 Widschwendter M, Daxenbichler G, Dapunt O, Marth C (1995) Effects of retinoic acid and γ-interferon on expression of retinoic acid receptor and cellular retinoic acid-binding protein in breast cancer cells. *Cancer Res* 55: 2135–2139
73 Gianni M, Zanotta S, Terao M, Rambaldi A, Garattini E (1996) Interferons induce normal and aberrant retinoic-acid receptors type α in acute promyelocytic leukemia cells: potentiation of the induction of retinoid-dependent differentiation markers. *Int J Cancer* 68: 75–83
74 Liu M, Lee MH, Cohen M, Bommakanti M, Freedman LP (1996) Transcriptional activation of the Cdk inhibitor p21 by vitamin D3 leads to the induced differentiation of the myelomonocytic cell line U937. *Gene Develop* 10: 142–153
75 Chin YE, Kitagawa M, Su WCS, You ZH, Iwamoto Y, Fu XY (1996) Cell growth arrest and induction of cyclin-dependent kinase inhibitor p21WAF1/CIP1 mediated by STAT1. *Science* 272: 719–722
76 Liu M, Iavarone A, Freedman LP (1996) Transcriptional activation of the human p21WAF1/CIP1 gene by retinoic acid receptor. *J Biol Chem* 271: 31 723–31 728
77 Shalinsky DR, Bischoff ED, Gregory ML, Lamph WW, Heyman RA, Hayes JS, Thomazy V, Davies PJA (1996) Enhanced antitumor efficacy of cisplatin in combination with ALRT 1057 (9-*cis* retinoic acid) in human oral squamous carcinoma xenografts in nude mice. *Clin Cancer Res* 2: 511–520
78 Aebi S, Kroning R, Cenni B, Sharma A, Fink D, Los G, Weisman R, Howell SB, Christen RD (1997) All-*trans* retinoic acid enhances *cis*-platin-induced apoptosis in human ovarian adenocarcinoma and in squamous head and neck cancer cells. *Clin Cancer Res* 3: 2033–2038

Vitamin A and retinoids: an update of biological aspects and clinical applications
M.A. Livrea (ed.)
© 2000 Birkhäuser Verlag Basel/Switzerland

Treatment of acute promyelocytic leukemia with all-*trans* retinoic acid

P. Fenaux[1] and L. Degos[2]

[1] *Service des Maladies du Sang, CHU, 1 place de Verdun, F-59037 Lille, France*
[2] *Service d'Hématologie, Hôpital Saint-Louis, F-75475 Paris, France*

Introduction

Acute promyelocytic leukemia (APL) is a specific type of acute myeloid leukemia (AML) characterized by the morphology of blast cells (M_3 in the French American British classification of AML) [1, 2], the t [15, 17] translocation [3] which fuses the promyelocytic leukemia (PML) gene on chromosome 15 to the retinoic acid receptor (RARα) α gene on chromosome 17 [4, 5], and by a coagulopathy combining disseminated intravascular coagulation (DIC) and fibrinolysis [6, 7]. Until recently, intensive chemotherapy, usually combining an anthracycline and cytosine arabinoside (AraC) was the only effective treatment of APL [8–22].

In recent years, discovery of the *in vitro* and *in vivo* differentiation of APL blasts by all-*trans* retinoic acid (ATRA) has modified the therapeutic approach of APL [23–25] but also lead to important advances in the biology of APL and opened new perspectives for differentiation therapy in cancer [26].

Background: results of chemotherapy alone in APL

Although complete remission (CR) rates of only 50 to 60% were generally reported ten years ago, results have subsequently improved, and CR rates of 70% to 80% have been reported in the most recent series [8–25]. Failure to achieve CR was due, in early reports, to central nervous system (CNS) bleeding during the first days of treatment or sepsis during the phase of aplasia whereas resistant leukemia was generally seen in less than 10% of the patients, probably reflecting the high sensitivity of APL cells to anthracyclines.

Several studies, including recent ones, have shown that total induction doses of daunorubicin (DNR) greater than 200 mg/m^2 to 250 mg/m^2 were required to obtain these results [8, 9, 23, 24]. Furthermore, it has not been demonstrated that anthracycline-AraC combinations are superior to anthracyclines alone if the latter are used at a high dose, e.g. at least 300 mg/m^2 during induction for DNR. High dose AraC and other induction drugs (6Thioguanine, VP 16) do not seem to bring any benefit during induction chemotherapy in APL and, in the case of high dose AraC, could increase toxicity [23, 25].

Significant coagulopathy, present at diagnosis in 80% of cases of APL, is worsened (or triggered in the remaining patients) by the onset of chemotherapy. Whereas the beneficial role of heparin is very controversial, the use of intensive platelet support during chemotherapy has probably been the major factor for reduction of the incidence of hemorrhagic deaths in APL, especially in patients presenting with hyperleukocytosis, who have an increased risk of early death due to bleeding [10].

Once CR has been achieved, several studies have shown that APL, even when treated with chemotherapy alone, is associated with a lower risk of relapse than other types of AML treated identically [11, 23]. Some studies have suggested that prolonged maintenance chemotherapy could prolong remissions.

Poor prognostic factors of CR achievement in newly-diagnosed APL treated with chemotherapy alone include: an age older than 50 [9, 10], hyperleukocytosis at diagnosis [13, 15], microganular APL variant [18], fever, severe bleeding at diagnosis (especially in fundus oculi) [13] or major thrombocytopenia [10]. Shorter remissions are seen in patients with hyperleukocytosis [11] and in patients with microgranular APL variant [9].

Thus, anthracycline-AraC regimens with sufficient anthracycline dosage, in combination with intensive platelet support during induction, yield CR in 75 to 80% of newly diagnosed APL patients. With anthracycline-based consolidation and possibly maintenance chemotherapy, median CR duration ranges from 11 to 25 months so that, overall, only 35 to 45% of the patients can be cured by chemotherapy alone. Patients presenting with high leucocyte counts have a particularly poor prognosis with chemotherapy alone, as their CR rate is only 50 to 60%, and the risk of relapse is high.

First results obtained with ATRA alone in APL

In the first reports of ATRA therapy, CR rates of about 90% were obtained in newly-diagnosed and first relapse APL, generally with a 45 mg/m^2 daily dose of ATRA, and it was demonstrated that response was not obtained by cytotoxicity but by differentiation of APL blasts into neutrophils, leading to progressive replacement of leukemic hematopoiesis by normal polyclonal hematopoiesis [26–31]. Rapid improvement of coagulopathy, instead of the initial worsening observed with conventional chemotherapy, was also seen.

However, two major drawbacks of ATRA treatment were observed. First, mainly in newly-diagnosed APL, a rapid rise in leukocytes was seen in one third to one half of the patients, accompanied by clinical signs of "ATRA syndrome" which proved fatal in some patients [28, 32, 33]. Low dose chemotherapy (with hydroxyurea or low dose AraC) did not succeed in lowering leukocyte counts and preventing the fatal outcome, whereas more intensive anthracycline-AraC chemotherapy was able to reduce leukocyte counts and allow most patients to enter CR [70]. The second drawback was that most of the patients who achieved CR with ATRA and received either ATRA alone or low dose chemotherapy for maintenance therapy relapsed within a few months of CR achievement [25, 34–37]. These findings prompted clinicians to administer treatment combining ATRA and intensive chemotherapy in APL.

ATRA combined to intensive chemotherapy in newly-diagnosed APL

Non-randomized studies and two randomized trials have demonstrated the superiority of combined treatment with ATRA and intensive chemotherapy over intensive chemotherapy alone in newly-diagnosed APL.

Non-randomized studies

A first pilot study treated 26 newly-diagnosed cases of APL with ATRA until CR, followed by three courses of daunorubicin (DNR)-AraC and 25 (96%) achieved CR, as compared to 76% in a historical control group treated by chemotherapy alone (difference not significant) [38, 39]. With a minimum follow up of 38 months from CR achievement, event free survival (EFS), disease free interval (DFI) and survival were significantly higher after ATRA followed by chemotherapy. Actual survival at four years was 40% after chemotherapy alone, and 77% after ATRA + chemotherapy. The combination of ATRA and chemotherapy clearly reduced the number of relapses occurring within 18 months of CR achievement, whereas the number of late relapses was similar to that seen after chemotherapy alone [38]. This suggested that combination therapy did not just delay relapses but truly reduced their incidence.

Warrell et al. [40], in 49 newly-diagnosed APL patients, obtained a CR rate of 85% with ATRA alone, and three anthraycline-AraC courses were subsequently administered. Sixty five percent of the patients who achieved CR remained in CR and median survival was not reached whereas, in a historical control group of 80 patients treated by CT alone, median CR duration was 14 months and median survival was 17 months (all those differences were significant). Likewise, Kanamaru et al. [41] obtained a CR rate of 89% in 109 newly-diagnosed APL cases with a combination of ATRA and CT. Actual EFS and disease free survival (DFS) at 23 months were 75% and 81%, respectively. In a historical control group treated with CT alone by the same group, the CR rate, actual EFS and DFS at 23 months had been 70%, 48% and 65%, respectively (all those differences were significant). Wiley et al. [42] obtained CR in 18 of 19 newly-diagnosed APL patients treated with ATRA followed by CT and with a median follow up of 13 months, only one had relapsed.

Finally, it was shown in one study that treatment with ATRA was associated with significantly fewer platelet and RBC transfusions, less days with fever and with antibiotics, and a shorter hospital stay than treatment with CT [43].

Randomized studies

A European trial (APL 91 trial) comparing chemotherapy alone (three intensive courses of DNR and AraC) and ATRA followed by the same chemotherapy in newly-diagnosed APL was performed between 1991 and 1992. In the ATRA group, the first chemotherapy course was rapidly added to ATRA if the WBC was greater than 5000/mm^3 at diagnosis, or increased during treatment. The trial was prematurely stopped after 18 months, because event free survival (EFS) was significantly better in the ATRA group [81, 82]. These results were confirmed in the last interim analysis, performed 61 months after the closing date of the study, which showed a significantly lower relapse rate, higher EFS and survival in the ATRA group. Survival at four years was 76% in the ATRA group, as compared to 53% in the chemotherapy group (Fenaux et al., unpublished observations).

A US intergroup study randomizing ATRA followed by chemotherapy and chemotherapy alone in newly-diagnosed APL, started in 1992, was closed at the beginning of 1995 and its results confirmed in terms of relapse and survival the superiority of ATRA followed by chemotherapy over chemotherapy alone.

Finally, results of two recently-closed European trials of ATRA + CT in newly-diagnosed APL, each of which have included about 500 patients, showed CR rates of 93 and 94%, respectively [46, 47]. These results, obtained on a multicenter basis, show that with better knowledge of the utilization of ATRA (and especially of the prophylaxis of its major side effect, the ATRA syndrome), very high CR rates can be achieved by combining this drug with chemotherapy in newly-diagnosed APL. Furthermore, analysis of minimal residual disease by RT-PCR, made in one of them (the Italian Gimema study) showed that molecular remission was achieved after ATRA and four cycles of chemotherapy in 98% of the patients who achieved hematological CR.

What is the optimal combination of ATRA and chemotherapy in APL? Is there a role for maintenance treatment?

Schedule of ATRA and chemotherapy for induction treatment

Most combinations of ATRA and chemotherapy, in APL, have used ATRA first, followed by chemotherapy after CR achievement. In an attempt to determine the optimal combination of ATRA and chemotherapy for induction treatment, the European APL group compared, in a randomized trial, ATRA followed by chemotherapy to ATRA plus chemotherapy (where chemotherapy was added on day three of ATRA treatment). Significantly fewer relapses were seen with ATRA plus chemotherapy (4% versus 17% at two years).

ATRA has also been used following chemotherapy in APL. This schedule has not been compared to other combinations between ATRA and chemotherapy. Furthermore, most clinicians would be reluctant not to use ATRA upfront in APL because of its favorable effect on coagulopathy.

Role of maintenance treatment in APL

In the first studies using ATRA in APL, it was shown that continuous maintenance with ATRA was unable to avoid relapses, and that patients who relapsed while under ATRA were unable to respond to ATRA, even at higher doses. A probable explanation was that prolonged ATRA treatment induced, after a few weeks, hypercatabolism of the drug, which reduced its plasma levels. *In vitro* studies also found that this hypercatabolism was reversible two to three months after discontinuation of the drug, providing a rationale for intermittent ATRA as maintenance in APL. In a randomized trial that was closed in September 1998, the European APL group showed that intermittent maintenance with ATRA, 15 days every three months, reduced the incidence of relapse in newly-diagnosed APL after CR achievement with ATRA and chemotherapy followed by consolidation chemotherapy.

This study also showed that continuous low dose chemotherapy with 6-mercaptopurine and methotrexate reduced the incidence of relapse and that this effect was additive to that of intermittent ATRA (Fenaux et al., unpublished observations)

Treatment of relapsing APL with ATRA

CR rates of 85 to 90% were obtained with ATRA in first relapsed patients previously treated with chemotherapy alone, [34, 40, 49, 50] but almost all the patients subsequently maintained on ATRA alone or mild chemotherapy relapsed again, generally within a year. On the other hand, long term results of the European APL 91 trial suggest that almost one half of those patients can be helped in the long term by consolidation with allogeneic or autologous stem cell transplantation, or with intensive consolidation chemotherapy (P. Fenaux, unpublished observations). Still, in the APL 91 trial, long term survival of patients who initially received no ATRA was significantly lower than that of patients who initially received ATRA. This shows that ATRA, in APL, should not be "kept" for relapse, but rather incorporated into the front line therapeutic regimen.

Relapses now mainly occur in patients who have already received ATRA. In relapses occurring within three to four months after discontinuation of ATRA, no response to ATRA, even at higher doses, can be observed [31, 51, 40]. On the other hand, most of the patients who relapse later, after ATRA discontinuation, achieve CR after retreatment with ATRA [40, 45]. In the APL 91 trial, all relapses occurred more than five months after ATRA discontinuation and all those patients responded again to ATRA (P. Fenaux, unpublished results.)

Treatment of APL at the second or subsequent relapse with ATRA is disappointing. In patients not previously exposed to ATRA some CR can be obtained, but patients generally relapse again rapidly, whereas in patients previously exposed to ATRA no response is generally seen [28, 51, 40]. As seen above, resistance to ATRA can be due to development of hypercatabolism of the drug, which is in principle reversible several months after drug discontinuation. Resistance to ATRA, in other patients, could be due to point mutations in the RARα region of the PML-RARα gene, which develop in patients already treated with ATRA.

Preliminary results with Arsenic trioxyde (As_2O_3) suggest that this drug can be effective in 80 to 90% of APL refractory to ATRA. It induces differentiation of APL cells at low doses, like ATRA, whereas at higher doses, the effect on those cells is induction of apoptosis [51, 52].

ATRA syndrome and other side effects of ATRA

ATRA therapy is usually well tolerated but a few major side effects can occur, mainly the ATRA syndrome.

Incidence and clinical signs

A progressive and symptomless rise in WBC counts is frequent with ATRA treatment, but in some cases a rapid rise of WBC counts associated with cardiopulmonary and renal failure were reported [28, 32]. This "ATRA syndrome" combines fever, respiratory distress, weight gain, lower extremity edema, pleural or pericardial effusions, hypotension and sometimes renal failure. These signs are preceded by increasing WBC counts in the majority of cases, but some patients develop symptoms at normal WBC counts [33, 40]. The ATRA syndrome occurred in 23% of the patients in the New York study and 5 of 13 patients with this complication died [50]. The incidence was somewhat lower in the French and Japanese studies, probably because of early addition of chemotherapy in case of rising WBC counts. On the other hand, by also taking into acount very moderate clinical or radiological signs, the European Group found an incidence of ATRA syndrome of 15% in more than 500 patients included in the recently closed APL 93 trial (Fenaux et al., unpublished observations).

Pathophysiology

It is still poorly understood. The ATRA syndrome is not due to leukostasis and/or thrombosis [33]. Because its clinical signs are reminiscent of those observed in the endotoxic shock and in the adult respiratory distress syndrome (ARDS), a possible stimulatory effect of ATRA on cytokine expression by APL cells has been envisaged. Induction of IL-1β and granulocyte colony stimulating factor (G-CSF) secretion by APL cells under ATRA may contribute to hyperleukocytosis *in vivo*. On the other hand, the secretion of IL-1β, IL6, TNFα and IL-8, which are involved in leucocyte activation and adherence, and are implicated in the development of ARDS, could have a pathogenetic role in the ATRA syndrome [51, 52].

Prophylaxis and treatment

When the ATRA syndrome has developed, addition of low dose chemotherapy and leucophereses are unable to reverse the symptoms. Two different approaches aimed at preventing and/or treating early the ATRA syndrome are currently proposed. One of them, used by the European and Japanese groups [38, 41, 44, 49], consists of adding chemotherapy from the onset of ATRA in patients presenting with high WBC counts (WBC > 5000/mm^3 in the European trial, or >3000/mm^3 in the Japanese trials) or when increases in the WBC are seen. This approach proved very effective on a multicenter basis, as the ATRA syndrome was seen in 64 of 413 patients treated in Europe [44], and seven of 109 patients treated by the Japanese group [41], and only six cases in the 522 patients proved fatal. A disadvantage of this approach is that about two thirds of the patients treated with ATRA also received early chemotherapy. However, several reports show that the period of neutropenia and thrombocytopenia is significantly shorter in patients who receive chemotherapy while already on ATRA, by comparison with chemotherapy alone [41].

The current US approach is to try to prevent the ATRA syndrome by high dose intravenous corticosteroids (Dexamethasone, 10 mg IV twice daily for three or more days) as soon as the first symptoms occur. This attitude proved very effective in the New York study, both for preventing the ATRA syndrome and reducing its mortality, from three in nine patients in the initial report to no fatalituies in the last two years [33, 40, 50]. It also proved efficient in a large US intergroup study [53]. However, as seen above, a recent European trial favours early introduction of chemotherapy in combination with ATRA, as it may reduce relapses.

Finally, there seems to be a consensus on the fact that patients presenting with high WBC counts (i.e. more than 15000 to 20000/mm^3) will very often develop severe ATRA syndrome with ATRA alone, and require chemotherapy and intravenous dexamethasone from the onset of treatment. Some of these patients even present with symptoms analogous to those of the ATRA syndrome at diagnosis [32].

Coagulopathy and thrombosis

In contrast to treatment with chemotherapy, no exacerbation of the bleeding tendency is observed in APL patients undergoing ATRA therapy. In the European APL 91 trial, median time to disappearance of significant coagulopathy was six days after chemotherapy alone and three days in the ATRA group (p = 0.001) [44]. ATRA therapy may be especially important in reducing the severity of the bleeding tendency in hyperleukocytic APL patients, a population still at a relatively high risk of early death with chemotherapy alone.

In APL patients treated with ATRA alone, primary fibrinogenolysis disappears during the first five days of treatment, while DIC and leukocyte-mediated proteolysis seem to persist during the first two or three weeks of ATRA therapy [54–56]. This could lead to a transient period of hypercoagulability, which could explain the few well documented cases of thromboembolic events in APL patients treated with ATRA [57–59].

Other side effects of ATRA

Dryness of lips and mucosae are usual but are reversible with symptomatic treatment. Increases in transaminases and triglycerides and bone pain are common, but they never required treatment discontinuation in our experience. Headache, due to intracranial hypertension, is generally moderate in adults but may be severe in children, and associated with signs of pseudotumor cerebri [60]. Lower ATRA doses (25 mg/m^2/day) reduce this side effect in children and seem as effective as conventional doses of 45 mg/m^2/day in inducing CR [60]. In our experience isolated fever, in the absence of other signs of the ATRA syndrome (or infection), frequently develops and is reversible within 48 h of ATRA discontinuation [41, 44].

A few other side effects, including hypercalcemia [61], erythema nodosum [62], marked basophilia [63, 64], severe myositis [65], and Sweet syndrome [66] have, on exceptions, been reported.

Conclusion

The advent of ATRA has been a major breakthrough in the treatment of APL. Optimal use of this drug is in combination with classical anthracycline-AraC chemotherapy during induction treatment. Recent results also suggest that ATRA—on an intermittent schedule—may also be beneficial during maintenance, possibly in combination with low dose chemotherapy.

References

1   Bennett JM, Catovsky D, Daniel MT, Flandrin G, Galton D, Gralnick M, Sultan C (1976) Proposals for the classification of the acute leukemias. *Brit J Haematol* 33: 451–460
2   Bennett JM, Catovsky D, Daniel MT, Flandrin G, Galton D, Gralnick M, Sultan C (1980) A variant form of hypergranular promyelocytic leukemia (M3). *Ann Int Med* 92: 280–287
3   Larson RA, Kondo K, Vardiman JW, Butler ARE, Golomb HM, Rowley JD (1984) Evidence for a 15;17 translocation in every patient with acute promyelocytic leukemia. *Amer J Med* 76: 827–835
4   De The H, Lavau C, Marchio A, Chomienne C, Degos L, Dejean A (1991) The PML-RARα fusion mRNA generated by the t(15;17) translocation in acute promyelocytic leukemia encodes a functionally altered RAR. *Cell* 66: 675–682
5   Kakizuka A, Miller WH, Umesono K, Warrell R, Frankel S, Dmitrovsky E, Evans R (1991) Chromosomal translocation t(15;17) in human acute promyelocytic leukemia fuses RARα with a novel putative transcription factor, PML. *Cell* 66: 663–673
6   Tallman MS, Kwaan HC (1992) Reassessing the hemostatic disorder associated with acute promyelocytic leukemia. *Blood* 79: 543–560
7   Dombret H, Sutton L, Duarte M, Daniel MT, Leblond V, Castaigne S, Degos L (1992) Combined therapy with all-*trans* retinoic acid and high-dose chemotherapy in patients with hyperleukocytic acute promyelocytic leukemia and severe visceral hemorrhage. *Leukemia* 6: 1237–1244
8   Bernard J, Weil B, Boiron M, Jacquillat C (1973) Acute promyelocytic leukemia. Results of treatment with daunorubicin. *Blood* 41: 489–496
9   Marty M, Ganem G, Fischer J, Degos L, Berger R, Boiron M (1984) Leucémie aiguë promyélocytaire: étude rétrospective de 119 malades traités par daunorubicine. *Nouv Rev Fr Hematol* 24: 371–378
10 Kantarjian H, Keating M, Walters R (1986) Acute promyelocytic leukemia. MD Anderson Hospital Experience. *Amer J Med* 80: 789–797
11 Cunningham I, Gee T, Reich L (1989) Acute promyelocytic leukemia: treatment results during a decade at Memorial Hospital. *Blood* 73: 1116–1122
12 Sanz M, Jarque I, Martin G (1988) Acute promyelocytic leukemia. Therapy results and prognostic factors. *Cancer* 61: 7–13
13 Cordonnier C, Vernant JP, Brun B (1985) Acute promyelocytic leukemia in 57 previously untreated patients. *Cancer* 55: 18–25
14 Fenaux P, Pollet JP, Vandenbossche L, Dupriez B, Jouet JP, Bauters F (1991) Treatment of acute promyelocytic leukemia: a report on 70 cases. *Leuk Lymphoma* 4: 249–256
15 Fenaux P, Tertian G, Castaigne S (1991) A randomized trial of amsacrine and rubidazone in 39 patients with acute promyelocytic leukemia. *J Clin Oncol* 9: 1556–1561
16 Rodighiero F, Avvisati G, Castaman G, Barbui T, Mandelli F (1990) Early deaths and anti-hemorrhagic treatments in acute promyelocytic leukemia. A GIMEmA retrospective study in 268 consecutive patients. *Blood* 11: 2112–2117
17 Hoyle C, Swirsky D, Freedman L, Hayhoe F (1988) Beneficial effect of heparin in the management of patients with APL. *Brit J Haematol* 68: 283–289
18 Goldberg MA, Ginsburg D, Mayer RJ et al (1987) Is heparin administration necessary during induction

chemotherapy for patients with acute promyelocytic leukemia? *Blood* 69: 187–191
19 Arlin Z, Kempin S, Mertelsmann R (1984) Primary therapy of acute promyelocytic leukemia: Role of Amsacrine and Daunorubicin based therapy. *Blood* 63: 211–212
20 Chan KW, Steinherz PG, Miller DR (1981) Acute promyelocytic leukemia in children. *Med Pediat Oncol* 9: 5–15
21 Carter M, Kalwinsky d Dahl G et al (1989) Childhood acute promyelocytic leukemia: a rare variant of non-lymphoid leukemia with distinctive clinical and biological features. *Leukemia* 4: 298–302
22 Thomas W, Archimbaud E, Treille-Ritouet D, Fiere D (1991) Prognostic factors in acute promyelocytic leukemia: a retrospective study of 67 cases. *Leuk Lymphoma* 4: 249–256
23 Head D, Kopecky KJ, WIillman CL, Appelbaum FR (1994) Aggressive daunomycin (DNR) therapy improves survival in acute promyelocytic leukemia (APL), a Southwest Oncology Group (SWOG trial). *Leukemia* 8, suppl2,: S38–S41
24 Willemze R, Suciu S, Mandelli F, De Witte Th Cadiou M, Castoldi GL, Liso V, Dardenne M, Solbu G, Zittoun R (1994) Treatment of patients with acute promyelocytic leukemia. The EORTC-LCG experience. *Leukemia* 8, suppl. 2: S48–S55
25 Bassan R, Battista R, Viero P, D'Emilio A, Buelli A, Montaldi A, Rambaldi A, Tremul L, Dini E, Barbui T (1995) Short-term treatment for adult hypergranular and microgranular acute promyelocytic leukemia. *Leukemia* 9: 238–243
26 Huang M, Yu-Chen Y, Shu-Rong C, Lu MX, Zhoa L, Gu LJ, Wang ZY (1988) Use of all-*trans* retinoic acid in the treatment of acute promyelocytic leukemia. *Blood* 72: 567–577
27 Chomienne C, Ballerini P, Balitrand N, Daniel MT, Fenaux P, Castaigne S, Degos L (1990) All-*trans* retinoic acid in promyelocytic leukemias. II. *In vitro* studies structure function relationship. *Blood* 76: 1710–1718
28 Castaigne S, Chomienne C, Daniel MT, Berger R, Fenaux P, Degos L (1990) All-*trans* retinoic acid as a differentiating therapy for acute promyelocytic leukemias. I. Clinical results. *Blood* 76: 1704–1709
29 Warrel RP, Frankel SR, Miller WH, Scheinberg DA, Itri IM, Hittelman WN, Vyas R, Andreeff M, Tafuri A, Jakubowski A et al (1991) Differentiation therapy of acute promyelocytic leukemia with tretinoin (all-*trans* retinoic acid). *New Engl J Med* 324: 1385–1395
30 Warrel RP, Frankel SR, Miller W, Itri L, Andreeff M, Jabukowski A, Gabrilove J, Gordon MS, Dmitrovsky E (1991) Differentiation therapy of acute promyelocytic leukemia with tretinoin (all-*trans* retinoic acid). *New Engl J Med* 324: 1385–1395
31 Ohashi H, Ichikawa A, Takagi N, Ohno R (1992) Remission induction of acute promyelocytic leukemia by all-*trans*-retinoic acid: Molecular evidence of restoration of normal hematopoiesis after differentiation and subsequent extinction of leukemic clone. *Leukemia* 6: 859–868
32 Fenaux P, Castaigne S, Chomienne C, Dombret H, Degos L (1992) All-*trans* retinoic acid treatment for patients with acute promyelocytic leukemia. *Leukemia* 6: 64–72
33 Frankel SR, Eardley A, Lauwers G, Weiss M, Warrell R (1992) The "retinoic acid syndrome" in acute promyelocytic leukemia. *Ann Int Med* 117: 292–299
34 Degos L, Chomienne C, Daniel MT, Berger R, Dombret H, Fenaux P, Castaigne S (1990) Treatment of first relapse in acute promyelocytic leukaemia with all-*trans* retinoic acid. *Lancet* 2: 1440
35 Wang ZY, Sun GL, Lu JX, Chen ZH (1990) Treatment of acute promyelocytic leukemia with all-*trans* retinoic acid in China. *Nouv Rev Fr Hematol* 32: 34–36
36 Sun G, Ouyang R, Chen S, Gu Y, Huang L, Lu J, Wang Z (1994) Follow up of 481 patients with APL after CR using ATRA. *Chinese J Hematol* 15 (8): 411–413
37 Sun GL, Zhou RF, Wu W, Jiang GS (1994) Follow up of 524 APL patients after CR induced by ATRA. *Leukemia* 8: 1082 (abstr)
38 Fenaux P, Castaigne S, Dombret H, Archimbaud E, Duarte M, Morel P, Lamy T, Tilly H, Guerci A, Maloisel F et al (1992) All-transretinoic acid followed by intensive chemotherapy gives a high complete remission rate and may prolong remissions in newly diagnosed acute promyelocytic leukemia: a pilot study on 26 cases. *Blood* 80: 2176–2181
39 Fenaux P, Wattel E, Archimbaud E, Sanz M, Hecquet B, Guerci A, Link H, Fegueux N, Fey M, Castaigne S et al (1994) Prolonged follow up confirms that all-transretinoic acid (ATRA) followed by chemotherapy reduces the risk of relapse in newly diagnosed acute promyelocytic leukemia (APL). *Blood* 84: 666–667
40 Warrell RP, Eardley A, Heller G, Miller WH, Frankel SR (1994) Treatment of acute promyelocytic leukemia with all-*trans* retinoic acid: an update of the New York experience. *Leukemia* 8: 926–933
41 Kanamaru A, Takemoto Y, Tanimoto M, Murakami H, Asou N, Kobayashi T, Kuriyama K, Ohmoto E, Sakamaki H, Tsubaki K et al (1995) All-*trans* retinoic acid for the treatment of newly diagnosed acute promyelocytic leukemia. *Blood* 85: 1202–1206
42 Wiley JS, Firkin FC (1995) Reduction of pulmonary toxicity by prednisolone prophylaxis during all-*trans* retinoic acid treamtent of acute promyelocytic leukemia. *Leukemia* 9: 774–778
43 Eardley AM, Heller G, Warrell RP (1994) Morbidity and costs of remission induction therapy with all-*trans* retinoic acid compared with standard chemotherapy in acute promyelocytic leukemia. *Leukemia* 8: 934–939
44 Fenaux P, Le Deley MC, Castaigne S, Archimbaud E, Chomienne C, Link H, Guerci A, Duarte M, Daniel MT et al (1993) Effect of all-transretinoic acid in newly diagnosed acute promyelocytic leukemia. Results of a multicenter randomized trial. *Blood* 82: 3241–3250

45 Fenaux P, Chastang C, Chomienne C, Degos L (1994) Tretinoin with chemotherapy in newly diagnosed acute promyelocytic leukaemia. *Lancet* 343:1033–1034
46 Degos L, Dombret H Chomienne C, Daniel MT, Miclea JM, Chastang C, Castaigne S, Fenaux P (1995) All-*trans*-retinoic acid as a differentiating agent in the treatment of acute promyelocytic leukemia. *Blood* 85: 2643–2653
47 Avvisati G, Baccarini M, Ferrara F, Lazzarino M, Resegotti L, Mandelli F (1994) AIDA protocol (all-*trans* retinoic acid + idarubicin) in the treatment of newly diagnosed acute promyelocytic leukemia (APL): A pilot study of the Italian cooperative group GIMENA. *Blood* 84: 380a (abstr, suppl)
49 Ohno R, Yoshida H, Fukutani H, Naoe T, Oshima T, Kyo T, Endoh N, Fujimoto T, Kobayashi T, Hiraoka A et al (1993) Multi-institutional study of all-*trans* retinoic acid as a differentiation therapy of refractory acute promyelocytic leukemia. *Leukemia* 7: 1722–1727
50 Frankel SR, Eardley A, Heller G, Berman E, Miller WH, Dmitrovsky E, Warrell RP (1994) All-*trans* retinoic acid for acute promyelocytic leukemia. Results of the New York study. *Ann Int Med* 120: 279–286
51 Delva L, Cornic M, Balitrand N, Guidez F, Miclea JM, Delmer A, Teillet F, Fenaux P, Castaigne S, Degos L, Chomienne C (1993) Resistance to All-*trans* retinoic acid therapy in relapsing acute promyelocytic leukemia. *Blood* 82: 2175–2181
52 Dubois C, Schlageter MH, De Gentile A, Guidez F, Balitrand N, Toubert ME, Krawice I, Fenaux P, Castaigne S, Najean Y et al (1994) Hematopoietic growth factor expression and ATRA sensitivity in Acute Promyelocytic Leukemia. *Blood* 83: 3264–3270
53 Tallman MS, Andersen J, Schiffer CA, Bloomfield C, Rowe J (1997) All-*trans* retinoic acid in acute promyelocytic leukemia. *New Engl J Med* 337: 1021–1030
54 Tapiovaara H, Matikainen S, Hurme M, Vaheri A (1994) Induction of differentiation of promyelocytic NB4 cells by retinoic acid is associated with rapid increase in urokinase activity subsequently downregulated by production of inhibitors. *Blood* 83: 1883–1891
55 Federici AB, Berkowitz SD, Lattuada A, Mannucci PM (1993) Degradation of von Willebrand factor in patients with acquired clinical conditions in which there is heightened proteolysis. *Blood* 81: 720–730
56 Dombret H, Scrobohaci ML, Ghorra P, Zini JM, Daniel MT, Castaigne S, Degos L (1993) Coagulation disorders associated with acute promyelocytic leukemia: Corrective effect of all-*trans* retinoic acid treatment. *Leukemia* 7: 2–9
57 Runde V, Aul C, Sudhoff T, Heyll A, Schneider W (1993) Retinoic acid in the treatment of acute promyelocytic leukemia: Inefficacy of the 13-*cis* isomer and induction of complete remission by the all-*trans* isomer complicated by thromboembolic events. *Ann Hematol* 64: 270–277
58 Forjaz De Lacerda J, Alves Do Carmo J, Lurdes Guerra M, Geraldes J, Forjaz De Lacerda JM (1993) Multiple thrombosis in acute promyelocytic leukaemia after tretinoin. *Lancet* 342: 114–115
59 Hashimoto S, Koike T, Tatewaki W, Seki Y, Sato N, Azegami T, Tsukada N, Takahashi H, Kimura H, Ueno M et al (1994) Fatal thromboembolism in acute promyelocytic leukemia during all-*trans* retinoic acid therapy combined with antifibrinolytic therapy for prophylaxis of hemorrhage. *Leukemia* 8: 1113–1115
60 Mahmoud HH, Hurwitz CA, Roberts WM, Santana VM, Ribeiro RC, Krance RA (1993) Tretinoin toxicity in children with acute promyelocytic leukaemia. *Lancet* 342: 1394–1395
61 Akiyama H, Nakamura N, Nagasaka S, Sakamaki H, Onozawa Y (1992) Hypercalcaemia due to all-*trans* retinoic acid. *Lancet* i: 308–309
62 Hakimian D, Tallman MS, Zugerman C, Caro WA (1993) Erythema nodosum associated with all-*trans*-retinoic acid in the treatment of acute promyelocytic leukemia. *Leukemia* 7: 758–759
63 Koik T, Tatewaki W, Aoki A (1992) Brief report: severe symptoms of hyperhistaminemia after the treatment of acute promyelocytic leukemia with tretinoin (all-*trans* retinoic acid). *New Engl J Med* 327: 385–387
64 Iwakiri R, Inokuchi K, Dan K, Nomura T (1994) Marked basophilia in acute promyelocytic leukaemia treated with all-*trans* retinoic acid: molecular analysis of the cell origin of the basophils. *Brit J Haematol* 86: 870–872
65 Miranda N, Oliveira P, Frade MJ, Melo J, Marques MS, Parreora A (1994) Myositis with tretinoin. *Lancet* 344: 1096–1105
66 Tomas JF, Escudero A, Fernandez-Ranada JM (1994) All-*trans* retinoic acid treatment and Sweett syndrome. *Leukemia* 8: 1596–1597

Vitamin A and retinoids: an update of biological aspects and clinical applications
M.A. Livrea (ed.)
© 2000 Birkhäuser Verlag Basel/Switzerland

Bioactivities of N-(4-hydroxyphenyl) retinamide

F. Formelli

Istituto Nazionale per lo Studio e la Cura dei Tumori, Department of Experimental Oncology, via Venezian 1, I-20133 Milan, Italy

Introduction

N-(4-hydroxyphenyl)retinamide (4HPR-fenretinide) is a synthetic amide of all-*trans* retinoic acid (RA) which was synthesised in 1978 [1]. After the important observation that RA reduced mouse skin carcinogenesis, many analogs, now termed retinoids, were synthesised and tested for their efficacy. The major problem that arose in testing retinoids as therapeutic agents was their toxicity. The ratio of efficacy to toxicity, i.e. the "therapeutic index" became an important consideration in assessing the potential clinical utility of different retinoids. Since the first *in vivo* studies on inhibition of experimentally induced carcinogenesis, 4HPR proved to be one of the most effective retinoids with relatively low toxicity. It has then been proven to be active in the prevention and treatment of a variety of tumors in animals and, by inducing apoptosis, it has shown *in vitro* growth inhibitory activity against a large number of human tumor cell lines from different tissue origins. For all these reasons it was selected to enter into clinical trials. 4HPR was first tested for the treatment of skin disease and it is currently under investigation as a chemopreventive and therapeutic agent.

Preclinical efficacy studies

In tracheal organs cultured *in vitro*, which was a common screening procedure for evaluating new retinoids, 4HPR was shown to be highly active in reversing keratinization [2]. The first studies mainly examined the efficacy of 4HPR as a chemopreventive agent of mammary tumors. In 1979 Moon et al. [2] first reported the efficacy of 4HPR against mammary carcinogenesis. Of all the retinoids evaluated 4HPR and retinyl acetate were the most efficacious, however, 4HPR was markedly less toxic than retinyl acetate. Its administration resulted in a potent antiproliferative effect on mammary epithelium, evidenced by decreased ductal branching and end bud proliferation. In carcinogen-induced mammary tumors, positive chemopreventive interaction has been reported between 4HPR and ovariectomy [3] and between 4HPR and tamoxifen [4, 5]. 4HPR was also effective in inhibiting mammary tumorigenesis in mice [6]. 4HPR can also prevent tumors in organs other than the mammary gland (Tab. 1). It was an effective inhibitor of urinary bladder [7, 8], skin [9, 10], colon [11], lung [12], prostate [13, 14] and seminal vesicle [14] carcinogenesis. 4HPR also prevented virally-induced T-lymphomas in mice [15]. In *in vivo* tumor experimental models 4HPR has been shown to be effective not only as a preventive agent, but also as a therapeutic agent (Tab. 1). A moderate number of complete and

Table 1. Preclinical activity of 4HPR

	Tumor (carcinogen/or origin)	Species	Reference
In vivo preventive effect	Mammary (MNU)	Rat	2
	Mammary (spontaneous)	Mouse	6
	Urinary Bladder (OH-BBN)	Rat	7
	Urinary Bladder (OH-BBN)	Mouse	8
	Skin (DMBA/TPA)	Mouse	9
	Skin (Spontaneous)	Rat	10
	Colon (DMH)	Rat	11
	Lung (DEN)	Hamster	12
	Prostate (MNU+TP)	Rat	13,14
	Seminal vesicle (MNU+TP)	Rat	14
	T-lymphoma (virus)	Mouse	15
In vivo therapeutic effect	Mammary (MNU)	Rat	16
	Prostate (rat tumor line)	Rat	17,18
	Ovary (human tumor line)	nu/nu mouse	19
In vitro antiproliferative effect	Breast carcinoma	Human	24,25,26,27
	Prostate carcinoma.	Human	28,29
	Leukemia and lymphoma	Human	30,31
	Neuroblastoma carcinoma	Human	32,33,34
	Ovarian carcinoma	Human	35,36
	Cervical carcinoma	Human	37,38
	Head and neck squamous carcinoma	Human	39
	Esophageal squamous carcinoma	Human	40
	Small-cell lung carcinoma	Human	41

MNU: N-methyl-N-nitrosourea; DMBA: dimethylbenz(a)anthracene; OH-BBN: N-butyl-N-(4-hydroxy-buthyl)nitrosamine, DMH: 1,2-dimethylhydrazine; DEN: diethylnitrosamine; TP: testosterone propionate; TPA,:12-0-tetradecanoyl-phorbol-13-acetate.

partial regressions of already established MNU-induced mammary tumors were reported in rats treated with high doses of 4HPR [16]. 4HPR was effective in inhibiting rat prostate tumor growth and metastases in the lung [17, 18]. The intracavitary administration of 4HPR had a therapeutic effect in a human ovarian carcinoma xenografted intraperitoneally into nude mice [19]. The same treatment potentiated the activity of cisplatin in the same tumor [19]. In rodents, after the administration of equivalent molar doses, 4HPR was less toxic than RA and retinyl acetate [2]. Unlike most natural and synthetic retinoids, 4HPR did not cause hepatic injuries and bone fractures [20, 21]. Moreover, 4HPR has proved to be less teratogenic than RA [22] and non-genotoxic [23].

The possible modes of action of 4HPR have been mostly investigated in already transformed tumor cells. 4HPR suppressed the proliferation of cell lines of different tumor types such as human breast carcinoma [24–27], prostate adenocarcinoma [28, 29], leukemias [30, 31], neuroblastoma [32, 33, 34], ovarian carcinoma [35, 36], cervical carcinoma [37, 38], head and neck squamous cell carcinoma [39], esophageal squamous carcinoma [40] and small-cell lung cancer [41] (Tab. 1). This effect occured at concentrations from 1 μM, a concentration which is pharmacologically achieavable in man [42], to 10 μM. Interestingly, 4HPR was shown to be much less effective in inhibiting proliferation of normal human cervical and ovarian cells than that of the respective tumor cells [36, 38]. In most of the tested cell systems, 4HPR was more potent than either RA or 13-*cis*-RA, and instead of having differentiative or cytostatic effects, as most retinoids do, 4HPR induced apoptosis. Thus apoptosis of DNA-damaged cells may constitute the mechanism of the 4HPR preventive effect.

The apoptotic-inducing effect of 4HPR has been associated with different mechanisms. 4HPR apoptosis was delayed by ectopic overexpression of bcl-2 in human leukemic cells [31] and, in breast cancer cells, 4HPR treatment resulted in decreased bcl-2 mRNA levels [43]. The apoptosis related genes, p53 and c-myc, do not seem to be involved in the apoptotic effect of 4HPR in human leukemic cell lines [31] although the sensitivity of human small-cell lung cancer cell lines to 4HPR was associated with a high expression of c-myc [41]. Induction of apoptosis by 4HPR was associated with induction of transforming growth factor β1 (TGFβ1) expression in prostate cancer cells [29] and in down-modulation of β1 integrin expression in neuroblastoma cells [34]. Studies conducted on myeloid and cervical cancer cells [31, 38] suggest that 4HPR induces apoptosis by eliciting oxidative stress and generating reactive oxygen species. It has been also suggested that one of the mechanisms by which 4HPR inhibits tumor cell growth is through induction of the enzyme that metabolizes retinol to the biologically active metabolite 4-oxoretinol [44]. In breast cancer cells, 4HPR, like RA, enhanced the metabolism of retinol to 4-oxoretinol in estrogen receptor positive (ER^+) but not in ER^- cell lines [44].

The interrelationship between 4HPR and retinoic acid receptors (RAR) is not clear. 4HPR has been shown to bind poorly to partially purified RARα, RARβ and RARγ *in vitro* [27, 45] and to only minimally activate retinoic acid receptor elements (RARE) and retinoid X receptor elements (RXRE) [27]. In cells obtained by cotransfection, expressing high RARs and RXRs levels, 4HPR was a higly selective activator of RARs and a potent activation protein 1 (AP-1) inhibitor [46]. Indirect evidence shows that RARs, and in particular RARβ, might be implicated in the 4HPR antiproliferative effects. Among four human ovarian cancer cell lines, the most sensitive cell line to 4HPR had the greatest expression of RARα and RARγ and it was the only one to constitutively express RARβ [35]. 4HPR, similarly to RA, induced the expression of RARβ mRNA levels in human normal mammary epithelial cells, but not in mammary tumor cells [47]. Since the RARβ gene promoter contains a RARE, these results suggest that 4HPR may mediate its effects via nuclear RARs. Similar results were obtained in a recent study [36] in which following 4HPR treatment, ovarian cancer cells that were sensitive to 4HPR contained higher levels of RARβ than more resistant cells. Thus, even if the binding of 4HPR to RARs and RXRs has not yet been proven, the activity of this retinoid seems to be associated, in same cases, with RAR expression.

Pharmacokinetics and pharmacological effects

In rodents, 4HPR was metabolized to more lipophilic compounds, including M-(4-methoxyphenyl)retinamide (4MPR), 4HPR-o-glucuronide and numerous other polar compounds, but was not detectably hydrolyzed to RA [48, 49]. Compared with rats, mice metabolized 4HPR more extensively and showed a larger percentage of 4MPR in their organs [50]. After an intravenous administration in rats, 4HPR reached the highest concentrations in the liver, but was not stored there [50]. Its elimination from the rat mammary gland was much slower (44 h) than from plasma and liver (12 h). Of the administrated 4HPR, 64% was excreted in the feces in five days [48].

In humans, the pharmacokinetcs of 4HPR have been studied both in healthy volunteers [51, 52] and cancer patients [42, 53, 54] (Tab. 2). After oral administration of a dose of 300 mg, peak plasma concentrations of 4HPR, whose MW is 391.55 kDa, were in the range of 198–596 ng/ml (i.e. 0.5–1.5 µM) and they were reached between five and seven hours. The concentrations of 4MPR were lower than those of the parent drug and were reached between eight and 12 h. The bioavailability of 4HPR, like that of other retinoids, is enhanced by administration with food and is influenced by meal composition [52]. After a single dose, the mean elimination half-life was 14–20 h for 4HPR and 22–28 h for 4MPR (Tab. 2).

At variance with RA, whose continuous administration results in a rapid decrease in its plasma concentration, the pharmacokinetics of 4HPR do not change appreciably during long-term treatment. After multiple daily doses for 28 days the half-life of 4HPR increased only slightly (27 h) whereas 4MPR showed evidence of accumulation with a half life of 45–51 h. Daily administration of a dose of 200 mg for five years resulted in steady state 4HPR plasma concentrations of approximately 400 ng/ml, i.e. 1 µM which remained constant during the whole period [42]. After interruption of the five year treatment, the half-lives of 4HPR and 4MPR were 27 and 54 h respectively [42], similar to those reported after 28 daily treatments. At six

Table 2. Pharmacokinetic parameters of 4HPR and 4MPR after single and multiple doses in humans

Dose mg/day	Treatment period	4HPR			4MPR			References
		Peak concentration (ng/ml)	Trough concentration (at 12–19 h) (ng/ml)	t1/2β (h)	Peak concentration (ng/ml)	Trough concentration (at 12–19 h) (ng/ml)	t1/2β (h)	
150[a]	1 day	99		20	35		28	51
300[a]	1 day	198		18	50		22	51
300[b]	1 day	500		14	100		23	53
300[c]	1 day	198			91			52
300[d]	1 day	596			211			52
150[a]	28 days	128		27	56		45	51
300[a]	28 days	257		27	116		51	51
200[b]	5 years		≅400	27		≅400	54	42

[a]10' after a standardard breakfast meal
[b]No detail is given about interval from drug and food intake
[c]During fasting
[d]After meal

and 12 months after discontinuation of the drug, the concentrations of 4HPR in plasma were at the limit of detectability (0.01 μM) whereas those of 4MPR were five-fold higher [42].

No 4HPR and no metabolite or conjugate were found in the urine of treated subjects [51]. 4HPR has been shown to accumulate in human breast tissues [42]. Its concentrations and those of 4MPR in nipple discharge, which is secreted by the breast gland, were 10–30 fold greater, respectively, than those found in plasma. 4HPR crosses the human placenta, but at variance with isotretinoin (13 cisRA), it is not converted to RA and it is not stored in the embryo [55].

A relevant pharmacological effect of 4HPR, found both in rats [56] and in humans [53, 54], is the rapid and marked decrease in the plasma concentrations of both retinol and its plasma transport protein RBP. This effect is reversible, it already occurs after a single dose (Tab. 3) and it is proportional to the dose of 4HPR administered [54]. This phenomenon and its dose dependency are associated with impaired dark adaptation, a side-effect reported in patients treated with 4HPR [57–61]. In order to minimise this side-effect in clinical trials, treatment with 4HPR is interrupted for three days at the end of each month. This increases plasma retinol concentrations and, as a consequence, the storage of retinol in the pigment epithelial cells of the eye is enhanced. During daily treatment with 200 mg 4HPR, baseline retinol levels were reduced by 71% and this reduction was steady during five years of continuous treatment [42]. After the three day drug interruption, the mean reduction was 38% [42] (Tab. 3). The mechanisms responsible for retinol level reduction by 4HPR seem to be due to the high binding affinity of 4HPR with RBP [62] and with the lack of binding of the 4HPR-RBP complex with transthyretin (TTR) [63]. This interferes with secretion of retinol-RBP from the liver [64].

4HPR has been shown to influnce the immune system. In vitamin A-deficient rats, 4HPR increased the number of lymphocytes and natural killer (NK) cells and it slightly increased NK cell lytic efficiency [65]. In breast cancer patients treated for six months with 200 mg 4HPR, the NK activity was augmented by 73% as compared to that of control patients [66] (Tab. 3).

Another interesting pharmacological effect of 4HPR which might be relevant for its chemopreventive activity is the reduction of circulating insuline-like growth factor 1 (IGF-1), a potent

Table 3. 4HPR pharmacological effects in humans

Parameter	Dose mg/day	Length of treatment	Effect	References
Plasma retinol levels	200	1 day	38% reduction	54
	300	1–2 weeks	60% reduction	53
	200	5 years	71% reduction	42
	200	after 3 days of drug interruption	38% reduction	42
NK activity of peripheral blood mononuclear cells	200	6 months	73% increase	66
Plasma IGF-1 levels	200	12 months	15.3% reduction	67

mitogen for transformed epithelial cells (Tab. 3). The administration of 4HPR to breast cancer patients, caused a significant decline in plasma IGF-1 levels [67] and this effect was particularly pronounced in premenopausal women.

Clinical studies

4HPR was first introduced into the clinic for treatment of skin diseases. Three out of eight patients with severe chronic plaque psoriasis were graded as being better after a 600 mg daily dose [58]. Subsequently in a phase II trial in patients with advanced breast cancer and melanoma who had been heavily pretreated with other drugs, 4HPR at doses of 300 and 400 mg/day, showed no beneficial effect [59]. The same doses were not effective in patients with myelodysplatic syndromes; on the contrary, disease acceleration was observed in some patients [68].

During that period a randomised chemoprevention trial with 4HPR in breast cancer was going to start. The impaired dark adaptation observed in patients treated with 800, 600 and 400 mg/day [57–59] suggested the running of a phase I trial with the aim of identifying the highest dose of 4HPR to be safely administered for a long period of time. Stage I breast cancer patients were randomised to receive 100, 200, 300, mg/day 4HPR and placebo. The 300 mg/day provoked alterations in the electroretinogram (ERG) in one out of 25 patients after six months of treatment [61]. The 200 mg/day dose proved to have minimal side effects after 6–12 months of administration [69]. In the meantime the relationship between 4HPR intake and the reduction in retinol levels reduction was observed [54]. Thus, the 200 mg/day dose was adopted with a three day drug holiday at the end of each month in order to allow plasma retinol recovery.

Three randomised chemoprevention trials in breast cancer, basal cell carcinoma and oral leukoplakia started at the Istituto Nazionale Tumori (INT)-Milan in 1987. The aim of the breast prevention trial is to evaluate the efficacy of 4HPR administered for five years in preventing controlateral primaries in women previously operated on for early breast cancer [70]. Final analysis of the study will be soon performed. Preliminary results from this study suggest a chemopreventive effect of 4HPR in ovarian cancer [71]. A significant reduction in ovarian cancer incidence has been observed during the treatment in patients belonging to the 4HPR group.

The aim of the basal cell carcinoma study is the prevention of recurrences or new occurrences after surgical resection [72]. In patients surgically treated for oral leukoplakia new occurrences were reduced by 75% in the 4HPR group vs the control group, although recurrences were the same in both groups [73]. Administered topically, 4HPR has been shown to be effective in patients with diffuse oral lichen planus or leukoplakia [74] and with actinic keratoses [75].

In patients with superficial bladder cancer, treatment with 200 mg/day 4HPR decreased the proliferative activity of cells and delayed the development of DNA aneuploidy evaluated in cells obtained by serial bladder washings [76]. A phase I/II trial in stage IV breast cancer patients, evaluating tamoxifen at 20 mg/day with 4HPR at doses up to 700 mg/day showed no synergistic toxicity and responses among receptor-positive previously untreated patients [77, 78].

A study was undertaken to determine the toxicity and feasibility for the use of 4HPR as a chemopreventive agent in men at risk for prostate cancer [79]. In this study the administered dose was 100 mg/day. 4HPR was well tolerated and it did not produce any side-effects. The number of subjects was too small for statistical analyses. In any case, no therapeutic benefit was found.

Since a potential role for 4HPR in the treatment of rheumatoid arthritis has been proposed [80], a pilot study has been run in patients with this disease [81]. No clinical improvement has been found.

In all the clinical studies reported, 4HPR has been found to be well tolerated. It does not give rise to ligament calcification, skeletal hyperostosis, and it does not influence bone density [69]. Its continuous administration does not modify lipoprotein levels [82] and is not associated with hepatotoxicity [69]. 4HPR causes dermatological side-effects [61] although its dermatological tolerabilty is good even after more than three years of administration [69]. The major adverse effect of 4HPR is decreased night vision which is associated with retinol level reduction as previously described. The dose of 200 mg/day with a three day drug interruption at the end of each month, produced mild changes in night vision measurable by dark adaptometry which were inversely correlated with plasma retinol levels [60, 83]. The cumulative incidence of visual complaints was nearly 20% at five years of treatment and the occurrence of this symptom was more frequent at the start of treatment [84]. All 4HPR adverse effects are readily reversible upon treatment interruption.

Thus, up to now, 4HPR has been shown to be well-tolerated in humans with no major toxicity associated with its use. Although not showing benefit in the treatment of some advanced cancers it has shown considerable promise as a preventive agent against epithelial neoplasms in several tissues.

Acknowledgements
The author wish to thank Ms. Elena Morittu for editorial assistance.

References

1   Gander RJ, Gurney JA (1978) All-*trans* retinoic acid esters and amides. U.S. patent 4,108,880, August 22, 1978. *Chem Abstr* 88: 89892
2   Moon RC, Thompson HJ, Becci PJ, Grubbs CJ, Gander RJ, Newton DL, Smith JM, Phillips SL, Henderson WR, Mullen LT et al (1979) N-(4-hydroxyphenyl)retinamide, a new retinoid for the prevention of breast cancer in the rat. *Cancer Res* 39: 1339–1346
3   McCormick DL, Mehta RG, Thompson CA, Dinger N, Caldwell JA, Moon RC (1982) Enhanced inhibition of mammary carcinogenesis by combination N-(4-hydrophenyl)retinamide and ovariectomy. *Cancer Res* 42: 508–512
4   McCormick DL, Moon RC (1986) Retinoid-tamoxifen interaction in mammry cancer chemoprevention. *Carcinogenesis* 7: 193–196
5   Ratko TA, Detrisac CJ, Dinger MN, Thomas CF, Kellof GJ, Moon RC (1989) Chemopreventive efficacy of combined retinoid and tamoxifen treatment following surgical excision of a primary mammary cancer in female rats. *Cancer Res* 49: 4472–4476
6   Welsch CW, Dehoog JV, Moon RC (1983) Inhibition of mammary tumorigenesis in nulliparous C3H mice by chronic feeding of the synthetic retinoid, N-(4-hydroxyphenyl)-retinamide. *Carcinogenesis* 4: 1185–1187
7   Moon RC, McCormick DL, Mehta RG (1983) Inhibition of carcinogenesis by retinoids. *Cancer Res* 43: 2469 s-2475 s
8   Moon RC, McCormick DL, Becci PJ, Shealy YF, Frickel F, Paust J, Sporn MB (1982) Influence of 15 retinoic acid amides on urinary bladder carcinogenesis in the mouse. *Carcinogenesis* 3: 1469–1472
9   McCormick DL, Moon RC (1986) antipromotional activity of dietary N-(4-hydroxyphenyl)retinamide in two-

stage skin tumorigenesis in CD-1 and sencar mice. *Cancer Lett* 31: 133–138

10 Ohshima M, Ward JM, Wenk ML (1985) Preventive and enhancing effects of retinoids on the development of naturally occurring tumors of skin, prostate gland, and endocrine pancreas in aged male aci/seghapbr rats. *J Nat Cancer Inst* 74: 517–524

11 Silverman J, Katayama S, Zelenakas K, Lauber J, Musser TK, Reddy M, Levenstein MJ, Weisburger JH (1981) Effect of retinoids on the induction of colon cancer in F344 rats by N-methyl-N-nitrosourea or by 1,2-dimethylhydrazine. *Carcinogenesis* 2: 1167–1172

12 Moon RC, Rao KVN, Detrisac CJ, Kellof GJ (1992) Hamster lung cancer model of carcinogenesis and chemoprevention. *In:* GR Newell, WK Hong (eds): *The biology and prevention of aerodigestive tract cancers.* Plenum Press, New York, 55–61

13 Pollard M, Luckert PH, Sporn MB (1991) Prevention of primary prostate cancer in Lobund-Wistar rats by N-(4-hydroxyphenyl)retinamide. *Cancer Res* 51: 3610–3611

14 Scott M, Anzano MA, Slayter MV, Anver MR, Green DM, Shrader MW, Logsdon DL, Driver CL, Brown CC, Peer CW et al (1995) Chemopreventive activity of tamoxifen, N-(4-hydroxyphenyl)retinamide, and the vitamin D analogue Ro24-5531 for androgen-promoted carcinomas of the rat seminal vesicle and prostate. *Cancer Res* 55: 5621–5627

15 Chan L-NL, Zhang S, Cloyd M, Chan T-S (1997) N-(4-hydroxyphenyl)retinamide prevents development of T-lymphomas in AKR/J mice. *Anticancer Res* 17: 499–504

16 Dowlatshahi K, Mehta RG, Thomas CF, Dinger NM, Moon RC (1989) Therapeutic effect of N-(4-hydroxyphenyl)-retinamide on N-methyl-N-nitrosourea-induced rat mammary cancer. *Cancer Lett* 47: 187–192

17 Pollard M, Luckert PH (1991) The inhibitory effect of 4-hydroxyphenyl retinamide (4-HPR) on metastasis of prostate adenocarcinoma-III cells in Lobund-Wistar rats. *Cancer Lett* 59: 159–163

18 Pienta KJ, Nguyen NM, Lehr JE (1993) Treatment of prostate cancer in the rat with the synthetic retinoid fenretinide. *Cancer Res* 53: 224–226

19 Formelli F, Cleris L (1993) Synthetic retinoid fenretinide is effective against a human ovarian carcinoma xenograft and potentiates cisplatin activity. *Cancer Res* 53: 5374–5376

20 Hixson EJ, Denine EP (1979) Comparative subacute toxicity of retinyl acetate and three synthetic retinamides in swiss mice. *J Nat Cancer Inst* 63: 1359–1364

21 McCarthy DJ, Lindamood C, III, Gundberg CM, Hill DL (1989) Retinoid-induced hemorrhaging and bone toxicity in rats fed diets deficient in vitamin K. *Toxicol Appl Pharmacol* 97: 300–310

22 Kenel MF, Krayer JH, Merz EA, Pritchard JF (1988) Teratogenicity of N-(4-hydroxyphenyl)-all-*trans*-retinamide in rats and rabbits. *Teratog Carcinog Mutagen* 8: 1–11

23 Paulson JD, Oldham JW, Preston RF, Newman D (1985) Lack of genotoxicity of the cancer chemopreventive agent N-(4-hydroxyphenyl)retinamide. *Fundam Appl Toxicol* 5(1): 144–150

24 Fontana JA (1987) Interaction of retinoids and tamoxifen on the inhibition of human mammary carcinoma cell proliferation. *Exp Cell Biol* 55: 136–144

25 Marth C, Bock G, Daxenbichler G (1985) Effect of 4-hydroxyphenylretinamide and retinoic acid on proliferation and cell cycle of cultured human breast cancer cells. *J Nat Cancer Inst* 75: 871–875

26 Pellegrini R, Mariotti A, Tagliabue E, Bressan R, Bunone G, Coradini D, Della Valle G, Formelli F, Cleris L, Radice P et al (1995) Modulation of markers associated with tumor aggressiveness in human breast cancer cell lines by N-(4-hydroxyphenyl)retinamide. *Cell Growth Differ* 6: 863–869

27 Sheikh MS, Shao Z-M, Li Z-S, Ordenez JV, Conley BA, Wu S, Dawson MI, Han Q-X, Chao W, Quick T et al (1995) N-(4-hydroxyphenyl)retinamide (4-HPR)-mediated biological actions involve retinoid receptor-independent pathways in human breast carcinoma. *Carcinogenesis* 16: 2477–2486

28 Igawa M, Tanabe T, Chodak GW, Rukstails DB (1994) N-(4-hydroxyphenyl)retinamide induces cell cycle specific growth inhibition in PC3 cells. *Prostate* 24: 299–305

29 Roberson KM, Penland SN, Padilla GM, Selvan RS, Kim C-S, Fine RL, Robertson CN (1997) Fenretinide: induction of apoptosis and endogenous transforming growth factor β in PC-3 prostate cancer cells. *Cell Growth Differ* 8: 101–111

30 Delia D, Aiello A, Lombardi L, Pelicci PG, Grignani F, Grignani F, Formelli F, Menard S, Costa A, Veronesi U et al (1993) N-(4-hydroxyphenyl)retinamide induces apoptosis of malignant hemopoietic cell lines including those unresponsive to retinoic acid. *Cancer Res* 53: 6036–6041

31 Delia D, Aiello A, Formelli F, Fontanella E, Costa A, Miyashita T, Reed JC, Pierotti MA (1995) Regulation of apoptosis induced by the retinoid N-(4-hydroxyphenyl)retinamide and effect of deregulated bcl-2. *Blood* 85: 359–367

32 Di Vinci A, Geido E, Infusini E, Giaretti W (1994) Neuroblastoma cell apoptosis induced by the synthetic retinoid N-(4-hydroxyphenyl)retinamide. *Int J Cancer* 59: 422–426

33 Ponzoni M, Bocca P, Chiesa V, Decensi A, Pistoia V, Raffaghello K, Rozzo C, Montaldo PG (1995) Differential effects of N-(4-hydroxyphenyl)retinamide and retinoic acid on neuroblastoma cells: apoptosis versus differentiation. *Cancer Res* 55: 853–861

34 Rozzo C, Chiesa V, Caridi G, Pagnan G, Ponzoni M (1997) Induction of apoptosis in human neuroblastoma cells by abrogation of integrin-mediated cell adhesion. *Int J Cancer* 70: 688–698

35 Supino R, Crosti M, Clerici M, Warlters A, Cleris L, Zunino F, Formelli F (1996) Induction of apoptosis by fenretinide (4HPR) in human ovarian carcinoma cells and its association with retinoic acid receptor expres-

sion. *Int J Cancer* 65: 1–7
36 Sabichi AL, Hendricks DT, Bober MA, Birrer MJ (1998) Retinoic acid receptor β expression and growth inhi-
    bition of gynecologic cancer cells by the synthetic retinoid N-(4-hydroxyphenyl)retinamide. *J Nat Cancer Inst*
    90: 597–605
37 Oridate N, Lotan D, Mitchell MF, Hong WK, Lotan R (1995) Induction of apoptosis by retinoids in human
    cervical carcinoma cell lines. *Int J Oncol* 7: 433–441
38 Oridate N, Higuchi M, Suzuki S, Shroot B, Hong WK, Lotan R (1997) Rapid induction of apoptosis in human
    C33A cervical carcinoma cells by the synthetic retinoid 6-[3-(1-adamantyl)hydroxyphenyl]-2-naphtalene car-
    boxylic acid (CD437) *Int J Cancer* 70: 484–487
39 Oridate N, Lotan D, Xu X-C, Hong WK, Lotan R (1996) Differential induction of apoptosis by all-*trans*-
    retinoic acid and N-(4-hydroxyphenyl)retinamide in human head and neck squamous cell carcinoma cell lines.
    Clin Cancer Res 2: 855–863
40 Muller A, Nakagawa H, Rustgi AK (1997) Retinoic acid and N-(4-hydroxy-phenyl)retinamide suppress
    growth of esophageal squamous carcinoma cell lines. *Cancer Lett* 113: 95–101
41 Kalemkerian GP, Slusher R, Ramalingam S, Gadgeel S, Mabry M (1995) Growth inhibition and induction of
    apoptosis by fenretinide in small-cell lung cancer cell lines. *J Nat Cancer Inst* 87: 1674–1680
42 Formelli F, Clerici M, Campa T, Di Mauro MG, Magni A, Mascotti G, Moglia D, De Palo G, Costa A, Veronesi
    U (1993) Five-year administration of fenretinamide: pharmacokinetics and effects on plasma retinol concen-
    trations. *J Clin Oncol* 11: 2036–2042
43 Wang TTY, Phang JM (1996) Effect of N-(4-hydroxyphenyl)retinamide on apoptosis in human breast cancer
    cells. *Cancer Lett* 107: 65–71
44 Chen AC, Guo X, Derguini F, Gudas LJ (1997) Human breast cancer cells and normal mammary epithelial
    cells: retinol metabolism and growth inhibition by the retinol metabolite 4-oxoretinol. *Cancer Res* 57:
    4642–4651
45 Sani BP, Shealy YF, Hill DL (1995) N-(4-hydroxyphenyl)retinamide: interactions with retinoid-binding pro-
    teins/receptors. *Carcinogenesis* 16: 2531–2534
46 Fanjul AN, Delia D, Pierotti MA, Rideout D, Qiu J, Pfahl M (1996) 4-hydroxyphenyl retinamide is a highly
    selective activator of retinoid receptors. *J Biol Chem* 271: 22 441–22 446
47 Swisshelm K, Ryan K, Lee X, Tsou HC, Peacocke M, Sager R (1994) Down-regulation of retinoic acid recep-
    tor β in mammary carcinoma cell lines and its up-regulation in senescing normal mammary epithelial cells.
    Cell Growth Differ 5: 133–141
48 Swanson BN, Newton DL, Roller PP, Sporn MB (1981) Biotransformation and biological activity of N-
    (4-hydroxyphenyl)retinamide derivatives in rodents. *J Pharmacol Exp Ther* 219: 632–637
49 Bunk MJ, Kinahan JJ, Sarkar NH (1985) Biotransformation and protein binding of N-(4-hydroxyphenyl)reti-
    namide in murine mammary epithelial cells. *Cancer Lett* 26: 319–326
50 Hultin TA, May CM, Moon RC (1986) N-(4-hydroxyphenyl)-*all-trans*-retinamide pharmacokinetics in female
    rats and mice. *Drug Metab Disposition* 14: 714–717
51 Professional staff Johnson and Johnson Co (1987) Fenretinide. *Drugs Future* 12: 305–306
52 Doose DR, Minn FL, Stellar S, Nayak RK (1992) Effects of meals and meal composition on the bioavail-
    ability of fenretinide. *J Clin Pharmacol* 32: 1089–1095
53 Peng Y-M, Dalton WS, Alberts DS, Xu M-J, Lim H, Meyskens FL (1989) Pharmacokinetics of N-4-hydrox-
    yphenyl-retinamide and the effect of its oral administration on plasma retinol concentrations in cancer patients.
    Int J Cancer 43: 22–26
54 Formelli F, Carsana R, Costa A, Buranelli F, Campa T, Dossena G, Magni A, Pizzichetta M (1989) Plasma
    retinol level reduction by the synthetic retinoid fentretinide: a one year follow-up study of breast cancer
    patients. *Cancer Res* 49: 6149–6152
55 Formelli F, De Palo G, Costa A, Veronesi U (1997) Human transplacental passage of the retinoid fenretinide
    (4HPR). *Eur J Cancer* 34: 428–429
56 Formelli F, Carsana R, Costa A (1987) N-(4-hydroxyphenyl)retinamide (4-HPR) lowers plasma retinol levels
    in rats. *Med Sci Res* 15: 843–844
57 Kaiser-Kupfer MI, Peck GL, Caruso RC, Jaffe MJ, DiGiovanna JJ, Gross EG (1986) Abnormal retinal func-
    tion associated with fenretinide, a synthetic retinoid. *Arch Ophthalmol* 104: 69–70
58 Kingston TP, Lowe NJ, Winston J, Heckenlively J (1986) Visual and cutaneous toxicity which occurs during
    N-(4-hydroxyphenyl)retinamide therapy for psoriasis. *Clin Exp Dermatol* 11: 624–627
59 Modiano MR, Dalton WS, Lippman SM, Joffe L, Booth AR, Meyskens FL Jr (1990) Phase II study of fen-
    retinide (N-[4-hydroxyphenyl]retinamide) in advanced breast cancer and melanoma. *Invest New Drugs* 8:
    317–319
60 Caruso RC, Zujewski J, Iwata F, Podgor MJ, Conley BA, Ayres LM, Kaiser-Kupfer MI (1998) Effects of fen-
    retinide (4-HPR) on dark adaptation. *Arch Ophthalmol* 116: 759–763
61 Costa A, Malone W, Perloff M, Buranelli F, Campa T, Dossena G, Magni A, Pizzichetta M, Andreoli C, Del
    Vecchio M et al (1988) Tolerability of the synthetic retinoid fenretinide (HPR). *Eur J Cancer Clin Oncol* 25:
    805–808
62 Berni R, Formelli F (1992) *In vitro* interaction of fenretinide with plasma retinol-binding protein and its func-
    tional consequences. *FEBS Lett* 1: 43–45

63 Berni R, Clerici M, Malpeli G, Cleris L, Formelli F (1993) Retinoids: *in vitro* interaction with retinol-binding protein and influence on plasma retinol. *FASEB J* 7: 1179–1184

64 Holven KB, Natarajan V, Gundersen TE, Moskaug JØ, Norum KR, Blomhoff R (1997) Secretion of N-(4-hydroxyphenyl) retinamide-retinol-binding protein from liver parenchymal cells: evidence for reduced affinity of the complex for transthyretin. *Int J Cancer* 71: 654–659

65 Zhao Z, Matsuura T, Popoff K, Ross AC (1994) Effects of N-(4-hydroxyphenyl)-retinamide on the number and cytotoxicity of natural killer cells in vitamin-A-sufficient and -deficient rats. *Nat Immun* 13: 280–288

66 Villa ML, Ferrario E, Trabattoni D, Formelli F, De Palo G, Magni A, Veronesi U, Clerici E (1993) Retinoids, breast cancer and NK cells. *Brit J Cancer* 68: 845–850

67 Torrisi R, Pensa F, Orengo MA, Calsafados E, Ponzani P, Boccardo F, Costa A, Decensi A (1993) The synthetic retinoid fenretinide lowers plasma insulin-like growth factor I levels in breast cancer patients. *Cancer Res* 53: 4769–4771

68 Garewal HS, List A, Meyskens F, Buzaid A, Greenberg B, Katakkar S (1989) Phase II trial of fenretinide [N-(4-hydroxyphenyl)retinamide] in myelodysplasia: possible retinoid-induced disease acceleration. *Leuk Res* 13: 339–343

69 Rotmensz N, De Palo G, Formelli F, Costa A, Marubini E, Campa T, Crippa A, Danesini GM, Delle Grottaglie M, Di Mauro MG et al (1991) Long-term tolerability of fenretinide (4-HPR) in breast cancer patients. *Eur J Cancer* 27: 1127–1131

70 De Palo G, Camerini T, Marubini E, Costa A, Formelli F, Del Vecchio M, Mariani L, Miceli R, Mascotti G, Magni A et al (1997) Chemoprevention trial of contralateral breast cancer with fenretinide. Rationale, design, methodology, organization, data management, statistics and accrual. *Tumori* 83: 884–894

71 De Palo G, Veronesi U, Camerini T, Formelli F, Mascotti G, Boni C, Fosser V, Del Vecchio M, Campa T, Costa A et al (1995) Can fenretinide protect women against ovarian cancer? *J Nat Cancer Inst* 87: 146–147

72 De Palo G, Formelli F (1995) Risks and benefits of retinoids in the chemoprevention of cancer. *Drug Safety* 13: 245–256

73 Chiesa F, Tradati N, Marazza M, Rossi N, Boracchi P, Mariani L, Clerici M, Formelli F, Barzan L, Carrassi A et al (1992) Prevention of local relapses and new localisations of oral leukoplakias with the synthetic retinoid fenretinide (4-HPR). Preliminary results. *Oral Oncol, Eur J Cancer* 28B: 97–102

74 Tradati N, Chiesa F, Rossi N, Grigolato R, Formelli F, Costa A, De Palo G (1994) Successful topical treatment of oral lichen planus and leukoplakias with fenretinide (4-HPR). *Cancer Lett* 76: 109–111

75 Moglia D, Formelli F, Baliva G, Bono A, Accetturi M, Nava M, De Palo G (1996) Effects of topical treatment with fenretinide (4-HPR) and plasma vitamin A levels in patients with actinic keratoses. *Cancer Lett* 110: 87–91

76 Decensi A, Bruno S, Costantini M, Torrisi R, Curotto A, Gatteschi B, Nicolò G, Polizzi A, Perloff M, Malone WF et al (1994) Phase IIa study of fenretinide in superficial bladder cancer, using DNA flow cytometry as an intermediate end point. *J Nat Cancer Inst* 86: 138–140

77 Cobleigh MA, Dowlatshahi K, Deutsch TA, Mehta RG, Moon RC, Minn F, Benson AB, Rademaker AW, Ashenhurst JB, Wade JL (1993) Phase I/II trial of tamoxifen with or without fenretinide, an analogue of vitamin A, in women with metastatic breast cancer. *J Clin Oncol* 11: 474–477

78 Cobleigh MA, Lincoln S, Mullane M, Benson AB, Minn F (1993) Phase I/II trial of tamoxifen (tam) + fenretinide (4HPR) in stage IV, receptor-positive, previously-untreated breast cancer. *Breast Cancer Res Treat* 21: 151–

79 Pienta KJ, Esper PS, Zwas F, Krzeminski R, Flaherty LE (1997) Phase II chemoprevention trial of oral fenretinide in patients at risk for adenocarcinoma of the prostate. *Amer J Clin Oncol* 20: 36–39

80 Haraoui B, Wilder RL, Allen JB, Sporn MB, Helfgott RK, Brinkerhoff CE (1985) Dose-dependent suppression by the synthetic retinoid, 4-hydroxyphenyl retinamide, of streptococcal cell wall-induced arthritis in rats. *Immunopharmacology* 7: 909–916

81 Gravallese EM, Handel ML, Coblyn J, Anderson RJ, Sperling RI, Karlson EW, Maier A, Ruderman EM, Formelli F, Weinblatt ME (1996) N-[4-hydroxyphenyl]retinamide in rheumatoid arthritis: a pilot study. *Arthritis Rheum* 39: 1021–1026

82 Pizzichetta M, Rossi R, Costa A, Guindani A, De Palo G (1992) Lipoproteins in fenretinide (4-HPR) treated patients. *Diabetes Nutr Metab* 5: 71–72

83 Decensi A, Torrisi R, Polizzi A, Gesi R, Brezzo V, Rolando M, Rondanina G, Orengo MA, Formelli F, Costa A (1994) Effect of the synthetic retinoid fenretinide on dark adaptation and the ocular surface. *J Nat Cancer Inst* 86: 105–110

84 Mariani L, Formelli F, De Palo G, Manzari A, Camerini T, Campa T, Di Mauro MG, Crippa A, Delle Grottaglie M, Del Vecchio M et al (1996) Chemoprevention of breast cancer with fenretinide (4-HPR): study of long-term visual and ophthalmologic tolerability. *Tumori* 82: 444–449

Therapeutic uses of retinoids in skin diseases

C.C. Geilen, B. Almond-Roesler and C.E. Orfanos

Department of Dermatology, University Medical Center Benjamin Franklin, The Free University of Berlin, Hindenburgdamm 30, D-12200 Berlin, Germany

Introduction

Since retinol became available in the mid 1940s, at least 2500 different retinoids, structurally related to retinol, were synthesized. Dermatologists have played a leading role in developing this new class of drugs. Topical retinoic acid was first successfully used for treatment of keratoses in 1962 [76]. In the beginning of the 1970s, the first systemic use of the natural retinoid, retinoic acid, [72] and the first synthetic retinoids, 13-*cis* retinoic acid and retinoic acid ethylamide, [61,69] were published. In recent years, a group of at least ten retinoid compounds gained their reputation as effective drugs in the therapeutic armament of the dermatologist. Topical as well as oral retinoids are used in a broad spectrum of dermatoses. Today, for more than 100 different dermatological disorders exists credible evidence of therapeutic efficacy.

In the following chapter an overview on the use of retinoids for therapy of dermatological diseases will be presented, focussing the interest on psoriasis and other disorders of keratinisation, acne, acneiform dermatoses and seborrhoea, epithelial tumors, and cutaneous lymphomas.

Psoriasis and other disorders of keratinisation

Retinoids exhibit significant effects on cell proliferation and epidermal differentiation, and are used for treatment of severe hyperkeratotic inflammatory diseases and genodermatoses. The topical and also the oral treatment of *psoriasis* with synthetic retinoids such as 13-*cis*-retinoic acid and retinoic acid ethylamide has been reported since the beginning of the 1970s by our group. Aromatic retinoids of the second generation, as etretinate and acitretin, represent today an established systemic antipsoriatic regimen [27, 30, 63]. Systemic retinoids may be administered as oral monotherapy or in combination with other therapeutic modalities (e.g. tar, dithranol, mild corticosteroids, ultraviolet light B [UVB] or psoralen ultraviolet light A [PUVA]) [58]. Clearly, the onset of the therapeutic effect is rather slow but it is reliable if the given dose is within a well established therapeutic range [49,55,70]. In plaque-type psoriasis, etretinate/acitretin should be used as monotherapy, or preferably in combination with topical dithranol, in dosages between 0.3–1.0 mg/kg/day over 4–12 weeks. In combination regimens of oral retinoids with UVB or with UVA (ReSUP or RePUVA), the dosage of etretinate/acitretin can be maintained by 0.3–0.5 mg/kg/day and the duration of treatment is 4–8 weeks. In *pus-*

tular types of psoriasis, e.g. type Zumbusch, palmoplantar pustulosis or acrolocalized suppurative psoriasis pustulosa Hallopeau, systemic etretinate or acitretin has been reported as the first line therapy [59, 48]. In these cases, etretinate/acitretin are given in a higher initial dosage up to 1.0 mg/kg/day and are slowly decreased to 0.5–0.6 mg/kg/day over 3–6 months. Maintenance treatment is required for 6–12 months after clearing. In contrast, in erythrodermic psoriasis it is preferred to start with a low initial dosage, slowly increasing up to 0.5–0.6 mg/kg/day over three months. Maintenance is required for six months for avoiding relapses.

A beneficial response of stationary, plaque-type psoriasis by third generation retinoids for topical use such as tazarotene (0.05–0.1% gel) has recently been reported [12, 25, 47]. However, topical retinoids are more potent in combination with phototherapy or with mild topical corticosteroids in order to avoid retinoid-induced irritation [60, 43].

Oral retinoids were known to normalize hyperkeratotic and dyskeratotic conditions and to reduce scaling in severe keratotic genodermatoses. Clearing is not complete but the overall improvement of skin appearance and function justifies their use. *Darier's disease* [59, 13], *ichthyosis vulgaris, congenital ichthyosis* (particularly the dry lamellar type), various types of *palmoplantar keratodermas,* and also *erythrokeratodermia figurata variabilis* (Mendes da Costa) respond well to etretinate/acitretin and today represent standard indications for systemic retinoid treatment [33, 4]. In most cases, treatment with low initial doses (0.3–0.6 mg/kg/day) is preferred in these indications for avoiding mucocutaneous side-effects such as retinoid dermatitis, intertriginous maceration, oozing and also increased bulla formation, e.g. in epidermolytic hyperkeratosis. In these genetic disorders life-long treatment has to be continued under minimal dosing for improving life quality.

Teratogenicity and bone toxicity of oral retinoids have to be controlled and supervised carefully mostly in young individuals and in females of child-bearing age. In mild cases of Darier's disease and ichthyoses, topical isotretinoin (0.05%) may be a helpful alternative to the systemic administration of the drug [10, 75] Other rare keratotic diseases such as *ichthyosis hystrix, hyperkeratotic verrucous naevi, keratosis lichenoides chronica* etc. may respond to standard oral retinoid doses to some degree, showing reduction of hyperkeratosis and skin smoothening. Due to the rarity of such entities, however, the overall experiences are still restricted.

Acne, acneiform dermatoses, and seborrhoea

The sebosuppressive effect of retinoids is being increasingly used for treatment of severe acne. Introduced in 1979, isotretinoin (13-*cis*-retinoic acid) remains a major advance in the therapy of acne conglobata and acne nodulocystica [66]. Oral isotretinoin is the only substance that cures acne while oral antibiotics and other treatments have only moderate, temporary effects [38, 71]. Isotretinoin influences the basic pathogenic factors for acne. It directly suppresses the sebum production and decreases the abnormal desquamation of the sebaceous follicle. This drug has also been shown to diminish the proliferation of Propionibacterium acnes and to improve inflammatory acne lesions. Other systemic retinoids such as natural all-*trans*-retinoic acid and 9-*cis*-retinoic acid, or synthetic retinoids like etretinate, acitretin and arotinoids were less active or completely inactive on sebosuppression and therefore on acne [27, 35, 62].

Conditions warranting oral isotretinoin therapy include severe acne and poorly responsive acne which improves less than 50% after six months of therapy with antibiotics. Furthermore, acne which relapses, scars or induces consequential psychological disturbance should be treated with oral isotretinoin [17, 64]. Different therapeutic dosage regimes were recommended with good clinical response. Isotretinoin can be given at 0.5 mg/kg/day or at a higher dose, e.g. 1 mg/kg/day, over a minimum of six months. To optimize the therapeutic outcome and minimize the dose-dependent side-effects, a scheme with 1 mg/kg/day for three months and subsequent reduction to 0.5 mg/kg/day and 0.2 mg/kg/day for an additional nine months is recommended. The relapse rate was lower in patients treated with 1 mg/kg/day (22 to 30%) vs. lower dosage treatment (39 to 82%) over a ten years follow up period. The current recommendation for isotretinoin therapy is a cumulative dose >150 mg/kg or 0.5 to 1 mg/kg/day over twelve months in severe acne [63]. A particular indication for choosing higher doses is severe involvement of the trunk [51].

Treatment with intermittent moderate dose isotretinoin was effective in mild (facial) acne resistant to conventional antibiotic therapy in adults and was recommended as a cost-effective alternative to full-dose isotretinoin [31]. Isotretinoin therapy normally results in complete clearing of nodulocystic acne, and many patients permanently remain free of disease. Only a few patients respond poorly or have a high rate of relapse, mainly young patients or women with elevated androgens. Patients with haemorrhagic or other severe type acne lesions may develop acne-fulminans-like eruptions under full-dose isotretinoin [52].

Isotretinoin in the treatment of acne is a safe drug, with no serious long-term side-effects. The symptoms reported were predominantly mild mucocutaneous (4.8%) or musculoskeletal (2%). Xeroderma, dry eyes, arthralgia, and possible exacerbation of eczema, were infrequently found side-effects [32]. It could be shown that the plasma levels of isotretinoin and its main metabolite 4-*oxo*-isotretinoin correlated with the number and frequency of side-effects [1]. Preliminary data showed that oral isotretinoin reduces the total numbers of antibiotic resistant Propionibacterium acnes on the skin of acne patients [15]. Contraception is essential one month before therapy and at least three months after discontinuation of isotretinoin intake in women of child-bearing age [63].

Today, several topical retinoids are used in the management of mild to moderate acne. More than 50% of acne patients belong to this group [29]. Tretinoin was the first retinoid used in the topical treatment more than 25 years ago. Topical isotretinoin, which became available later, is less irritating, but is probably less effective than tretinoin [42]. A group of retinoids that include adapalene, tazarotene, and reformulations of tretinoin represent new and forthcoming agents for topical treatment. Studies indicate that several of these agents are associated with less skin irritation than previous formulations while they retain potent comedolytic activity. Adapalene, a naphthoic acid derivative with retinoid-like activity, possesses significant anti-inflammatory activity. It binds to retinoic acid receptors which are found predominantly in the terminal differentiation zone of the epidermis and is probably more active than tretinoin in modulating cellular differentiation. The cutaneous tolerability of adapalene aqueous gel was better than that of tretinoin gel [9, 16, 73]. In a large study, side-effects occurred in 5.1% of those in the ada-palene-treated groups, compared with 9.1% in the tretinoin-treated groups [14].

Isotretinoin was also found to be highly effective in the clearing of refractory *rosacea* lesions. The anti-inflammatory action of isotretinoin is considered to be the responsible mechanism for efficacy in *rosacea*. A daily dose of 0.4 to 1 mg/kg/day for up to six months is beneficial and was reported by many authors [36, 63]. Remissions of up to two years after discontinuation have been documented. Mucocutaneous side-effects were predictable and dose-dependent and systemic side-effects were rarely problematic. A study with 22 patients with severe or recalci-trant *rosacea* revealed that low-dose oral isotretinoin (10 mg/d), topical applied tretinoin (0.025% cream), and the combined treatment were beneficial in the treatment of severe or recal-citrant *rosacea* [22].

Other indications for isotretinoin are *rhinophyma* and *Gram-negative folliculitis*. Rhinophyma, in early inflammatory stages, improves after a daily dose of 0.5 to 1 mg/kg body weight for three to six months. Gram-negative folliculitis responds to an oral dose of 0.5 to 2 mg/kg/day and usually does not show relapses. The growth conditions of the bacteria were impaired by reducing sebaceous gland volume.

Isotretinoin is an extremely effective drug if given systemically in severe forms of *sebor-rhoea*, being the only retinoid with potent sebostatic properties. It is the most effective drug in reducing sebaceous gland size (up to 90%) by decreasing proliferation of basal sebocytes, sup-pressing sebum production and inhibiting sebocyte differentiation *in vivo* [67, 74, 80]. At very low doses of 0.1 mg/kg daily it reduces sebocyte lipid synthesis by 75%. Patients who have received oral isotretinoin therapy for seborrhoea do not usually experience a relapse for months. In treatment of seborrhoea, a maintenance dose of 5 to 10 mg/day over years is effective. The duration of the sebosuppressive effect lasts for about eight weeks after discontinuation of ther-apy (0.1–0.3 mg/kg/day) and seems to be dose-dependent [63].

Epithelial tumors

Epithelial tumors of the skin are the most common malignancies in humans and a worldwide increased incidence of all forms has been recognized. In recent years, organ transplant recipi-ents have an especially increased risk of developing skin cancer due to prolonged immunosup-pressive treatment. Although distant metastasis from light-induced squamous cell carcinoma is uncommon, it occurs in 3 to 5% of all patients [45] and transplant recipients have a ten-fold higher mortality from skin cancer, thus suggesting increased metastatic potential [23, 40]. With this background, the development of cancer preventional and anticancer strategies appears nec-essary, especially for this group of patients.

Prevention of epithelial tumors

Synthetic retinoids have been examined in several chemoprevention trials in the skin. The data indicate that topical and systemic application of retinoids have a significant effect in reversing premalignant skin lesions [5, 53].

A successful prevention of basal cell carcinomas and squamous cell carcinomas in patients with *Xeroderma pigmentosum* has been reported for systemic isotretinoin [46] and etretinate therapy [3]. Over a period of two years, isotretinoin was shown to reduce the occurrence of basal cell carcinomas by 80% and of squamous cell carcinomas by 60%. For etretinate, a reduction of 75% was reported with respect to the prevention of squamous cell carcinomas. Both retinoids have also been shown to be effective in preventing basal cell carcinomas in organ transplant recipients [40] and in the basal cell nevus syndrome [28, 34]. Furthermore, a combination of topical tretinoin and low-dose oral etretinate (10 mg/d) has been suggested for chemoprevention with reduced side-effects [69].

Therapy of epithelial tumors

Basal cell carcinomas have been shown to be retinoid-sensitive, but as well as in the therapy of keratoacanthomas, discontinuation of the retinoid therapy led to relapse of these malignancies [63].

In contrast to the benefit of chemoprevention by retinoids, no satisfactory therapeutic results have been obtained in *squamous cell carcinomas* either with etretinate/acitretin or isotretinoin monotherapy. Recently, an effective combination therapy with isotretinoin and interferon α-2a in advanced squamous cell carcinomas was reported [54, 77]. Of a total of 34 patients with advanced squamous cell carcinomas, a remission of 65% (eight complete remissions, 14 partial remissions) has been observed.

Cutaneous lymphomas

Successful monotherapy of cutaneous T cell lymphoma has been reported with etretinate [56] as well as with isotretinoin [24, 41, 57, 79] and with arotinoid Ro 13-6298 [78]. However, a combination of etretinate/PUVA therapy (RePUVA) appeared to be more effective [59]. Jones and coworker [39] reported on the successful treatment of mycosis fungoides and Sezary syndrome with a combination of etretinate and electron beam therapy.

A promising new approach for treatment of cutaneous T cell lymphoma is the combination of etretinate and interferon α-2b [2, 19, 20] or interferon α-2b [8]. Remission rates of 77% and 53% were reported, respectively. Combination of systemic isotretinoin and interferon α-2b also showed good therapeutic results with remission rates between 57% [44] and 82% [21].

Miscellaneous cutaneous disease

In a series of other dermatoses oral retinoids have been applied with varying clinical efficacy. Etretinate/acitretin were found effective in: (i) *lichen planus* [50] including oral manifestations of *lichen mucosae oris* with papillomatous and erosive/bullous lesions, (ii) cutaneous variants of *lupus erythematosus* [11], particularly the hyperkeratotic lesions of chronic-discoid lupus erythematosus, and (iii) *lichen sclerosus et atrophicans* [7]. Both *lichen planus* and *lichen mucosae oris* were also shown to respond to low-dose oral tretinoin. Complete remission was observed in 13/18 patients [65]. The beneficial effect of retinoids in these entities underlines their immunomodulatory dermal action. *Prurigo nodularis* may be another entity responding well to systemic retinoid therapy.

From *in vitro* studies it has been shown that systemic retinoids may decrease endothelial proliferation [37]. Recently, the beneficial effect of systemic tretinoin in patients with *HIV-related Kaposi's* sarcoma was reported in first pilot clinical studies [6]. As shown in cutaneous T cell lymphomas, it seems that retinoids and interferons may act synergistically. The mechanisms underlying this synergistic effect were thought to be an induction of the expression of RARs and, *vice versa*, the expression of interferon-receptors by retinoids. Therefore, combination therapy of HIV-related Kaposi's sarcoma with interferon α and systemic tretinoin may be a promising concept.

Conclusions

Retinoids are remarkable drugs in that they show multiple pharmacological effects. They act in different ways on a broad spectrum of unrelated disorders and up to now, no explanation for these diverse biological effects is available. The concept of nuclear retinoid receptors has powered a high amount of basic research dealing with the mechanism of action and led to the development of receptor-specific retinoids. The development of *in vitro* screening assays is neccessary to test the biological activity of these hundreds of retinoids. From these studies, new retinoids will become available for clinical use in the near future.

Reference

1   Almond-Roesler B, Blume-Peytavi U, Bisson S, Krahn M, Rohloff E, Orfanos CE (1998) Monitoring of isotretinoin therapy by measuring the plasma levels of isotretinoin and 4-*oxo*-isotretinoin. a useful tool for management of severe acne. *Dermatology* 196: 176–181

2   Altomare GF, Capella GL, Pigatto PD, Finzi AF (1993) Intramuscular low dose alpha-2b interferon and etretinate for treatment of mycosis fungoides. *Int J Dermatol* 32: 138–141

3   Berth-Jones J, Cole J, Lehmann AR, Arlett CF, Graham-Brown RA (1993) Xeroderma pigmentosum variant: 5 years of tumour suppression by etretinate. *J Royal Soc Med* 86: 355–356

4   Blanchet-Bardon C, Nazzaro V, Rognin C, Geiger J-M, Puissant A (1991) Acitretin in the treatment of severe disorders of keratinization. Results of an open study. *J Am Acad Dermatol* 24: 982–986

5   Bollag W, Holdener EE (1992) Retinoids in cancer prevention and therapy. *Ann Oncol* 3: 513–526

6   Bonhomme L, Fredj G, Ecstein E, Maurisson G, Farabas C, Misset JL, Jasmin C (1994) Treatment of AIDS-associated Kaposi's sarcoma with oral tretinoin. *Am J Hosp Pharm* 51: 2417–2419.

7   Bousema MT, Romppanen U, Geiger JM, Baudin M, Vaha-Eskeli K, Vartiainen J, Vupala S (1994) Acitretin in the treatment of severe lichen sclerosus et atrophicus of the vulva: a double-blind, placebo-controlled study.

J Am Acad Dermatol 30: 225–231

8 Braathen LR, McFadden N (1989) Successful treatment of mycosis fungoides with the combination of etretinate and human recombinant interferon alpha-2a. *J Dermatol Treat* 1: 29–32

9 Brogden RN, Goa KE (1997) Adapalene. A review of its pharmacological properties and clinical potential in the management of mild to moderate acne. *Drugs* 53: 511–519

10 Burge SM, Buxton PK (1995) Topical isotretinoin in Darier's disease. *Br J Dermatol* 133: 924–928

11 Callen JP (1994) Treatment of cutaneous lesions in patients with lupus erythematosus. *Dermatol Clin* 2: 201 –206

12 Chandraratna RAS (1996) Tazarotene – first of a new generation of receptor-selective retinoids. *Br J Dermatol* 135: 18–25

13 Christophersen J, Geiger J-M, Danneskiold-Samsoe P, Kragballe K, Larsen FG, Lauberg G, Serup J, Thomsen K (1992) A double-blind comparison of acitretin and etretinate in the treatment of Darier's disease. *Acta Derm Venereol* (Stockh) 72: 150–152

14 Clucas A, Verschoore M, Sorba V, Poncet M, Baker M, Czernielewski J (1997) Adapalene 0.1% gel is better tolerated than tretinoin 0.025% gel in acne patients. *J Am Acad Dermatol* 36: 116–118

15 Coates P, Adams CA, Cunliffe WJ, McGinley KT, Eady EA, Leyden JJ, Ravenscroft J, Vyakrnam S, Vowels B (1997) Does oral isotretinoin prevent Propionibacterium acnes resistance? *Dermatology* 195 Suppl 1: 4–9

16 Cunliffe WJ, Caputo R, Dreno B, Forstrom L, Heenen M, Orfanos CE, Privat Y, Robledo Auguilar A, Meynadier J, Alirezai M et al. (1997) Clinical efficacy and safety comparison of adapalene gel and tretinoin gel in the treatment of acne vulgaris: Europe and U.S. multicenter trials. *J Am Acad Dermatol* 36: 126–134

17 Cunliffe WJ (1987) Evolution of strategy for the treatmentof acne. *J Am Acad Dermatol* 16: 591–599

18 Dai S, Morrè DJ, Geilen CC, Almond-Roesler B, Orfanos CE, Morrè DM (1997) Inhibition of plasma membrane NADH oxidase activity and growth of HeLa cells by natural and synthetic retinoids. *Mol Cell Biochem* 166: 101–109.

19 Dreno B, Celerier P, Litoux P (1993). Roferon-A in combination with Tigason in cutaneous T-cell lymphomas. *Acta Haematol* 89: 28–32

20 Dreno B, Claudy A, Meynadier J, Verret JL, Souteyrand P, Ortonne JP, Kalis B, Godefroyw Y, Beerblock K, Thill L (1991) The treatment of 45 patients with cutaneous T-cell lymphoma with low doses of interferon-alpha 2a and etretinate. *Brit J Dermatol* 125: 456–459

21 Duvic M, Lemak NA, Redman JR, Eifel PJ, Tucker SL, Cabanilles FF, Kurzrock R (1996) Combined modality therapy for cutaneous T-cell lymphoma. *Am Acad Dermatol* 34: 1022–1029

22 Ertl GA, Levine N, Kligman AM (1994) A comparison of the efficacy of topical tretinoin and low-dose oral isotretinoin in rosacea. *Arch Dermatol* 130: 319–324

23 Euvrard S, Kanitakis J, Pouteil-Noble C, Disant F, Dureau G, Finaz de Villaine J, Claudy A, Thivolet J (1995) Aggressive squamous cell carcinoma in organ transplant recipients. *Transpl Proc* 27: 1767–1768.

24 Fitzpatrick JE, Mellette JR (1986) Treatment of mycosis fungoides with isotretinoin. *J Dermatol Surg Oncol* 12: 626–629

25 Foster RH, Brogden RN, Benfield P (1998) Tazarotene. *Drugs* 55: 705–711

26 Geiger J-M, Saurat J-H (1993) Acitretin and etretinate. How and when they should be used. *Dermatol Clin* 11: 117–129

27 Geiger JM, Hommel L, Harms M, Saurat JH (1996) Oral 13-*cis* retinoic acid is superior to 9-*cis* retinoic acid in sebosuppression in human beings. *J Am Acad Dermatol* 34: 513–515

28 Goldberg L, Hsu S, Alcalay J (1989) Effectiveness of isotretinoin in preventing the appearance of basal cell carcinomas in basal cell nevus syndrome. *J Am Acad Dermatol* 21: 144–145

29 Gollnick H, Schramm M (1998) Topical drug treatment in acne. *Dermatology* 196: 119–125

30 Gollnick HPM (1996) Oral retinoids – Efficacy and toxicity in psoriasis. *Br J Dermatol* 135 (suppl 49): 6–17

31 Goulden V, Clark SM, McGeown C, Cunliffe WJ (1997) Treatment of acne with intermittent isotretinoin. *Br J Dermatol* 137: 106–108

32 Goulden V, Layton AM, Cunliffe WJ (1994) Long-term safety of isotretinoin as a treatment for acne vulgaris. *Br J Dermatol* 31: 360–363

33 Happle R, Van de Kerkhof PCM, Traupe H (1987) Retinoids in disorders of keratinization: Their use in adults. *Dermatologica* 175 (suppl 1): 107–124

34 Hodak E, Ginzburg A, David M, Sadbank M (1987) Etretinate treatment of the nevoid basal cell carcinoma syndrome. *Int J Dermatol* 26: 606–609

35 Hommel L, Geiger JM, Harms M, Saurat JH (1996) Sebum excretion rate in subjects treated with oral all-*trans*-retinoic acid. *Dermatology* 193: 127–130

36 Hoting E, Paul E, Plewig G (1986) Treatment of rosacea with isotretinoin. *Int J Dermatol* 25: 660–663

37 Imcke E, Ruszczak Zb, Mayer-da-Silva A, Detmar M, Orfanos CE (1991) Cultivation of human dermal microvascular endothelial cells in vitro: Immunocytochemical and ultrastructural characterization and effect of treatment with three synthetic retinoids. *Arch Dermatol Res* 283: 149–157

38 Jansen T, Plewig G (1997) Advances and perspectives in acne therapy. *Eur J Med Res* 28: 321–334

39 Jones G, McLean J, Rosenthal D, Roberts J, Sander DN (1992) Combined treatment with oral etretinate and electron beam therapy in patients with cutaneous T-cell lymphoma (mycosis fungoides and Sezary syndrome). *J Am Acad Dermatol* 26: 960–967

40 Kelly JW, Sabto J, Gurr FW, Bruce F (1991) Retinoids to prevent skin cancer in organ transplant recipients. *Lancet* 338: 1407
41 Kessler JF, Jones SE, Levine N, Lynch PJ, Booth AR, Meyskens FL Jr. (1987) Isotretinoin and cutaneous helper T-cell lymphoma (mycosis fungoides). *Arch Dermatol* 123: 201–204
42 Kligman AM (1997) The treatment of acne with topical retinoids: one man's opinions. *J Am Acad Dermatol* 36: 92–95
43 Kligman AM (1998) The growing importance of topical retinoids in clinical dermatology: a retrospective and prospective analysis. *J Am Acad Dermatol* 39: S2-S7
44 Knobler RM, Trautinger F, Radaszkiewicz T, Kokoschka EM, Micksche M (1991) Treatment of cutaneous T cell lymphoma with a combination of low-dose interferon alfa-2b and retinoids. *J Am Acad Dermatol* 24: 247 –252
45 Ko CB, Walton S, Keczkes K, Bury HP, Nichloson C (1994) The emerging epidemic of skin cancer. *Brit J Dermatol* 130: 269–272
46 Kraemer KH, Di Giovanna JJ, Moshell AN, Tarone RE, Peck GL (1988) Prevention of skin cancer in xeroderma pigmentosum with the use of oral isotretinoin. *N Engl J Med* 318: 1633–1637
47 Krueger GG, Drake LA, Elias PM, Lowe NJ, Guzzo C, Weinstein GD, Lew-Kaya DA, Lue JC, Sefton J, Chandraratna RA (1998) The safety and efficacy of tazarotene gel, a topical acetylenic retinoid, in the treatment of psoriasis. *Arch Dermatol* 134: 57–60
48 Lassus A, Geiger JM (1988) Acitretin and etretinate in the treatment of palmoplantar pustulosis: A double-blind comparative trial. *Br J Dermatol* 119: 755–759
49 Lauharanta J, Geiger JM (1989) A double-blind comparison of acitretin and etretinate in combination with bath PUVA in the treatment of extensive psoriasis. *Br J Dermatol* 121: 107–112
50 Laurberg G, Geiger JM, Hjorth N, Holm P, Hou-Jensen K, Jacobsen KU, Nielsen AO, Pichard J, Serup J, Sparre-Jorgensen A et al. (1991) Treatment of lichen planus with acitretin. A double-blind placebo controlled study in 65 patients. *J Am Acad Dermatol* 24: 434–437
51 Layton AM, Knaggs H, Taylor J, Cunliffe WJ (1993) Isotretinoin for acne vulgaris – 10 years later: A sfe and successful treatment. *Br J Dermatol* 12: 292–296
52 Leyden JJ (1997) Oral isotretinoin. How can we treat difficult acne patients? *Dermatology* 195 (Suppl 1): 29 –33
53 Lippman SM, Brenner SE, Hong WK (1994) Cancer chemoprevention. *J Clin Oncol* 12: 851–873
54 Lippman SM, Parkinson DR, Itri LM, Weber RS, Schantz SP, Ota DM, Schustermann MA, Krakoff IH, Guttermann JU, Hong WK (1992) 13-cis retinoic Acid anf interferon alpha-2a: Effective combination therapy for advanced squamous cell carcinoma of the skin. *J Natl Cancer Inst* 84: 241–245
55 Lowe NJ, Prystowsky JH, Bourget T, Edelstein J, Nychays, Armstrong R (1991) Acitretin plus UVB therapy for psoriasis. Comparisons with placebo plus UVB and acitretin alone. *J Am Acad Dermatol* 24: 591–594
56 Molin L, Thomsen K, Volden G, Aronsson A, Hammar H, Helbe L, Wantzin GL, Roupe G (1987) Oral retinoids in mycosis fungoides and Sezary syndrome: A comparison of isotretinoin and etretinate. *Acat Derm Venerol* (Stockh) 67: 232–236
57 Neely SM, Mehlmauer M, Feinstein DI (1987) The effect of isotretinoin in six patients with cutaneous T-cell lymphoma. *Arch Intern Med* 147: 529–531
58 Orfanos CE, Ehlert R, Gollnick H (1987). The retinoids: A review of their clinical pharmacology and therapeutic use. *Drugs* 34: 459–503
59 Orfanos CE, Landes E, Bloch PH (1978) Traitément du psoriasis pustuleux par un nouveaux rétinoide aromatique (Ro 10-9359). *Ann Dermatol Vénéréol* 103: 807–811
60 Orfanos CE, Schmidt HW, Mahrle G, Gartmann H, Lever WF (1973) Retinoic acid in psoriasis: Its value for topical therapy with and without corticosteroids. Clinical, histological and electron microscopical studies on forty-four hospitalized patients with extensive psoriasis. *Br J Dermatol* 88: 167–182
61 Orfanos CE, Schmidt HW, Mahrle G, Runne U (1972) Die Wirksamkeit von Vitamin-A-Säure bei Psoriasis. Topische Kombinationsbehandlung mit Corticoiden. Zwei neue VAS-Präparate zur peroralen Therapie. *Arch Dermatol Forsch* 144: 424–426
62 Orfanos CE, Zouboulis CC (1998) Oral retinoids in the treatment of seborrhoea and acne. *Dermatology* 196: 140–147
63 Orfanos CE, Zouboulis ChC, Almond-Roesler B, Geilen CC (1997) Current use and future potential role of retinoids in dermatology. *Drugs* 53: 358–388
64 Ortonne JP (1997) Oral isotretinoin treatment policy. Do we all agree? *Dermatology* 195 (Suppl 1): 34–37
65 Ott F, Bollag W, Geiger JM (1996) Efficacy of oral low-dose tretinoin (all-*trans*-retinoic acid) in lichen planus. *Dermatology* 192: 334–336
66 Peck GL, Olsen TG, Yoder FW, Strauss JS, Downing DT, Pandya M, Butkus D, Arnaud-Battandier J (1979) Prolonged remission of cystic and conglobate acne with 13-cis retinoic acid. *N Engl J Med* 300: 329–333
67 Ridden J, Ferguson D, Kealey T (1990) Organ maintenance of human sebaceous glands: In vitro effects of 13-cis retinoic acid and testosterone. *J Cell Sci* 95: 125–136
68 Rook AH, Jaworsky C, Nguyen T, Grossmann RA, Wolfe JT, Witmer WK, Kligman AM (1995) Beneficial effect of low-dose systemic retinoid in combination with topical tretinoin for the treatment and prophylaxis of premalignant and malignant skin lesions in renal transplant recipients. *Transplantation* 59: 714–719.

69 Runne U, Orfanos CE, Gartmann H (1973) Perorale Applikation zweier Derivate der Vitamin A-Säure zur internen Psoriasis-Therapie. 13-cis-beta-Vitmain-A-Säure und Vitamin A-Säure-aethylamid. *Arch Derm Forsch* 247: 171–180

70 Ruzicka T, Sommerburg C, Braun-Falco O, Koster W, Lengen W, Lensing W, Letzel H, Meigel WN, Paul E, Przybilla B et al. (1990) Efficiency of acitretin in combination with UV-B in the treatment of severe psoriasis. *Arch Dermatol* 126: 482–486

71 Saurat JH (1997) Oral isotretinoin. Where now, where next? *Dermatology* 195 (Suppl 1): 1–3

72 Schuhmacher A, Stüttgen G (1971) Vitamin-A-Säure bei Hyperkeratosen, epithelialen Tumoren und Akne. Dtsch Med Wochenschr 96: 1547–1551

73 Shalita A, Weiss JS, Chalker DK, Ellis CN, Greenspan A, Katz HJ, Kantor J, Milikan LE, Swinehart T, Swinyer L, Whitemore C, Baker M, Czernielewski J (1996) A comparison of the efficacy and safety of adapalene gel 0.1% and tretinoin gel 0.025% in the treatment of acne vulgaris: a multicenter trial. *J Am Acad Dermatol* 34: 482–485

74 Shapiro SS, Hurley J, Vane FM, Doran T (1989) Evaluation of potential therapeutic entities for the treatment of acne. In: Reichert U, Shroot B, eds. Pharmacology of Retinoids in the Skin. Basel: Karger 104–112

75 Steijlen PM, Reifenschweiler DO, Ramaekers FC, van Muijen GN, Happle R, Link M, Ruiter DJ, van de Kerkhof PC (1993) Topical treatment of ichthyoses and Darier's disease with 13-*cis*-retinoic acid. A clinical and immunohistochemical study. *Arch Dermatol Res* 285: 221–226

76 Stüttgen G (1962) Zur Lokalbehandlung von Keratosen mit Vitamin A-säure. Dermatologica 124: 65–71

77 Toma S, Palumbo R, Vincenti M, Aitini E, Paganini G, Pronzato P, Grimaldi A, Rosso R (1994) Efficiency of recombinant alpha-interferon 2a and 13-cis-retinoic acid in the treatment of squamous cell carcinoma. *Ann Oncol* 5: 463–465

78 Tousignant J, Raymond GP, Light MJ (1987) Treatment of cutaneous T-cell lymphoma with the arotinoid Ro 13-6298. *J Am Acad Dermatol* 16: 167–171

79 Zachariae H, Thestrup-Pedersen K (1990) Interferon alpha and etretinate combination treatment of cutaneous T-cell lymphoma. *J Invest Dermatol* 95 (Suppl): 206–208

80 Zouboulis ChC, Krieter A, Gollnick H Mischke D, Orfanos CE (1994) Progressive differentiation of human sebocytes in vitro is characterized by increased cell size and altering antigenic expression and is regulated by culture duration and retinoids. *Exper Dermatol* 3: 151–160

Retinoid treatment of photoaged skin

A.A. Bajoghli and B.A. Gilchrest

Department of Dermatology, Boston University School of Medicine, Boston, MA 02118, USA

Introduction

Photoaging comprises the adverse effects of chronic ultraviolet radiation (UVR) on the skin, superimposed on the intrinsic aging process [1]. The magnitude of the problem of photoaging is large and expanding [2] as a result of factors such as increasing ultraviolet exposure from ozone layer depletion, increasing popularity of tanning salons, and "sun-worshipping" behavior of the public, combined with demographic shifts that have tripled the proportion of older persons in society during the past century and are projected again to double in the next century [3]. Ongoing public awareness of the consequences of chronic sun exposure has not yet diminished the population's desire to tan or motivation to use adequate sun protection measures. Consequently, photoaging is a common reason for consulting dermatologists and other physicians, and the public interest in products intended to reverse its effects is enormous. This chapter reviews evidence gathered from human and animal studies regarding the ability of topical retinoid application to reverse many aspects of the photoaging process.

Photoaging vs. chronological aging

By definition photoaging involves UVR (200–400 nm) [4] exposure, mainly from the sun but also from therapeutic sources of UVR and tanning salons. Ultraviolet C (UVC, 200–290 nm) is the most biologically damaging portion of the UVR spectrum but is almost entirely absorbed by the ozone layer of the earth's atmosphere. The strongest ultraviolet light to reach the earth's surface is ultraviolet B (UVB, 290–320 nm). Ultraviolet A (UVA, 320–400 nm) has greater penetration into the skin but is much weaker than UVB.

Intrinsic aging of the skin refers to the changes in the sun-protected areas of elderly patients, analogous to changes that occur in internal organs. Histologically, there is progressive and fairly uniform cell loss and atrophy; physiologically, there is loss of maximal function, reserve capacity, and environmental responsiveness [5]. Chronologically aged skin has a smooth and unblemished appearance with fine wrinkling and loss of elasticity. Histological changes are subtle, with flattening of the dermo-epidermal junction being most apparent [6]. In contrast, photoaged skin is a variable and non-uniform combination of hypertrophic injury responses, atrophic changes resulting from destruction of specific cell types, and "atypia" associated with the multi-step process of UVR carcinogenesis leading from initiation through progression to frank malignancy [7].

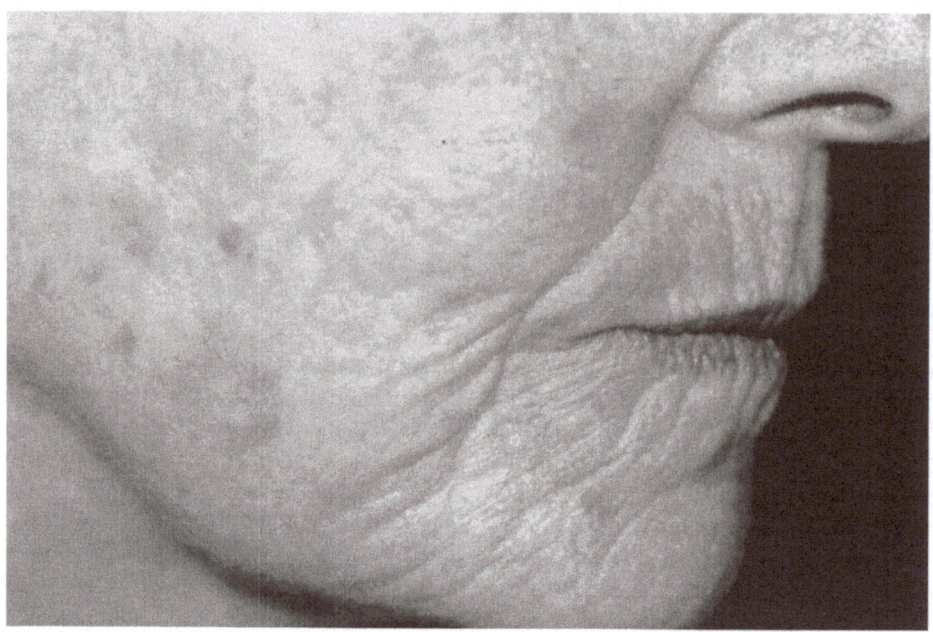

Figure 1. Moderate to severe photodamage. Note fine and coarse wrinkling, lentigenes, and actinic keratoses.

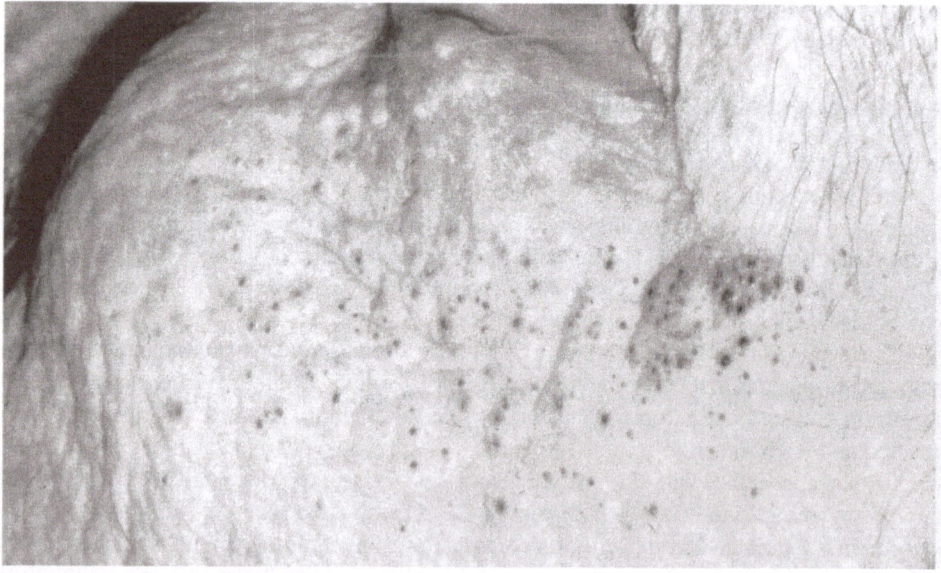

Figure 2. Solar comedones of Favre-Racouchout syndrome, although not formally evaluated in the controlled Renova trials because such lesions are a relatively uncommon manifestation of photoaging, comedones are known to respond well to retinoic acid therapy.

Clinical features of photoaging include fine and coarse wrinkles, skin roughness, sallowness, mottled hyper- and hypo-pigmentation, lentigines, telangiectasias, and skin laxity (Fig. 1). Actinic damage may cause a leathery texture, solar comedones and cysts termed Favre-Racouchout syndrome (Fig. 2). On the neck, deep furrows may assume a rhomboid pattern termed cutis rhomboidalis nuchae (Fig. 3). The most significant clinical manifestations include pre-malignant lesions (actinic keratosis and lentigo maligna) and skin cancers (basal cell carcinoma, squamous cell carcinoma, and melanoma), all of which have increased in recent years [8–10]. Histological changes include epidermal thinning and thickening with disordered maturation, variable basal keratinocyte atypia, and large, irregularly grouped melanocytes [11]. Photoaged dermis has a prominent band of elastic tissue with hallmark changes in tinctorial properties and ultra-structure, termed elastosis [12–14] (Fig. 4). The elastosis is separated from a relatively normal-appearing area (Grenz zone) just below the epidermis that stains positively with reticulin, indicating new collagen formation [15]. Therefore, the Grenz zone may be analogous to a "microscar" resulting from decades of repair from ultraviolet insults [16, 17]. There is also a perivascular deposition of basement membrane-like material in the micro-vasculature.

Photoaging accounts for the vast majority of the skin's age-related cosmetic changes. In addition, it is a major contributor to skin cancer and other medical problems. Only in the past three decades have investigators recognized that photoaging differs from intrinsic (chronological) aging and in 1992 the term photoaging ("chronic solar skin damage") was first listed in the International Classification of Disease 9th Revision Clinical Modification (ICD 9 CM; code 692.74) [18, 19].

Treatment of photoaging: a revolutionary concept

Beginning in the 1970s, Dr. Albert Kligman, of the University of Pennsylvania, who had earlier pioneered the use of topical tretinoin for acne [20], observed that his middle-aged acne patients experienced benefits from the therapy beyond resolution of comedones, papules and pustules. Many women reported, and Dr. Kligman agreed, that their facial skin was smoother, more evenly pigmented and had a "rosy glow", changes that conferred a younger and less sun-damaged appearance [21] (Tab. 1). Using the hairless mouse described above, he and Dr. Lorraine Kligman then reported that topical tretinoin could reverse histological changes associated with chronic UV irradiation and improve the appearance of the mouse skin [25]. This work received little attention from the academic dermatology community, however. In 1986, impressed by the animal data and growing anecdotal experience that tretinoin benefited photoaged skin, the Ortho Pharmaceutical Corporation, manufacturer of Retin-A ™ (topical tretinoin, FDA approved for acne treatment) sponsored a conference and subsequent publication of proceedings on innovative uses for tretinoin ranging from treatment of flat warts to xerophthalmia to photoaging [26]. Publication of Dr. Kligman's clinical experience in a widely read journal certainly heightened dermatological awareness of tretinoin as a proposed therapy for photoaging, but many remained unconvinced of this revolutionary concept in the absence of well-controlled trials.

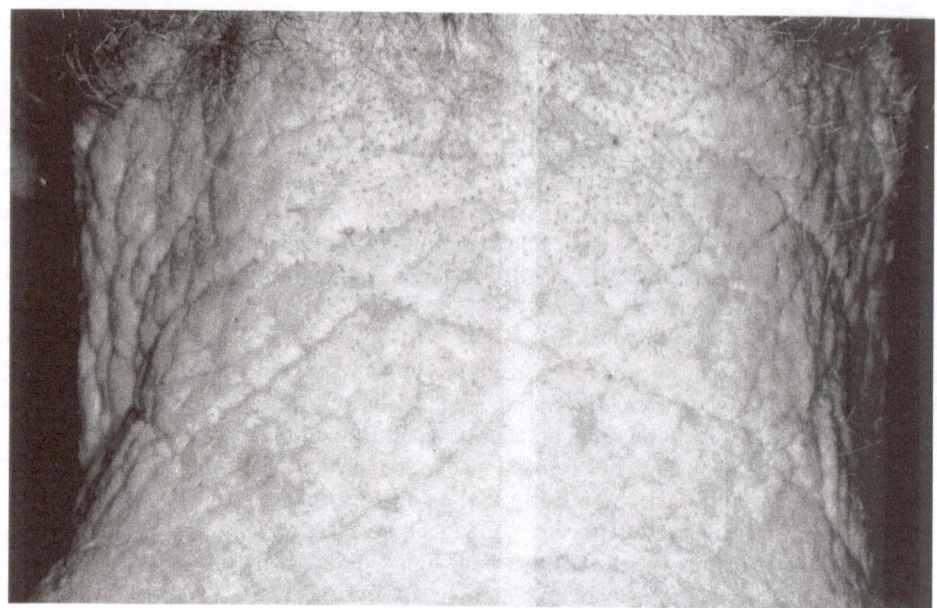

Figure 3. *Cutis rhomboidalis nuchae*. Deep and coarse furrows occurring on the neck as a result of photoaging.

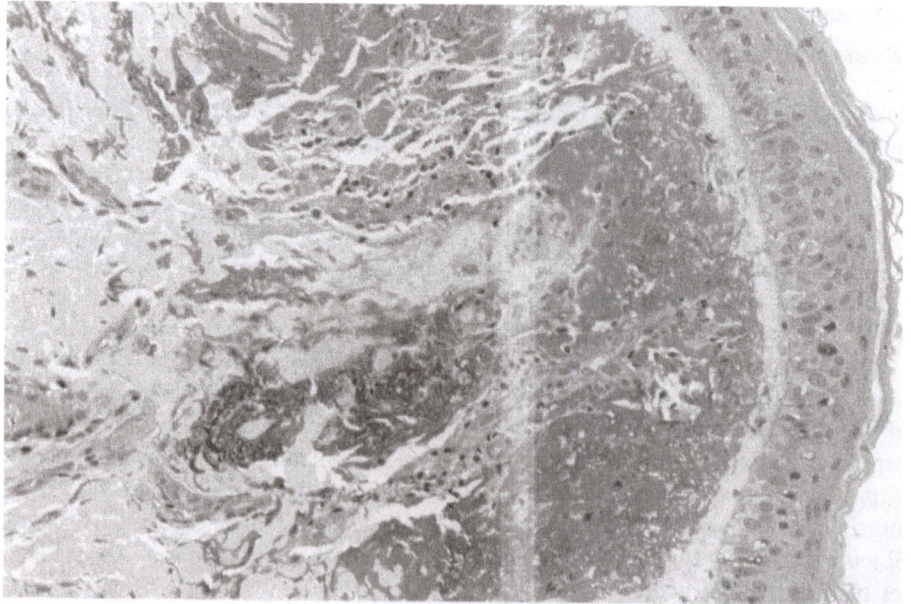

Figure 4. Light micrograph of a biopsy cross section from photodamaged facial skin revealing prominent elastosis. (PMS stain, original magnification × 40).

Table 1. Chronology of tretinoin therapy for photoaging

Date	Event	Reference
1986	First report that tretinoin improves photoaging	Kligman et al. [21]
1988	First controlled trial of tretinoin use in photoaging	Weiss et al. [27]
1990	Scientific community concludes that there is no definitive evidence that tretinoin decreases wrinkles	NIH Consensus Panel [22]
1991–1992	Multi-center trials establish safety and efficacy of tretinoin emollient cream for treatment of photoaging	Weinstein et al. [39] Olsen et al. [40] Bhawan et al. [47]
1991–1996	Continued publication of controlled trials demonstrating safety and efficacy of tretinoin for photoaging	Grove et al. [43] Griffiths et al. [45, 51] Rafal et al. [53].
1992	FDA Advisory Panel unanimously recommends approval of Renova™ for photoaging	FDA Online [23]
1996	FDA approves tretinoin (Renova™) for photoaging	FDA Online [24]

In 1988, Dr. John Voorhees and his colleagues [27] at the University of Michigan published the first randomized double-blind vehicle-controlled clinical trial of tretinoin as a treatment for photoaged skin, confirming the earlier controlled studies. Not surprisingly, the first demonstration that a topically applied cream could reduce or reverse unwanted age-associated skin changes within four months, published in a peer-reviewed prestigious medical journal, generated enormous press coverage and captured the public's imagination. At least at the time, however, the report was met with frank disbelief and open hostility from many dermatological "authorities" who in essence publicly accused the investigators of falsifying data and/or misrepresenting their findings for self-aggrandizement or financial gain [28]. Why? Personal animosities and petty jealousy are always candidate factors. However, in retrospect it is apparent that in the 1980s it was simply beyond belief that a topical preparation could alter aging changes in skin. If tretinoin appeared to decrease wrinkling, there had to be a trick [29–34]. Only the publication of multiple well-controlled trials involving many hundreds of patients and dozens of investigations from around the world, yielding uniformly positive and consistent findings, finally overcame the initial disbelief (Tab. 1). It is a testimony to the work summarized below that only one decade later, both physicians and patients take for granted effective medical treatment of aging and photoaging.

Animal models

Systematic study of the process of photoaging is difficult or impossible using human subjects. Animal models in which the process is accelerated are therefore desirable models. There is no naturally occurring animal model of photodamage, as diurnal animals are covered with fur, feathers, or scales that protect them from UVR [35]. However, the normally nocturnal laboratory generated albino hairless mouse (Skh-HR1) has been widely utilized because the animals wrinkle in response to UVR within the relatively short murine life expectancy. Mice were first UV irradiated using an established protocol for ten weeks to induce photodamage, and topical tretinoin was then applied in concentrations ranging from 0.005% to 0.05% for an additional 5–10 weeks. New dermal collagen formed only in the treatment group, and the levels were significantly higher with long term use as well as higher concentrations of tretinoin. The clinical improvement of photoaged skin in this animal model was demonstrated by Kligman et al [25]. in 1982. Later studies by Bryce [36] in 1988, Schwartz [37] in 1991, and Chen [38] in 1992 demonstrated that wrinkle effacement following topical tretinoin application was associated with new intracellular and extracellular dermal type I collagen synthesis.

Clinical efficacy

The extensive clinical trials done to date establish the clinical efficacy of tretinoin. In 1986 Albert Kligman [21] first published his observations regarding the potential of tretinoin in the treatment of photodamage. He reported that patients with acne developed a smoother skin and subjectively more "youthful" skin after application of topical tretinoin (Retin-A cream™), which he had pioneered two decades earlier in the management of acne vulgaris. Kligman and his colleagues then conducted an open study of eight elderly patients receiving tretinoin cream (0.05%) on the face for six to twelve months [21]. There were six age-matched patients who received the vehicle alone. Although there were only slight clinical differences, the histological outcomes in the treatment arm were impressive. The treatment group showed thickening of the previously atrophic epidermis, elimination of keratinocyte dysplasia and atypia, more uniform dispersion of melanin, and the formation of new dermal collagen and blood vessels.

The first double blind, vehicle controlled, randomized study was done by Weiss, Voorhees, and colleagues [27]. Patients applied tretinoin (0.1%) to one forearm and vehicle cream to the other for 16 weeks. Half of the patients received tretinoin cream to the face, and the other half received the vehicle cream alone. All 30 patients who completed the study showed a statistically significant improvement in photoaging on the tretinoin treated forearms, but not on the vehicle-treated forearms ($P < 0.0001$). Fourteen of the 15 patients who received tretinoin to the face had an improvement in photoaging, whereas none of the vehicle treated patients' faces improved, a highly statistically significant difference in response between the two groups. The greatest improvement was in fine wrinkling, which was statistically significant after two months of therapy (Fig. 5). Other features such as coarse wrinkles, roughness, and sallowness of the skin also showed an improvement but to a lesser degree.

In 1991 Weinstein et al. [39] repeated a double-blind, randomized, and vehicle-controlled multi-center trial using three different concentrations of topical tretinoin cream ranging from 0.01 to 0.05% on a daily basis for six months. The study involved 251 healthy white patients, predominately women aged 29 to 50 years (mean, 41 years) with mild to moderate facial photodamage. After 24 weeks there was an overall 79% clinical improvement in the 0.05% group compared to 48% in the vehicle control group ($p = 0.002$). The most remarkable clinical improvements were in fine wrinkling, skin laxity and roughness, and pigmentary changes.

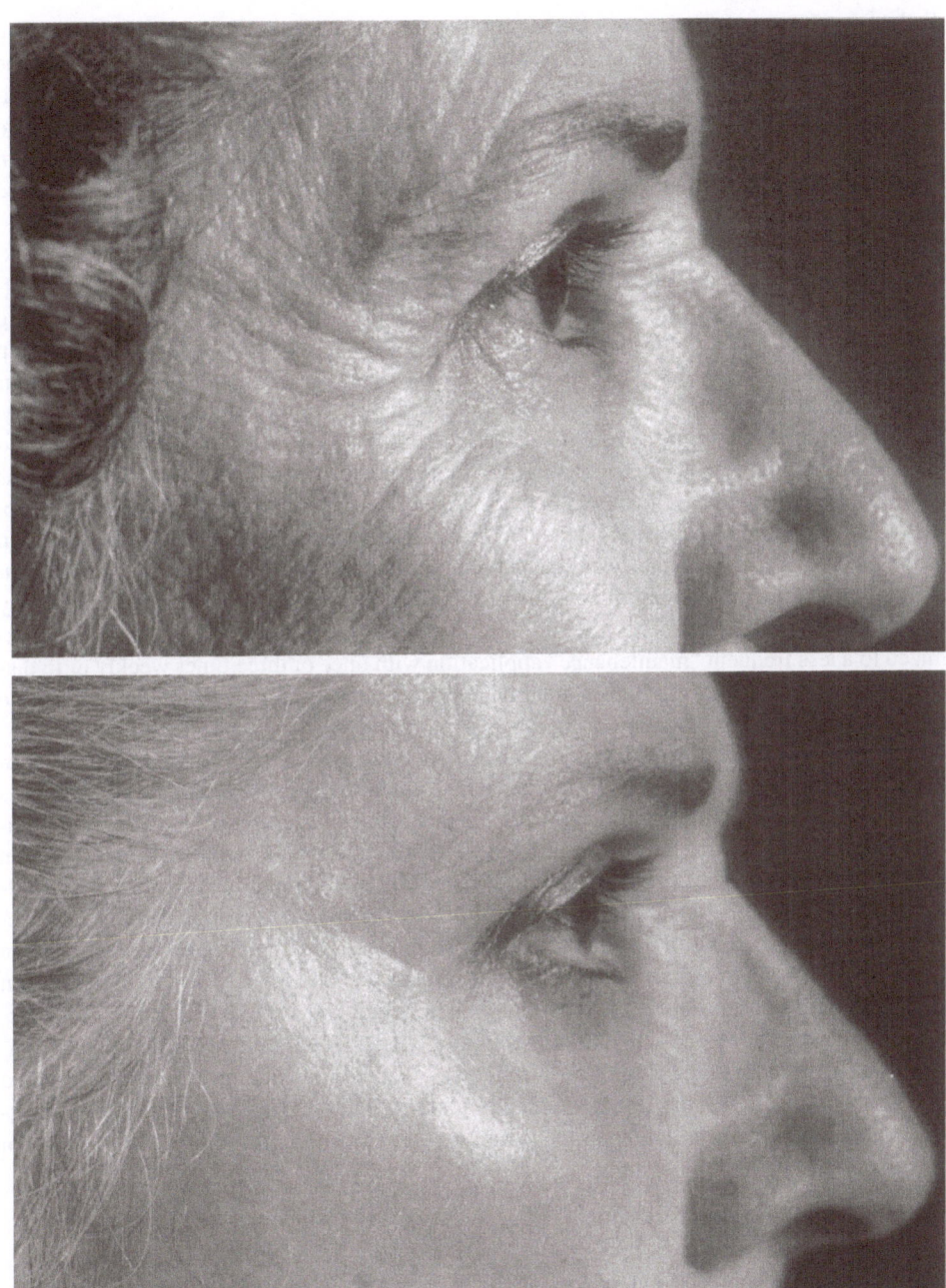

Figure 5. Photomicrographs at baseline (upper panel) and following treatment for 18 months (lower panel) with once daily application of 0.05% tretinoin emollient cream (Renova). Reprinted from Ellis et al. [41].

Patients' self-assessments were consistent with investigator ratings in overall response to therapy. Patients using 0.05% tretinoin responded most favorably to therapy, with 87% of the subjects judging their skin to be improved relative to baseline compared with 43% of the vehicle group. In addition, histological changes of increased epidermal thickness, decreased melanin content, and stratum corneum compaction were again noted. The high response rate in the control group was attributed to the sunscreen use and careful skin care regimen mandated for all trial participants.

Olsen et al. [40] evaluated 296 patients with mild to moderate photodamage in a 24 week, double blind, randomized, multi-center study. Tretinoin emollient cream (0.05%) gave a significantly better global response to therapy than vehicle, with 68% of patients exhibiting improvement at the end of therapy, compared with 43% of subjects in the vehicle group (p < 0.001). Studies beyond six months were initially done by Ellis et al. [41]. After a four month topical tretinoin (0.1% cream), randomized, double-blind, vehicle-controlled study, patients were asked to continue for up to 22 months on an open label basis. Patients sustained clinical improvement even when the dosage or frequency of application of tretinoin was reduced. The greatest improvement was in the reduction of lentigines (71% in the treated group, P = 0.003). Another long-term study done by Thorne et al. [42] using 0.01% or 0.05% tretinoin initially for six months and then continued for a year confirmed the clinical benefits, most notably mottled hyperpigmentation and fine wrinkling. An important component of this study was the follow-up of 149 patients for an additional six months during which they were randomized to a less frequent "maintenance" application with once or three times weekly or to discontinuation of treatment. Nearly 80% maintained their improvement and 47% maintained the full benefit despite discontinuation of therapy for six months.

In order to have more objective criteria for patient evaluation and to eliminate the criticism that "blinded" investigators could identify tretinoin-treated patients by their retinoid dermatitis and thus consciously or unconsciously bias their evaluations, Grove et al. [43] devised computerized image analysis of pre-delineated sites with silicone rubber replicas to measure skin topography in addition to clinical measures in two multi-center, double-blind, randomized studies of tretinoin emollient cream. This optical profilometry technique is both sensitive and objective for measuring fine wrinkles and confirmed improved skin topography in the tretinoin (0.05%) emollient cream users compared to controls. This technology was also used by Leyden et al. [44] to demonstrate significant reduction in severity of fine wrinkles in the tretinoin versus the vehicle-treated group (p = 0.024) in another randomized, double blind, vehicle controlled trial. In 1992 Griffiths et al. [45] introduced a method in which investigators compared patient photographs to standard photographs to quantify photodamage and demonstrated significantly greater agreement among graders than was previously achieved using only a written descriptive scale.

Histological data

Like the clinical changes, histological alternations resulting from chronic sun exposure were once considered irreversible. Histological evaluation of patients in the clinical trials discussed above has revealed both transient epidermal retinoid changes and changes which parallel the clinical benefits, and provides clinical-pathological correlation for the observed improvements. The findings are summarized in Table 2 [46].

Table 2. Morphological effects of tretinoin on photodamaged skin

Histologic parameter	Early, 3–6 months	Late, 12–48 months
Stratum corneum morphology	Compact (altered)	Basket weave (normal)
Granular layer thickness	Increased	Normal
Epidermal thickness	Increased	Normal
Keratinocytic atypia	None or decreased	None or decreased
Melanocytic atypia	None or decreased	None or decreased
Epidermal mucin	No change to increased	Increased
Epidermal melanin	Decreased and more evenly distributed	Decreased and more evenly distributed
Papillary dermalcolagen	No change	improved organization and new synthesis
Anchoring fibrils	No change	Increased
Elastosis	No change	Decreased
Vascularity	No change to slight increase	Increased

Modified from Bhawan J, Gilchrest BA [46].

The largest histological study examined 533 patients before and six months after treatment with 0.05%, 0.01% tretinoin emollient cream or vehicle [47]. Two millimeter punch biopsies were obtained from the periorbital area and computer assisted image analysis was used to quantify selected features from coded (blinded) histopathological slides. The two higher dose tretinoin groups showed statistically significant dose-dependent differences compared to the vehicle group. The most striking differences were an increased epidermal thickness and greater granular cell layer thickness (Fig. 6). The stratum corneum was also transformed from its normal basket weave pattern to a compact morphology in the treatment groups. A reduction of epidermal melanin content, increased epidermal mucin, and persistent absence of keratinocyte or melanocyte atypia were also noted. These "retinoid effects" had also been noted in the first controlled tretinoin trial after four months and peaked well before clinical improvements in photoaging [48].

The initial double blind placebo controlled trials were extended for an additional six months, but further histological assessments performed in a similar fashion showed reversal of some of the earlier changes [47]. The stratum corneum compaction decreased from 90% at 24 weeks to 51% at 48 weeks, the epidermal thickness decreased to 6% less than pre-treatment levels, and the granular layer thickness was reduced after 48 weeks to 25% over baseline compared to 60% increase at 24 weeks. Results of periorbital biopsies performed in 27 patients at baseline, 6, 12, and 48 months [44] were available and showed a slightly increased prevalence of stratum corneum compaction compared with baseline and 12 months and a normal granular layer and epidermal thickness. Further ultra-structural studies [49, 50] showed increased anchoring fib-

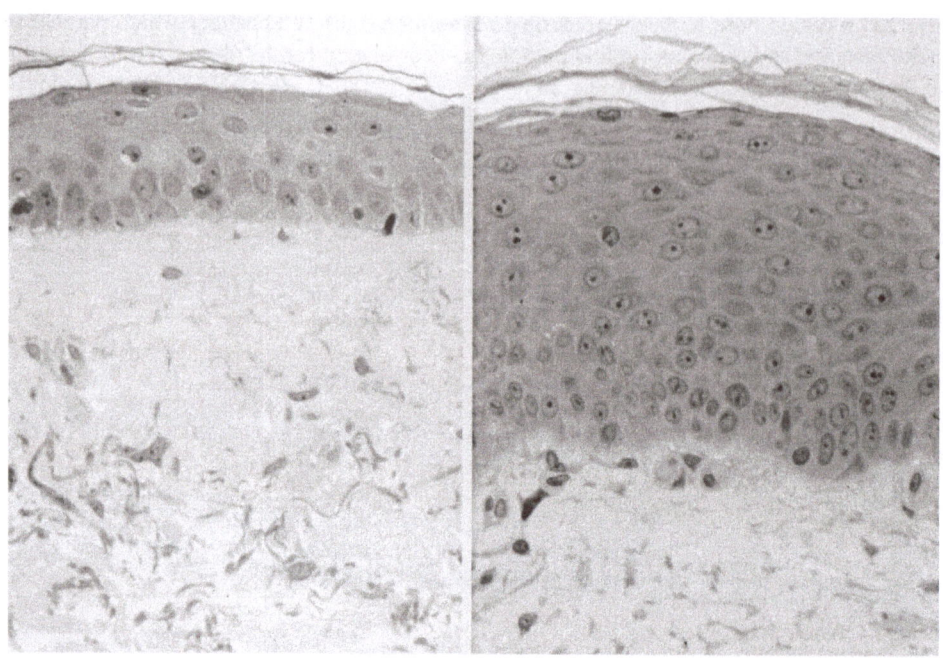

Figure 6. Light micrographs of a biopsy cross section from photodamaged facial skin of the same patient before (left panel) and 24 weeks after (right panel) applying 0.05% tretinoin emollient cream. (PMS stain, original magnification × 40). Micrographs courtesy of Dr. Jag Bhawan.

rils in the forearm biopsy specimens as well as decreased elastosis and new synthesis of papillary dermal collagen (type I collagen). The most significant histological changes in the long term studies (four years) include a decrease in melanin content and melanocyte numbers (81% decrease; p < 0.001), and an increase in the epidermal mucin (65% increase; p < 0.001). These findings along with an increase in dermal collagen most likely account for the clinical improvements in wrinkling [51]. No increase in keratinocyte or melanocyte atypia was noted in the short or long term studies of tretinoin use.

Recent work by Murphy et al. [52] has shown tretinoin to reverse Langerhans cell depletion in chronically photodamaged skin. Eight volunteers ranging in age from 38 to 51 with severe actinic damage were studied. Immunohistochemical studies done at baseline showed a profound depletion of CD1a-positive Langerhans cells in the epidermis. All patients demonstrated replenishment of inter-follicular epidermis by CD1a-positive Langerhans cells after six months of daily application of tretinoin. Thus, retinoid-induced enhancement of epidermal immunity in photodamaged skin may reflect restoration of antigen presenting Langerhans cells [52].

Other studies have begun to explore the benefits of topical tretinoin therapy in patients with predominantly pigmentary manifestations of photoaging. Topical tretinoin has been shown to have beneficial effects in the treatment of hyper-pigmentation such as "liver spots" (also termed actinic lentigines) in white persons (skin types I and II) [53]. Griffiths et al. [54] conducted a double-blind, placebo controlled, randomized study of 0.1% tretinoin cream for 40 weeks in 45 patients with skin type V (23 Chinese, 22 Japanese) in whom lentigines and irregular hyper-pigmentation were dominant components of photoaging. By the end of the treatment, hyper-pigmented lesions of the face were lighter or much lighter in 90% of patients receiving tretinoin compared with 33% receiving vehicle (p < 0.0001) [54].

Molecular mechanisms of retinoid action in photodamage improvement

Work in many laboratories over the past decade has established that tretinoin (retinoic acid, RA) exerts its multiple physiological and pharmacological effects principally, if not exclusively, by binding to a family of nuclear RA receptors (RARs) [55–57] .The RA/RAR complexes then bind RA response elements in the promoter region of target genes to modulate their transcription [58–60]. Unfortunately, to date the genes whose modulation (presumably by tretinoin therapy) improves photoaged skin have not been identified. Of interest, however, tretinoin appears to exert similar effects on both intrinsically aged and photoaged skin [61–63], suggesting target genes are involved in both processes.

Increased amounts of procollagen type I are deposited in tretinoin treated photoaged dermis, both intracellulary and extracellulary within 10–12 months [51] and may result from transcription up-regulation. Moreover, at an ultra-structural level, collagen fibers are both far more numerous and more orderly in treated skin [49, 50, 64].

A possible additional role for tretinoin has recently been suggested by Fisher et al. [65], who demonstrated that tretinoin-treated skin is protected from metalloproteinase-mediated destruction of dermal matrix following UV irradiation. Metalloproteinases (a family of 14 different enzymes) specifically degrade collagen and elastin [66–69]. A single exposure to even modest doses of UVR increases the expression and activity of three matrix metallo- proteinases: collagenase, a 92 kDa gelatinase, and stromelysin, leading to degradation of type I collagen and elastin [65]. Pre-treatment with topical tretinoin inhibits this induction of matrix metalloproteinases by 70 to 80 percent by blocking activation of the AP-1 transcription factor that otherwise results from UV exposure, but does not affect expression of the tissue inhibitor of metalloproteinases (TIMP) that is not regulated by AP-1 [65].

Tretinoin is thus postulated to shift the balance between proteolytic enzymes and their inhibitors in favor of the inhibitors, conserving dermal collagen and elastin. The molecular mechanisms by which tretinoin improves epidermal maturation and pigmentation or effects the other changes discussed above remain unknown.

Safety considerations

Irritation

Tretinoin formulations available today include the traditional topical tretinoin cream (0.025%, 0.05%, 0.1%), gel (0.025%, 0.01%) or liquid (0.025%) developed for acne therapy (Retin-A™), a "microsponge" slow release preparation of topical tretinoin 0.1% (Retin-A Micro™) [70], and the emollient cream 0.05% (Renova™, Retinova™) which was developed and approved specifically for photodamage therapy [71].

Retinoids cause a variable amount of irritation, more properly a heightening of cutaneous reactivity termed retinoid dermatitis, early in therapy [39–42, 44]. In the first few weeks approximately 80% of patients experience stinging, hyperesthesia or pruritus, fine scaling, dryness and mild erythema (Fig. 7). These effects are dose-dependent and peak within the first month of therapy. Weiss et al. reported 92% incidence with 0.1% tretinoin and Leyden et al. a 70% incidence using 0.05% tretinoin. However, in a double blind, vehicle controlled study of 179 patients, only 4% of patients using Renova™ discontinued the medication as a result of adverse reactions [57]. Retinoid dermatitis usually improves markedly during the course of treatment but may not completely disappear. Other side effects are rare.

Although pharmacists, physicians, and the lay press regularly warn that topical tretinoin therapy produces photosensitivity with easy sunburning, no data support this dogma. In the one published study experimentally examining the effect of tretinoin on UV-induced erythema threshold in human subjects, there was no effect [72]. Very possibly, tretinoin therapy increases the tendency to vasodilate following exposure to heat, including solar infrared energy, but this has never been documented.

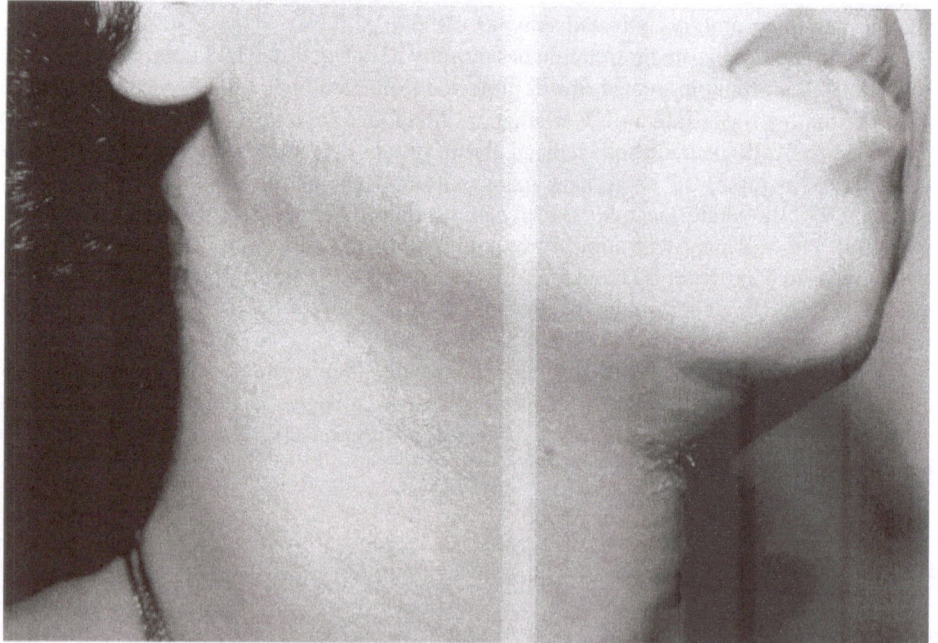

Figure 7. Face and neck 48 h following two daily applications of tretinoin 0.05% cream. Note mild erythema and scaling along the jaw line and submental area. The dermatitis cleared after two weeks of continued daily application.

Teratogenic considerations

Oral isotretinoin administration during early pregnancy causes a characteristic embryopathy (cardiac, cranio-facial and central nervous system abnormalities) and is contraindicated in pregnant women [73]. There are individual case reports of congenital defects associated with maternal use of topical tretinoin during the first trimester of pregnancy [74, 75], but no data to support a causal relationship.

For topical tretinoin to cause a fetal malformation, a sufficient quantity would need to be absorbed from the skin into the maternal blood stream. However, systemic absorption of this compound is negligible and insufficient to raise the endogenous plasma levels, as demonstrated by Worobec et al. [76] using radiolabeled tretinoin and by Chiang and colleagues [77] using a twice daily application of 0.025% tretinoin cream to greater than 4% of body surface area for one month. Jick et al. [78] reviewed 215 case histories of women who used tretinoin for acne in the first trimester and compared their pregnancy outcomes to 430 age-matched, non-exposed women. No increase in major congenital anomalies was noted and there were no instances of retinoid embryopathy. The overall prevalence of anomalies among infants born to tretinoin-exposed women was 1.9% compared with 2.6% for the non-exposed group, both of which are within the expected rates. Shapiro et al. [79] evaluated 94 cases of tretinoin use during the first trimester and a matched group of 133 control women, the most comprehensive clinical data yet available. Pregnancy outcomes did not differ between the two groups. There was no difference in the rate of malformations, gestational age, birth weight, or method of delivery. In rats and rabbits topical tretinoin is non-teratogenic when given in doses of 100 and 320 times the topical human dose, respectively, but offspring have lower average birth-weights presumably due to severe skin irritation of the pregnant animals [71].

On the basis of these data and further in-house review, the FDA has classified topical tretinoin as "Category C" [71], implying at least one animal study has shown teratogenic effects at higher doses than used in humans and no studies in humans have shown a teratogenic effect or human studies have not been done.

Usage guidelines

In the tremendous rush of patients seeking "rejuvenating" preparations, many have received little background information and even less instruction from their physicians regarding the optimal way to apply retinoids. Tretinoin is not a cosmetic and patients must be advised of its expected irritancy (Tab. 3). Patients should also be given realistic expectations and realize that retinoids are not a substitute for a facelift, blepharoplasty, liposuction, chemical peels or dermabrasion [80]. In addition, patients should be counseled regarding a comprehensive skin-care and sun-avoidance program including sunscreens and protective clothing. Finally, treatment must be individualized because patients seeking therapy represent a wide range of skin types, ages, and retinoid responsiveness [81]. Patients most likely to experience irritation include those with fair skin, freckles and blue eyes who sunburn easily (Fitzpatrick skin types I and II); co-existent dermatological disorders such as xerosis, atopic dermatitis, and acne rosacea; and skin hypersen-

Table 3. Guidelines for patient selection

Minimum criteria

    Access to adequate physician counseling.
    Dedicated, compliant and willing to either avoid sunlight or use sunscreen.
    Understands that improvements are often subtle and take several weeks to months.
    Aware of alternative surgical treatment options for photoaging.

Factors predicting a good response[*]

    Older age (>50 yr)
    Skin type III, IV, or V
    Prominent dyspigmentation
    Sallowness

Factors predicting little response and/or treatment intolerance[*]

    Younger age
    Skin type I or II
    Atopy
    Rosacea-prone complexion
    History of "sensitive skin"

---

[*] There are many exceptions in both groups.

sitive to commonly used topical products such as sunscreens, cosmetics, or perfumes. Lower sensitivity has been noted anecdotally in older patients with advanced photoaging and darker skin types with oilier skin. In those with sensitive skin, it is best to begin treatment with every other night applications or even every third night if irritation is encountered. After accommodation, daily applications can cautiously be restarted. Excessive use of toiletries and potentially irritating astringents, toners, cleansers, abrasives, and scrubbing solutions should be avoided.

Conclusion

Topical tretinoin is effective in reducing fine wrinkling, mottled hyperpigmentation and skin sallowness and roughness in patients with photodamaged skin. Histological changes such as increased mucin, decreased melanocytes and melanin redistribution and increased papillary collagen deposition correlate with the observed clinical benefits. With regard to molecular mechanism, tretinoin is known to bind nuclear RA receptors and the RA/RAR complexes in turn bind RA response elements in target genes to alter their transcription rate. To date, the genes relevant to improvement of aging/photoaging changes in skin have not been identified, although procollagen type I is strongly implicated. Recent data suggest that tretinoin may also work at the molecular level through inhibition of matrix metallo-proteinases, such as collagenase, by blocking induction of the UV-induced AP-1 transcription factor that regulates these enzymes.

    Patient selection and adequate counseling are paramount when using topical tretinoin in the management of photodamage. The compound is well tolerated and although tretinoin frequently induces a mild dermatitis, the problem can be solved by dosage or frequency reduction of the product. In several studies to-date, topical tretinoin therapy has not been shown to have any teratogenic risk. Moreover, topical application appears incapable of increasing endogenous circulating tretinoin levels.

References

1   Gilchrest BA (1994) Turning back the clock: retinoic acid modifies intrinsic aging changes. *J Clin Invest* 94: 1711–1712
2   Kaminer M (1995) Magnitude of the Problem. *In*: BA Gilchrest (ed): *Photodamage*. Blackwell Science, Cambridge, MA, 2–3
3   Wilmoth JR (1988) The future of human longevity: a demographer's perspective. *Science* 280(5362): 395–397
4   Ohnaka T (1993) Health effects of ultraviolet radiation. *Ann Physiol Anthropol* 12(1): 1–10
5   Lavker RM (1995) Cutaneous Aging: Chronologic versus Photoaging. *In*: BA Gilchrest (ed): *Photodamage*. Blackwell Science, Cambridge, MA, 123–125
6   Taylor CR, Stern RS, Leyden JJ, Gilchrest BA (1990) Photoaging/photodamage and photoprotection. *J Amer Acad Dermatol* 22: 1–5
7   Gilchrest BA, Blog FB, Szabo G (1979) Effects of aging and chronic sun exposure on melanocytes in human skin. *J Invest Dermatol* 73: 141–143
8   Weinstock MA (1994) Epidemiology of non-melanoma skin cancer: clinical issues, definitions, and classification. *J Invest Dermatol* 102(6): 4S–5S
9   Kwa RE, CampanaK, Moy RL (1992) Biology of cutaneous squamous cell carcinoma. *J Amer Acad Dermatol* 26: 1–26
10 Miller DL, Weinstock MA (1992) Non melanoma skin cancer in the United states: incidence. *J Amer Acad Dermatol* 30: 774–778
11 Uitto J (1986) Connective tissue biochemistry of the aging dermis: age-related alternations in collagen and elastin. *Dermatol Clin* 4: 443–446
12 Braverman IM, Fonferko E (1982) Studies in cutaneous aging: The elastic fiber network. *J Invest Dermatol* 78: 434–444
13 Kligman AM (1969) Early destructive effects of sunlight on human skin. *JAMA* 210: 2377–2380
14 Lavker RM, Kligman AM (1988) Chronic heliodermatitis: a morphologic evaluation of chronic actinic dermal damage. *J Invest Dermatol* 90: 325–330
15 Serri F, Tosti A, Cerimle D (1977) Studies on the patho-mechanics of chronic actinic dermatitis. *In*: A Castellani (ed): *Research in photo-biology*. Plenum Press, New York, 547–553
16 Baso CR, Longacre JJ, Unterthiner RA (1976) Electron microscopic study of fibrillogenesis and hypertrophic scar formation. *In*: JJ Longrace (ed): *The ultrastructure of collagen*. Charles C. Thomas, Springfield, 294–314
17 Linares HA (1976) Granulation tissue and hypertrophic scars. *In*: JJ Longrace (ed.): *The ultrastructure of collagen*. Charles C. Thomas, Springfield, 93–107
18 International Classification of Diseases 9th Revision (1991) US Deptartment of Health and Human Services, 578
19 International Classification of Disease 9th Revision (1992) US Department of Health and Human Services, 580
20 Kligman AM, Fulton JE, Plewig G (1969) Topical vitamin A acid in acne vulgaris. *Arch Dermatol* 99: 469–476
21 Kligman AM, Grove GL, Hirose R (1986) Topical tretinoin for photoaged skin. *J Amer Acad Dermatol*;15: 836–59
22 The role of tretinoin in photodamage (1990) NIH Consensus Panel. National Institutes of Health, Bethesda, MD
23 FDA approves Renova to Assist in Reducing Skin Damage (1998) (January 2, 1996 Report). Homepage of the Food and Drug Administration. Online. WWW.FDA.GOV
24 Dermatology Drugs Advisory Committee (1998) (April 9, 1992 Report). Homepage of the Food and Drug Administration. Online. WWW.FDA.GOV
25 Kligman LH, Akin FJ, Kligman AM (1982) Prevention of ultraviolet damage of the dermis of hairless mice by sunscreen. *J Invest Dermatol* 78: 181–189
26 Topical retinoids: an update (1986) Proceedings of a symposium. *J Amer Acad Dermatol* 15: 735–916
27 Weiss J, Ellis CN, Headington JT, Tincoff T, Hamilton TA, Voorhees JJ (1988) Topical tretinoin improves photoaged sk*In*: a double blind vehicle–controlled study. *JAMA* 259: 527–532
28 Roberts L (1988) Questions raised about anti-wrinkle cream. *Science* 239(4840): 564–565
29 Fackelman K (1988) Tretinoin: lasting results, lingering doubts. *Science News* 134(24): 375–376

30 Okun MR (1988) Topical Tretinoin for Photoaged skin. [letter]. *JAMA* 259: 3271–3272
31 Kanigsberg ND (1988) Topical tretinoin for photoaged skin. [letter]. *JAMA* 29: 3272
32 Tavelli BG, Storss FJ (1988) Topical tretinoin for photoaged skin. [letter]. *JAMA* 29: 3272
33 Pazmino P (1988) Topical tretinoin for photoaged skin. [letter]. *JAMA* 29: 3272
34 Marshall E (1990) Penn charges Retin-A inventor with conflict. *Science* 247: 1028–1029
35 Bissett DL, Hillebrand GG, Hannon DP (1989) The hairless mouse as model of skin photoaging: its use to evaluate photoptotective material. *Photodermatology* 6: 228–233
36 Bryce CF, Bogdan NJ, Brown CC (1988) Retinoic acids promote the repair of the dermal damage and the effacement of wrinkles in the UVB irradiated mouse. *J Invest Dermatol* 91: 175–180
37 Schwartz E, Cruickshank FA, Perlish JS, Fleischmajer R (1989) Alternations in dermal collagen in ultraviolet irradiated hairless mice. *J Invest Dermatol* 93: 142–145
38 Chen VL, Fleishmajer R, Schwartz E et al (1986) Immunochemistry of elastotic material in sun damaged skin. *J Invest Dermatol* 87: 334–337
39 Weinstein GD, Nigra TP, Pochi PE, Savin RC, Allan A, Benik K, Jeffes E, Lufranto L, Thorne EG (1991) Topical tretinoin for treatment of photodamaged skin. *Arch Dermatol* 127: 659–665
40 Olsen EA, Katz HI, Levine N, Shupack J, Billys MM, Praer S, Gold J, Stiller M, Lufrano L, Thorne EG (1992) Tretinoin emollient cream: a new therapy for photodamaged skin. *J Amer Acad Dermatol* 26: 215–224
41 Ellis CN, Weiss JS, Hamilton TA, Headington JT, Zelickson AS, Voorhees JJ (1990) Sustained improvement with prolonged topical tretinoin for photoaged skin. *J Amer Acad Dermatol;*23: 629–637
42 Thorne EG, Lufrano L Boateng F Sampson AR (1992) Long-term clinical experience with a topical retinoid. *Brit J Dermatol* 127(Suppl 41): 31–36
43 Grove GL, Grove MJ, Leyden JJ, Lufrano L, Schwab B, Perry BH, Thorne EG (1991) Skin replica analysis of photodamaged skin after therapy with tretinoin emollient cream. *J Amer Acad Dermatol* 25: 231–237
44 Leyden JJ, Grove GL, Grove MJ, Thorne G, Lufrano L (1989) Treatment of photodamaged facial skin with tropical tretinoin. *J Amer Acad Dermatol* 21: 638–44
45 Griffiths EM, Wang TS, Hamilton TA, Voorhees JJ (1992) A photonumeric scale for the assessment of cutaneous photodamage. *Arch Dermatol* 128: 347–351
46 Bhawan J, Gilchrest BA (1995) Morphological Effects of Tretinoin on Photodamaged Skin. *In:* BA Gilchrest (ed.): *Photodamage.* Blackwell Science, Cambridge, MA, 244
47 Bhawan J, Gonzalez-Serva A, Nehal K, Labadie R, Lufrano L, Thorne EG, Gilchrest BA (1991) Effects of tretinoin on photodamaged sk*In*: a histologic study. *Arch Dermatol* 127: 666–672
48 Bhawan J, Palko MJ, Lee J (1995) Reversible histologic effects of tretinoin on photodamaged skin. *J Geriatr Dermatol* 3: 62–67
49 Yamamoto O, Bhawan J, Solares G, Tsay AW, Gilchrest BA (1995) Ultra-structural effects of topical tretinoin on dermo-epidermal junction and papillary dermis in photodamaged skin. *Exp Dermatol* 4: 146–154
50 Yamamoto O, Bhawan J, Hara M, Gilchrest BA (1995) Keratinocyte degeneration in human facial sk*In*: documentation of new ultra-structural markers for photodamage. *Exp Dermatol* 4: 9–19
51 Griffiths EM, Russman AN, Majmudar G, Singer RS, Hamilton TA, Ellis CN, Voorhees JJ (1993) Restoration of collagen formation in photodamaged human skin by tretinoin. *New Engl J Med* 329: 530–535
52 Murphy GF, Katz S, Kligman A (1998) Topical tretinoin replenishes CD1a-positive epidermal Langerhans cells in chronically photodamaged human skin. *J Cutaneous Pathol* 25: 30–34
53 Rafal ES, Griffith EM, Ditre CM, Finkel LJ, Hamilton TA, Ellis CN, Voorhees JJ (1992) Topical tretinoin treatment for liver spots associated with photodamage. *New Engl J Med* 326: 368–372
54 Griffiths CE, Goldfarb MT, Finkel LJ, Roulia V, Bonawitz M, Hamilton TA, Ellis CN, Voorhees JJ (1994) Topical tretinoin treatment of hyperpigmented lesions associated with photoaging in Chinese and Japanese patients: A vehicle-controlled trial. *J Amer Acad Dermatol* 30: 76–84
55 Manglessdorf DJ, Ong ES, Dyck JA, Evans RM (1990) Nuclear receptor that identifies a novel retinoic acid response pathway. *Nature* 345: 224–229
56 Eller MS, Oleksiak MF, McQuaid TJ, McAfee SG, Gilchrest BA (1992) The molecular cloning and expression of two CRABP cDNAs from human skin. *Exp Cell Res* 199: 328–336
57 Nagpal S, Saunders M, Kastner P, Durand B, Nakshatri H, Chambon P (1992) Promoter context-and response element-dependent specificity of the transcriptional activation and modulating functions of retinoic acid receptors. *Cell* 70: 1007–1019
58 Duell EA, Astrom A, Griffiths CEM, Chambon P, Voorhees JJ (1992) Human skin levels of retinoic acid and cytochrome P-450 derived 4-hydrocyretinoic acid after topical application of retinoic acid *in vivo* compared to concentrations required to stimulate retinoic acid receptor-medicated transcription *in vitro*. *J Clin Invest* 90: 1269–1274
59 Lala DS, Mukherjee R, Schulman IG et al (1996) Activation of specific RXR heterodimers by an antagonist of RXR homodimers. *Nature* 383: 450–453
60 Chen JY, Clifford J, Zusi C, Starrett J, Tortolani D, Ostorowski J, Reczek PR, Chambon P, Gronemeyer H (1996) Two distinct actions of retinoid-receptor ligands. *Nature* 382: 819–822
61 Kligman AM, Dogadkina D, Lavker RM et al (1993) Effects of topical tretinoin on non-sun-exposed protected skin of the elderly. *J Amer Acad Dermatol* 29: 25–33
62 Varani J, Perone P, Griffiths CEM, Inman DR, Fligiel SEG, Voorhees JJ (1994) All-*trans* retinoic acid stim-

ulates events in organ cultured human skin that underlie repair: Adult skin from sun-protected and sun exposed sites responds in an identical manner to RA while neonatal foreskin responds differently. *J Clin Invest* 94: 1747–1756

63 Chen S, Kiss I, Tramposch KM (1992) The priming effect of ultraviolet B radiation on retinoic acid-stimulated collagen synthesis in the mouse photo-damage model. *Photoderm Photoimmunol Photomed* 9: 104–108

64 Feldman D, Bryce GF, Shapiro SS (1991) Ultrastructural effects of UVB radiation and subsequent retinoic treatment on the skin of hairless mice. J. Cut Pathol 18: 46–55

65 Fisher GJ, Datta SC, Talwzr HS, Wang ZQ, Varani J, Kang S, Voorhees JJ (1997) Pathophysiology of premature skin aging induced by ultraviolet light. *New Engl J Med* 337: 1419–1428

66 Birkedal-Hansen H, Moore WG, Bodden MK, Windsor LJ, Birkedal-Hansen B, DeCarlo A, Engler JA (1993) Matrix metalloproteinases: a review. *Crit Rev Oral Biol Med* 4: 197–250

67 Kahari VM Saarialho-Kere U (1997) Matrix metalloproteinases in skin. *Exp Dermatol* 6: 199–213

68 Denhardt DT, Feng B, Edwards et al (1993) Tissue inhibitor of metalloproteinases (TIMP): structure, control of expression and biological functions. *Pharmacol Ther* 59: 329–341

69 Gomez DE Alonso DF Yoshiji H Thorgeirsson UP (1997) Tissue inhibitors of metalloproteinases: structure, regulation and biological functions. Euro *J Cell Biol* 74:111–122

70 Retin-A prescribing information (1998) Physicians' desk reference. 52nd ed. Medical Economics Company, Inc. Montvale, NJ. 1986–1988

71 Renova prescribing information (1998) Physicians' desk reference. 52nd ed. Medical Economics Company Inc. Montvale, NJ. 1984–1986

72 Fisher GJ, Data SC, Talwar HS (1996) Molecular basis of sun-induced premature skin aging and tretinoin antagonism. *Nature* 279: 335–339

73 Lammer EJ, Chen DT, Hoar RM, Agnish ND, Benke PJ, Braun JT, Curry CJ, Fernhoff PM, Grix AW Jr, Lott IT (1985) Retinoic acid embryopathy. *New Engl J Med* 313: 837–841

74 Lipson AH, Collins F, Webster WS (1993) Multiple congenital defects associated with maternal use of topical tretinoin. *Lancet* 341: 1353–1355

75 Camera GPregliasco P (1992) Ear malformation in a baby born to mother using tretinoin cream. *Lancet* 339: 687

76 Worobec SM, Wond FGA, Tolamn EL Abrams LS, Parker GR, Thorne EG (1991) Percutaneous absorption of 3H-tretinoin in normal volunteers. *J Invest Dermatol*;96: 574A

77 Chain TC (1980) Gas chromatographic-mass spectrometric assay for low levels of retinoic acid in human blood. *J Chromatogr* 182: 335–340

78 Jick SS, Terris BZ, Jick H (1993) First trimester topical tretinoin and congenital disorders. *Lancet* 341: 1181–1182

79 Shapiro L, Pastuszak A, Curto G et al (1997) Safety of first-trimester exposure to topical tretinoin: Prospective cohort study. *Lancet* 350: 1143–1144

80 Weiss JS, Ellis CN, Goldfarb MT, Voorhees JJ (1990) Tretinoin therapy: practical aspects of evaluation and treatment. *J Int Med Res* 18(Suppl 3): 41C–48C

81 Kligman A (1989) Guidelines for the use of topical tretinoin for photoaged skin *J Amer Acad Dermatol* 21: 650–654

Vitamin A and retinoids: an update of biological aspects and clinical applications
M.A. Livrea (ed.)
© 2000 Birkhäuser Verlag Basel/Switzerland

Therapeutic uses of retinoic acid receptor antagonists and inverse agonists

E.S. Klein[1] and R.A.S. Chandraratna[1,2]

Retinoid Research, Departments of Biology[1] and Chemistry[2], Allergan Pharmaceuticals, Irvine, CA 92715, USA

Introduction

Retinoic acid receptors (RARs), members of the nuclear receptor superfamily, are transcription factors which mediate changes in gene expression in a ligand regulated manner. Ligands which bind to, and activate, the RARs (i.e. agonists) have proven useful in the treatment of a variety of dermatological disorders such as acne [1] and psoriasis [2] as well the treatment of photo-damaged skin [3]. RAR agonists have also been successful in the treatment of neoplastic disorders such as head and neck squamous cell carcinoma [4] and promyelocytic leukemia [5]. Further, an extensive amount of pre-clinical data suggest the possibility that RAR agonists may also be useful in treating disorders such as breast cancer [6, 7] and emphysema [8]. Agonists of the retinoid X receptor (RXR), which function as heterodimeric partners with RARs as well as other members of the nuclear receptor superfamily, have also been demonstrated to have promise as therapeutic agents in metabolic disease [9]. While antagonists of other members of the nuclear receptor superfamily, such as the estrogen receptor and progesterone receptor, have found significant applications in clinical medicine [10, 11], the utility of RAR antagonists is only now being considered. Formal possibilities might include pathogenic states caused either by inordinately high levels of retinoid signalling or which are dependent on endogenous retinoids for their maintenance. Progress in this area is now possible as potent and effective RAR antagonists have recently become available. Further, the recent identification of nuclear receptor cofactor proteins which interact with the RAR in an agonist or antagonist dependent manner, together with the description of a spectrum of RAR ligand activities from agonist to neutral antagonist to inverse agonist, now offers the possibility that ligand-regulated movement of these cofactors between the network of receptors which utilize them may offer a productive avenue for drug discovery.

RAR Antagonist/inverse agonist structures

Although the biology associated with retinoid agonists has been extensively investigated, the effects of retinoid antagonism have not been studied to a great extent. A major reason for this was that effective retinoid antagonists were not available until recently. The compounds 1 [12–14], 2 [12–14], 3 [15, 16] and 4 [17, 18] (Fig. 1) are relatively low affinity RAR antagonists and require that they be used in great excess to block retinoid activity. Compounds 5 [19] and 6 [20] are high affinity RAR antagonists that significantly inhibit all-*trans* retinoic acid (RA) function at equimolar concentrations. Interestingly, compound 6 functions as an inverse agonist at RARγ while the corresponding phenyl analog 7 is a neutral antagonist [21]. A systematic investigation of the effects of changing the position and nature of the substituent on the C-1 phenyl substituent of analogs of the type 6 and 7 has been carried out [22]. In terms of binding affinity to the RARs, it was determined that a para substituent was preferred over meta and ortho substituents and that a substituent of the size equivalent to a methyl or ethyl was optimal. It was also determined that the RAR binding affinities were rather insensitive to the electronic nature of the para substituent but a hydroxyl group was not well tolerated. The introduction of heteroatoms into the lipophilic region of the diaryl acetylene framework as in the general structure 8 gave RAR antagonists that were equivalent to or more potent than compound 6 [22]. It has also been possible to synthesize antagonists that are selective for the RARα subtype. Thus, compound 3 is moderately selective for RARα while compound 9 [23] is RARα-specific antagonist. The availability of this wide variety of structural types will greatly facilitate the search for therapeutic applications for RAR antagonists.

Agonists, neutral antagonists and inverse agonists

The concept of inverse agonism or negative agonism was developed from observations that certain antagonists of various G-coupled membrane receptors elicited an inhibitory effect upon unliganded basal receptor activity [24–26]. While it has been argued that such phenomena are the result of blockade of endogenous ligands [27], the identification of ligands (ie. neutral antagonists) which competitively antagonized the effects of agonists but did not exhibit inverse agonist activity [26, 28] suggested that inverse agonists, neutral antagonists and agonists have the ability to induce or stabilize receptor conformations which correlate with inactive, neutral and active receptor function, respectively. The ability of a ligand to read out as an inverse agonist therefore depends on the set point for the basal state (unoccupied receptor). This basal activity varies with receptor type [29] and is likely to be dependent upon the abundance [30] of not only the receptor in question but also the cofactors which communicate with the receptor in the signal transduction process. Therefore, a central paradigm of this concept is that inverse agonists function in an active manner, not by merely mediating competitive blockade of an agonist.

A theoretical model to explain the occurrence of inverse agonists was put forward by Lefkowitz and colleagues [26]. In this two-state receptor model, the receptor is proposed to exist in two conformations: R, denoting the inactive conformation and R*, denoting the active conformation which is capable of interacting with G-protein and initiating the cascade of signal transduction. The model further proposes that there is an inherent equilibrium between R and R* in the absence of ligand and that agonists shift the equilibrium towards R* while inverse agonists shift the equilibrium towards R. The position of the equilibrium in the absence of ligand will determine whether ligands will have the opportunity to exhibit inverse agonist as well as agonist activity (see below). In addition, this model proposes that neutral antagonists themselves will not have an inherent effect on the equilibrium, but will have the ability to antagonize the equilibrium shifts mediated by either agonist or inverse agonist.

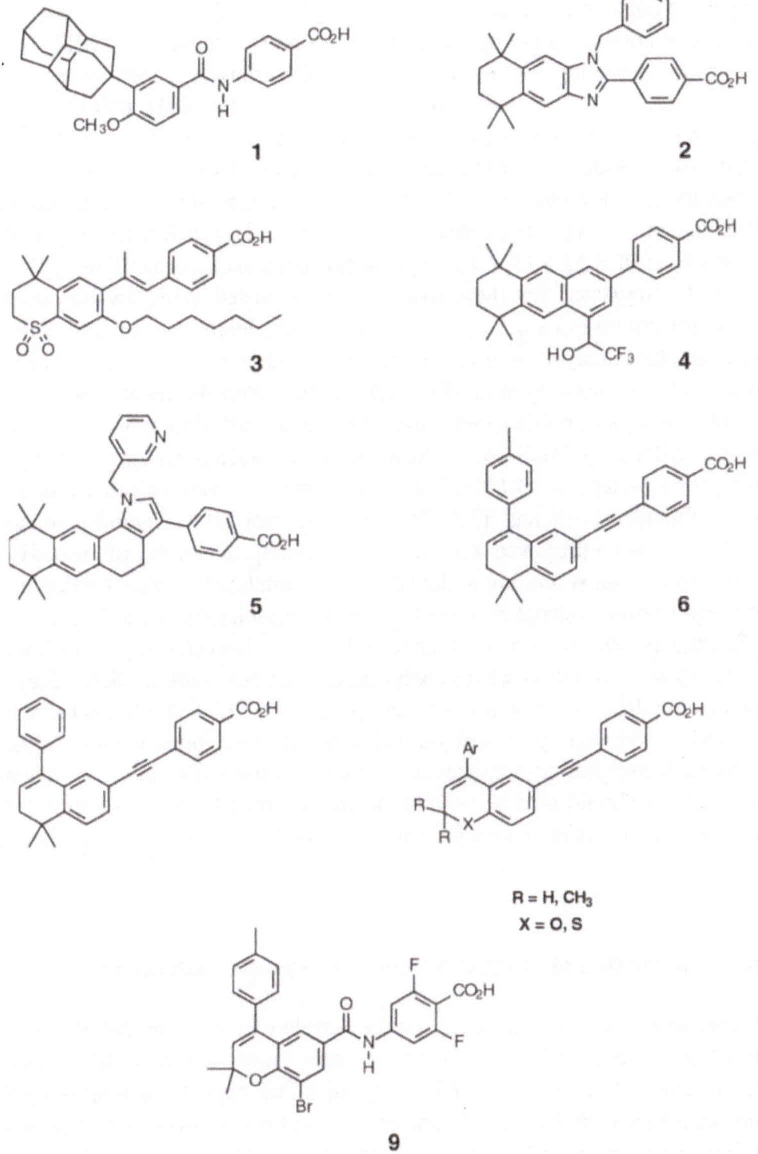

Figure 1. Structures of retinoic acid receptor antagonists.

The recent discovery of various cofactors which interact with nuclear receptor family members provides a conceptual framework to propose that an analogous spectrum of ligands will be possible for these DNA-bound, hormone-regulated transcription factors. Nuclear receptor cofactors appear to come in two basic varieties, positive and negative. Examples of positive acting, transcriptional co-activators such as steroid receptor coactivator 1 (SRC-1) [31], glucocorticoid receptor interacting protein 1 (GRIP-1) [32], transcriptional intermediary factor 1 (TIF-1) [33], activator of the thyroid and RA receptor (ACTR) [34], p300/CBP co-integrator associate protein (p/CIP) [35], receptor-associated coactivator 3 (RAC3) [36] were the first to be discovered. These proteins do not bind DNA but interact in an agonist-dependent manner with the transactivation domains located within the C-terminal ligand binding domains (LBDs) of the nuclear receptors. Crystalographic analysis of the LBD of RXRα [37], without ligand, as well as the LBDs of RARγ [38] and thyroid hormone receptor α (T3Rα) [39], with their respective ligands, suggested that these structurally conserved transactivation domains undergo a similar conformation change upon binding ligand, leading to co-activator interaction. Recent X-ray structure analysis [40] of both the unliganded and liganded LBD of peroxisome proliferator-activated receptor gamma (PPARγ) together with the nuclear receptor interaction domain of SRC-1 has proven this general model to be correct. Examples of negative cofactors include SMRT (Silencing Mediator of Retinoid and Thyroid receptors) [41] and N-CoR (Nuclear receptor Co-Repressor) [42]. These co-repressors were originally demonstrated to interact with unliganded RAR and T3R. Thus, a simple model for ligand-mediated transcriptional activation has been proposed which involves a reciprocal swap of co-activator in place of co-repressor upon agonist binding to the LBD of the nuclear receptor. Analogous to the two-state receptor equilibrium concept proposed for inverse agonist activity at G-coupled membrane receptors [26], the above model of nuclear receptor co-activator/co-repressor interactions can be extended to invoke a situation where such interactions are similarly defined by equilibrium constants describing these functionally opposite processes, which result in a basal level of transcriptional activity mediated by the unliganded receptor. An agonist would of course shift the equilibrium towards receptor-co-activator interaction resulting in an increase in gene transcription. Conceptually, it should also be possible to devise ligands that can increase receptor-co-repressor interactions (i.e. inverse agonists) and thereby decrease the basal level of gene expression.

Inverse agonist regulation of nuclear receptor-co-repressor interaction

Recent investigations have demonstrated that the simple reciprocal model of agonist-mediated replacement of co-repressor with co-activator is indeed more complex. Klein and co-workers [21] have demonstrated that certain RAR antagonists are capable of repressing the elevated basal activity of a transfected RARγ which has the additional *trans*-activation domain of the Herpes simplex virus viral protein VP-16 (RARγ-VP-16). These compounds are referred to as RAR inverse agonists. Structurally related ligands which do not exhibit this *trans*-repressive effect, but competitively inhibit the *trans*-repressive effect of inverse agonists, are referred to as neutral antagonists. Consistent with their *trans*-repressive effect, RAR inverse agonist binding to RARγ results in an increased association between the nuclear receptor and N-CoR (Klein et al., submitted). Such recruitment of co-repressor, demonstrated both in *in vitro* glutathione-S-transferase (GST) pull down experiments as well as in yeast two-hybrid experiments *in vivo*, suggest that the interaction of the receptor with the co-repressor may be more dynamic than originally proposed. That is, RAR inverse agonists may recruit co-repressor molecules to the receptor, lowering the basal activity relative to unliganded RAR. Such a scenario therefore may allow for not only repression of a RAR responsive gene but may modulate the expression of genes which are under the control of other transcription factors which also interact with co-repressor (see below).

This concept of inverse agonist-mediated recruitment of co-repressor has recently been extended to another nuclear receptor. The estrogen receptor (ER) has been demonstrated to interact with N-CoR *in vitro* [43] and co-immunoprecipitation of N-CoR from human mammary tumor MCF7 cells indicated the interaction between the ER and N-CoR was increased after treatment with the estrogen receptor antagonist 4-*trans*-hydroxytamoxifen (TOT) [44]. Thus, the mechanism by which TOT functions as an ER antagonist involves recruitment of a co-repressor to the ER and as such demonstrates that TOT is an ER inverse agonist. Hence, nuclear receptor inverse agonists appear to mediate their *trans*-repression via the modulation (i.e. recruitment) of receptor- co-repressor interactions.

Multiple functions of the co-repressor

Recent findings on the functions associated with nuclear receptor co-repressors are providing a useful insight into the potential uses of inverse agonists. N-CoR was originally described as a thyroid hormone receptor interacting protein in a yeast two-hybrid interaction screen [42]. Interestingly, binding of thyroid hormone resulted in a decreased interaction between N-CoR and the nuclear receptor. N-CoR and a related molecule, SMRT, have been proposed to be responsible for the transcription-suppressive effects of T3R and RAR in various transfection experiments [45, 46]. A domain immediately C-terminal of the DNA binding domain of RAR has been shown to be required for efficient RAR-N-CoR interaction *in vitro*. Mutation of this domain, referred to as the "AHTm" mutant (the single letter amino acid code for the residues altered), results in abrogation of RAR-N-CoR interaction [42]. Transfection of RAR-AHTm does not however result in elevated basal activity from a RAR responsive reporter compared to wild type RAR [47]. Thus, the transcription-suppressive effect of unliganded RAR does not correlate with the ability of this receptor to interact with N-CoR. However, RAR-AHTm does have an altered ligand responsiveness. Unlike wild type RAR/RXR heterodimers bound to direct repeat spaced by 1 basepair (DR-1) RAREs, which are not activated upon binding RAR ligands [48], those containing RAR-AHTm are activated by these ligands [47]. Thus, N-CoR appears to be involved in regulation of the receptor's response to bound ligands.

There has been a recent explosion of studies regarding the functional activities associated with co-repressor molecules. While originally defined as nuclear receptor co-repressors, we now know that these proteins interact with a variety of factors. N-CoR and SMRT have been characterized as part of a multi-subunit repression complex [49, 50]. Independent of nuclear receptor interaction, N-CoR interacts with mSin3A and B, mammalian homologues of the yeast transcriptional repressor Sin3 [51]. Also in this complex are proteins which exhibit histone deacetylase activity. It has been demonstrated that co-repressor and co-activator molecules are associated with distinct enzymatic activities which have opposing effects upon the acetylation state of core histones H3 and H4 [49, 52–54]. These histone proteins are part of the nucleosome, a structure consisting of DNA wrapped around complexed histone proteins. Co-activator proteins have inherent histone acetyltransferase (HAT) activity. Hormone-mediated recruitment of these co-activators to DNA-bound nuclear receptors has been proposed to lead to acetylation of core histones H3 and H4, resulting in a change of nucleosome structure. This elongation of the nucleosome is thought to allow access of transcription factors to their regulatory elements with a consequent increase in gene expression [55]. In contrast, co-repressor proteins are associated with protein complexes which exhibit histone deacetylase (HDAC) activity [49, 50]. Thus, interaction of co-repressor with RAR is proposed to result in contraction of the nucleosome structure leading to decreased transcription factor access with a consequent decrease in gene expression. Therefore, inverse agonists may have intrinsic biological activities which reflect their ability to modulate nucleosome structure through recruitment of co-repressor and associated HDAC activity.

In addition to nuclear receptors such as RAR, ER and T3R, N-CoR has been shown to interact directly or indirectly with transcription factors from other classes. The c-Myc proto-oncogene, a member of the helix-loop-helix (HLH) class of transcription factors, functions as a transcriptionally active DNA binding protein through its heterodimeric partnering with another HLH protein termed Max [56]. Competition between Max and other HLH proteins such as Mad or Mxi1 for heterodimerization with c-Myc results in c-Myc/Mad and c-Myc/Mxi1 heterodimers which are transcriptionally repressive [57]. While the repression activity of these c-Myc partners is mediated by their interaction with the aforementioned mSin3 proteins [51, 58], an interaction between mSin3 and N-CoR has been demonstrated to be necessary for mSin3-mediated repression of basal transcription [59]. The POU-Homeodomain transcription factor, Pit1, also interacts with N-CoR [44]. This pituitary specific transcription factor, responsible for the specification of lineage-specific gene expression of somatotroph, lactotroph and thyrotroph cell types in the anterior pituitary [60], is also regulated by cAMP responsive signal transduction pathways [61]. The apparent interaction of N-CoR with multiple transcription factors, both inside and outside of the nuclear receptor superfamily, suggests that the pathways mediated by these transcription factors may be functionally linked.

Inverse agonist modulation of gene expression

The shared interaction between the co-repressor N-CoR and a variety of nuclear receptors as well as non-receptor transcription factors suggests the possibility of modulation of gene expression due to the movement of co-repressor molecules, in a ligand dependent manner, from one set of factors to another. Cassanova and co-workers [62] used cotransfected cells to demonstrate that the transcription-suppressive effect of unliganded T3R could be relieved by co-transfection of RAR. This experiment predated the identification of co-repressor molecules and provided the first evidence for their existence. Recently, Xu and co-workers [44] have demonstrated that Pit1-dependent reporter gene activity can be activated by the treatment of cells with the ER inverse agonist tamoxifen, suggesting that recruitment of co-repressor to a nuclear receptor

(ER) away from another transcription factor (Pit1) can increase the transcriptional activity of that factor. It is tempting to speculate on an analogous scenario for explaining some curious pharmacological effects of certain nuclear receptor antagonists referred to as either mixed function agonists or selective estrogen receptor modulators (SERMs) [63]. Raloxifene is such a ligand, it exhibits ER antagonist activity in breast tissue and the uterus but has partial agonist activity in bone [64]. Thus, raloxifene is expected to provide the therapeutic benefits of estrogen replacement therapy to prevent bone loss but without the dangerous side effects of estrogen such as increased incidence of breast and uterine cancers. A raloxifene responsive element (RRE) has been mapped to the 5'-untranslated region of the TGFβ3 gene [65]. Raloxifene-induced activity through this RRE does not require the DNA binding domain of the ER. It is possible that this agonist-like action of raloxifene may result in recruitment of a co-repressor to the ER, away from an as yet unidentified transcription factor binding to the RRE of the TGFβ3 gene. However, the ability of raloxifene to recruit a co-repressor to the ER has yet to be demonstrated.

As mentioned above, the set point or basal activity of a particular nuclear receptor in the absence of ligand will determine if this receptor will theoretically allow for the existence of cognate inverse agonists. A recent finding by Forman and co-workers [66] suggests that natural examples of nuclear receptor inverse agonists may exist. The orphan receptor CARβ (constitutively active receptor) which binds as a heterodimeric partner with RXR to natural RAREs [67], exhibits constitutive transcriptional activity. In other words, its set point for the equilibrium between R and R* favors R*. Interestingly, the presence of androstene metabolites, androstenol and androstanol, results in a decreased interaction between CARβ and the receptor co-activator molecule SRC-1 and effective repression of CARβ-driven reporter gene activity. Whether these ligands recruit a co-repressor to CARβ has yet to be demonstrated. These androstane metabolites are the first examples of naturally occurring nuclear receptor inverse agonists.

Therapeutic uses of RAR antagonists and inverse agonists

Antagonism

Blockade of oral retinoid induced mucocutaneous irritation
Extended use of oral retinoid agonists (e.g. Accutane or 13-*cis* RA for nodular cystic acne) is accompanied by a broad spectrum of side effects ranging from the benign (dry mouth, chapped lips and skin) to the more serious (headache, bone spurs, hypertryglyredemia, pancreatitis) [68, 69]. Accutane is the preferred treatment for nodular cystic acne because of outstanding efficacy. However, mucocutaneous toxicities are universally associated with this drug resulting in decreased patient compliance and sub-optimal cumulative dosing. AGN 194310 and 193109 are potent retinoic acid receptor (RAR) antagonists/inverse agonists which inhibit the retinoid activation of RARs in *in vitro* receptor transactivation assays [20, 22] and the cutaneous toxicities associated with retinoid agonists in animal models. Topical application of RAR inverse agonist on hairless guinea pigs which received oral 13-*cis* RA resulted in amelioration of

retinoid-induced cutaneous irritation [70]. Thus, topical use of RAR antagonist/inverse agonist as an antidote to block the cutaneous toxicities of Accutane should greatly improve patient compliance.

While the blockade of the cutaneous side effects of oral retinoids may appear to be merely a cosmetic issue, the concurrent use of a topical RAR antagonist/inverse agonist together with oral Accutane may lead to significant therapeutic benefit. A significant percentage of nodular cystic acne patients experience a relapse within the first 18 months after the first course of Accutane treatment [71]. Interestingly, there appears to be a correlation between the cumulative dose over the course of therapy and the relapse rate [71]. Below a cumulative dose of 100 mg/kg, patients have a higher failure rate which approaches 90%. A 20 week treatment of 0.5 mg/kg/day, a common initial dose [72], is equivalent to a cumulative dose of 70 mg/kg. Patients who receive greater than 100 mg/kg over 20 weeks of treatment have only a 50% relapse rate [71]. As patient tolerance of associated mucocutaneous irritation can limit the cumulative dose of Accutane, concommitant AGN 194310 topical treatment to the sites of cutaneous irritation should allow an increase in the tolerable dose of Accutane and result in a decreased incidence of acne relapse. Such an increase in the tolerable dose of Accutane is also of potential importance to women of child bearing age. As the teratogenic potential of Accutane contraindicates its use in pregnant women [73], reaching the maximally effective cumulative dose in a shorter period of time should decrease the window of time that women will be exposed to the teratogenic risk of Accutane during the course of treatment.

Inhibition of retinoid augmented viral activation
While retinoid agonists have been demonstrated to exhibit antiviral properties [74] such as inhibition of reactivation of latent Epstein-Barr virus in infected lymphoid cells [75] and inhibition of cellular proliferation of human pappiloma virus (HPV) transformed human cervical epithelial cells [76], two examples of retinoid augmented viral activation suggest the possibility that antagonism of endogenous circulating retinoids may be effective in inhibiting viral replication. HIV viral gene expression appears to be sensitive to retinoids although the response appears to be cell type specific as evidence for both inhibition as well as induction of viral gene expression has been documented in response to RAR agonists [77, 78]. Analysis of the HIV Long Terminal Repeat (LTR), the initiation point of viral gene transcription, has led to the identification of a RARE capable of binding RARs and transfection of a HIV LTR-regulated reporter gene has demonstrated the ability of these viral regulatory sequences to be activated by RAR agonists [79, 80]. Interestingly, this regulatory element can be silenced by RAR antagonists [17] suggesting that HIV gene expression in some infected cell types might be inhibited by a RAR antagonist.

Human cytomegalovirus (hCMV) is a common latent infection which often goes unnoticed in immunocompetent individuals. However, immunocompromised individuals such as organ transplant recipients, cancer patients receiving chemotherapy and AIDS patients can exhibit clinical symptoms due to CMV reactivation. Interestingly, RA differentiation of the human embryonal carcinoma cell line NT-2/D2 into the neuronal lineage converts these cells from hCMV replication non-permissive to permissive [81]. Further, the promoter of the hCMV major immediate early (MIE) gene, the product of which is thought to be necessary for non-latent viral replication, is activated by physiological concentrations of RA [82]. Again, this promoter has been characterized as containing *bona-fide* RAREs which mediate induction of viral gene products by RAR ligands [83]. Angulo and co-workers [84] have demonstrated that the MIE gene of murine cytomegalovirus (MCMV) also contains multiple RAREs and can be activated by RA. Further, oral administration of RA exacerbated MCMV infection in immunocompetent mice and concomitant administration of an RAR antagonist/inverse agonist provided protection from the adverse effects of RA on MCMV infection [84]. Thus, RAR antagonists/inverse agonists may have therapeutic potential in inhibiting viral replication in immunocompromised hCMV hosts.

Inverse agonism

As mentioned above, while RAR inverse agonists and neutral antagonists both function as antagonists of RAR transactivation by RA or other RAR agonists, they can be distinguished according to their ability to repress RAR basal activity. The initial differentiation of RAR inverse agonists from neutral antagonists using a receptor based assay with elevated basal activity demonstrated that these two classes of retinoid ligand compete with one another [21]. Specifically, the *trans*-repression mediated by an inverse agonist can be competitively reversed by addition of a neutral antagonist. Similarly, a RAR agonist competes with these ligands as well. Thus, these three types of RAR ligand exhibit the characteristics of their analogous namesakes of G-coupled receptor ligands. Analysis of gene expression in cultured human keratinocytes treated with these ligands again distinguishes RAR neutral antagonists from inverse agonists. Cultured primary human keratinocytes grown to confluence or treated with either serum or elevated calcium leads to the expression of a variety of differentiation markers [85, 86]. MRP-8 (calgranulin A) is a marker of the abnormal differentiation associated with psoriasis which is not expressed in normal epidermis but is highly upregulated in psoriatic skin and in cultured keratinocytes differentiated with serum [87–89]. Since it exhibits considerable identity at the amino acid level (73%) to murine cytokine CP-10, a chemotactic factor for polymorphonuclear cells and macrophages [90], MRP-8 may function as a chemokine. Treatment of psoriatic skin with the RAR agonist tazarotene resulted in improvement of the disease with concurrent down regulation of abnormally expressed psoriatic differentiation markers, including MRP-8 [91]. RAR agonist treatment of serum-differentiated primary human keratinocytes similarly repressed the expression of MRP-8 [91]. Surprisingly, treatment of serum-differentiated human keratinocytes with a RAR inverse agonist also resulted in repression of MRP-8 expression while use of a neutral antagonist did not [21]. Consistent with their competitive activities upon RAR-VP-16 activity, RAR neutral antagonists provided a dose responsive antagonism of MRP-8 repression mediated either by RAR inverse agonists or agonists. The simultaneous administration of both a RAR inverse agonist and a RAR agonist, retinoids which as single agent treatments provide repression of MRP-8, resulted in mutual antagonism of MRP-8 repression [21]. That is, MRP-8 expression remains elevated when both ligands are present at approximately equivalent concentrations. Thus, RAR inverse agonists and agonists are both capable of mediating repression of MRP-8 through apparently distinct mechanisms that are mutually exclusive in this system. Again, as originally proposed for the G-coupled receptor system, RAR agonists, neutral antagonists and inverse agonists exhibit apparent competitive activities in their effects upon MRP-8 gene expression in cultured keratinocytes. Although the biological endpoint of MRP-8 repression is the same for both inverse agonist and agonist, the mechanism(s) underlying this effect is not. Transfection experiments using a DNA construct containing a reporter gene fused to the 5'-flanking region of the MRP-8 gene have identified a DNA response element which can be activated by the transcription factor NF-IL6 [92]. Activation of the MRP-8 promoter by cotransfected NF-IL6 is repressed by RAR agonist treatment of such transfected cells, suggesting that the mechanism by which RAR agonists repress MRP-8 expression involves antagonism of NF-IL6 activity. In contrast, RAR inverse agonists fail to repress the NF-IL6-activated MRP-8 promoter in similarly transfected cells [93]. Thus, a retinoid inverse agonist and a retinoid agonist are both capable of mediating repression of MRP-8 through apparently distinct mechanisms. The effects of RAR inverse agonists on MRP-8 gene expression in keratinocytes may be the result of recruitment of co-repressor to RARs situated near to the MRP-8 gene. Such an increased concentration of the associated repression complexes, with their histone deacetylase activity, may result in a localized compaction of the chromatin associated with the MRP-8 gene. Elucidation of this mechanism awaits further experimentation. However, the agonist-like effect of RAR inverse agonists upon MRP-8 and other cultured human keratinocyte differentiation markers, such as stromelysin [93], suggests that these agents may have therapeutic value in treating disorders of abnormal keratinocyte differentiation, such as psoriasis.

Concluding remarks

While classical pharmacology considered the binding of ligands and the resulting receptor activation, or lack thereof, as the distinguishing characteristics of agonists and antagonists, respectively, classical biochemistry provided detailed information on the regulation of protein conformation, quaternary structure, and biological activity. Recent structural analysis of members of the nuclear receptor superfamily, as well as the identification of their interacting co-activator/co-repressor proteins, has now provided insight into the molecular mechanisms by which these ligand-regulated receptors function. We now have an enormous amount of information to consider when pondering the activities of various nuclear receptor ligands. As a result, our ability to envision the effect of RAR ligands upon receptor activity has grown from a simple model of agonist versus antagonist to a wider spectrum of agonists, neutral antagonists and inverse agonists. As these ligands appear to delineate distinct types of receptor interactions with co-activator/co-repressor proteins which themselves are functionally integrated with a variety of transcription factor pathways, it is likely that the full utility of RAR inverse agonists has yet to be taken advantage of.

References

1 Bollag W (1983) The development of retinoids in experimental and clinical oncology and dermatology. *J Amer Acad Dermatol* 9: 797–805
2 Weinstein GD, Krueger GG, Lowe NJ, Duvic M, Friedman DJ, Jegasothy BV, Jorizzo JL, Shmunes E, Tschen EH, Lew-Kaya DA et al (1997) Tazarotene gel, a new retinoid, for topical therapy of psoriasis: vehicle-controlled study of safety efficacy and duration of therapeutic effect. *J Amer Acad Dermatol* 37: 85–92
3 Griffiths CE, Russman AN, Majmudar G, Singer RS, Hamilton TA, Voorhees JJ (1993) Restoration of collagen formation in photodamaged human skin by tretinoin (retinoic acid). *New Engl J Med* 329: 530–535
4 Lotan R (1997) Retinoids and chemoprevention of aerodigestive tract cancers. *Cancer Metastasis Rev* 16: 349–356
5 Huang ME, Ye YC, Chen SR, Chai JR, Lu JX, Zhoa L, Gu LJ, Wang ZY (1988) Use of all-*trans* retinoic acid in the treatment of acute promyelocytic leukemia. *Blood* 72: 567–72
6 Li X-S, Shao Z-M, Sheikh MS, Eiseman JL, Sentz D, Jetton AM, Chen J-C, Dawson MI, Aisner S, Rishi AK et al (1995) Retinoic acid nuclear receptor β inhibits breast carcinoma anchorage independent growth. *J Cell Physiol* 165: 449–458
7 Liu Y, Lee M-O, Wang H-G, Li Y, Hashimoto Y, Klaus M, Reed JC, Zhang X-K (1996) Retinoic acid receptor β mediates the growth inhibitory effect of retinoic acid by promoting apoptosis in human breast cancer cells. *Mol Cell Biol* 16: 1138–1149
8 Massaro GDC, Massaro D (1997) Retinoic acid treatment abrogates elastase-induced pulmonary emphysema in rats. *Nat Med* 3: 675–677
9 Mukherjee R, Davies PJ, Crombie DL, Bischoff ED, Cesario RM, Jow L, Hamann LG, Boehm MF, Mondon CE, Nadzan AM et al (1997) Sensitization of diabetic and obese mice to insulin by retinoid X receptor agonists. *Nature* 386: 407–410
10 Cadepond F, Ulmann A, Baulieu EE (1997) RU486 (mifepristone): mechanisms of action and clinical uses. *Annu Rev Med* 48: 129–156
11 MacGregor JI, Jordan VC (1998) Basic guide to the mechanisms of antiestrogen action. *Pharmcol Rev* 50: 151–196
12 Kaneko S, Kagechika H, Kawachi E, Hashimoto Y, Shudo K (1991) Retinoid antagonists. *Med Chem Res* 1: 220–225
13 Eyrolles L, Kawachi E, Matsushima Y, Nakajima O, Kagechika H, Hashimoto Y, Shudo K (1992) Retinoid antagonists: molecular design based on the ligand superfamily concept. *Med Chem Res* 2: 361–367
14 Eyrolles L, Kagechika H, Kawachi E, Fukasawa H, Iijima T, Matsushima Y, Hashimoto Y, Shudo K (1994) Retinobenzoic acids 6 Retinoid antagonists with a heterocyclic ring. *J Med Chem* 37: 1508–1517
15 Apfel C, Bauer F, Crettaz M, Forni L, Kamber M, Kaufmann F, LeMotte P, Pirson W, Klaus M (1992) A retinoic acid receptor α antagonist selectively counteracts retinoic acid effects. *Proc Natl Acad Sci USA* 89: 7129–7133
16 Klaus M, Mohr P (1993) Preparation and formulation of benzotheipins,-thiopyrans, and -thiophenes as immunomodulators. European Patent Appl EP 568898 A1
17 Lee M-O, Hobbs PD, Zhang X-K, Dawson MI, Pfahl M (1994) A synthetic retinoid antagonist inhibits the human immunodeficiency virus type 1 promoter. *Proc Natl Acad Sci USA* 91: 5632–5636
18 Lee M-O, Dawson MI, Picard N, Hobbs PD, Pfahl M (1996) A novel class of retinoid antagonists and their mechanism of action. *J Biol Chem* 271: 11 897–11 903
19 Yoshimura H, Nagai M, Hibi S, Kikuchi K, Abe S, Hida T, Higashi S, Hishinuma I, Yamanaka T (1995) A novel type of retinoic acid receptor antagonist: synthesis and structure-activity relationships of heterocyclic ring-containing benzoic acid derivatives. *J Med Chem* 38: 3163–3173
20 Johnson AT, Klein ES, Gillett SJ, Wang L, Song TK, Pino ME, Chandraratna RAS (1995) Synthesis and characterization of a highly potent and effective antagonist of retinoic acid receptors. *J Med Chem* 38: 4764–4767
21 Klein ES, Pino ME, Johnson AT, Davies PJA, Nagpal S, Thacher SM, Krasinski G, Chandraratna RAS (1996) Identification and functional separation of retinoic acid receptor neutral antagonists and inverse agonists. *J Biol Chem* 271: 22 692–22 696
22 Johnson AT, Wang L, Standeven AM, Escobar M, Chandraratna RAS (1999) Synthesis and biological activity of high affinity retinoic acid receptor antagonists. *Bioorg Med Chem* 7: 1321–1338
23 Teng M, Duong TT, Johnson AT, Klein ES, Wang L, Khalifa B, Chandraratna RAS (1997) Identification of highly potent retinoic acid receptor α-selective antagonists. *J Med Chem* 40: 2445–2451
24 Costa T, Herz A (1989) Antagonists with negative intrinsic activity at delta opioid receptors coupled to GTP-binding proteins. *Proc Natl Acad Sci USA* 86: 7321–7325
25 Chidiac P, Hebert TE, Valiquette M, Dennis M, Bouvier M (1994) Inverse agonist activity of β-adrenergic antagonists. *Mol Pharmacol* 45: 490–499
26 Bond RA, Leff P, Johnson TD, Milano CA, Rockman HA, McMinn TR, Apparsundaram S, Hyek MF, Kenakin TP, Allen LF et al (1995) Physiological effects of inverse agonists in transgenic mice with myocardial overexpression of the β2-adrenoreceptor. *Nature* 374: 272–275
27 Baxter GS, Tilford NS (1995) Endogenous ligands and inverse agonism. *Trends Pharmacol Sci* 16: 258–259

28 Mullaney I, Carr IC, Milligan G (1996) Analysis of inverse agonism at the delta opioid receptor after expression in Rat 1 fibroblasts. *Biochem J* 315: 227–234
29 Tiberi M, Caron MG (1994) High agonist-independent activity is a distinguishing feature of the dopamine D1B receptor subtype. *J Biol Chem* 269: 27 925–27 931
30 Samama P, Cotecchia S, Costa T, Lefkowitz RJ (1993) A mutation-induced activated state of the β 2-adrenergic receptor extending the ternary complex model. *J Biol Chem* 268: 4625–4636
31 Onate SA, Tsai SY, Tsai M-J, O'Malley BW (1995) Sequence and characterization of a coactivator for the steroid hormone receptor superfamily. *Science* 270: 1354–1357
32 Hong H, Kohli K, Trivedi A, Johnson DL, Stallcup MR (1996) Grip1, a novel mouse protein that serves as a transcriptional coactivator in yeast for the hormone binding domains of the steroid receptors. *Proc Natl Acad Sci USA* 93: 4948–4952
33 Le Douarin B, Zechel C, Garnier J-M, Lutz Y, Tora L, Pierrat B, Heery D, Gronemeyer H, Chambon P, Losson R (1995) The N-terminal part of TIF-1, a putative mediator of the ligand-dependent activation function (AF-2) of nuclear receptors, is fused to B-raf in the oncogenic protein T18. *EMBO J* 14: 2020–2033
34 Chen H, Lin RJ, Schiltz RL, Chakravarti D, Nash A, Nagy L, Privalsky ML, Nakatani Y, Evans RM (1997) Nuclear receptor coactivator ACTR is a novel histone acetyltransferase and forms a mutimeric activation complex with P/CAF and CBP/p300. *Cell* 90: 569–580
35 Torchia J, Rose DW, Inostroza J, Kamei Y, Westin S, Glass CK, Rosenfeld MG (1997) The transcriptional co-activator p/CIP binds CBP and mediates nuclear- receptor function. *Nature* 387: 677–684
36 Li H, Gomes PJ, Chen JD (1997) RAC3, a steroid/nuclear receptor-associated coactivator that is related to SRC-1 and TIF2. *Proc Natl Acad Sci USA* 94: 8479–8484
37 Bourguet W, Ruff M, Chambon P, Groemeyer H, Moras D (1995) Crystal structure of the ligand-binding domain of the human nuclear receptor RXRα. *Nature* 375: 377–382
38 Renaud J-P, Rochel N, Ruff M, Vivat V, Chambon P, Gronemeyer H, Moras D (1995) Crystal structure of the RARγ ligand-binding domain bound to all-*trans* retinoic acid. *Nature* 378: 681–689
39 Wagner RL, Apriletti JW, McGrath ME, West BL, Baxter JD, Fletterick RL (1995) A structural role for the hormone in the thyroid hormone receptor. *Nature* 378: 690–697
40 Nolte RT, Wisely GB, Westin S, Cobb JE, Lambert MH, Kurokawa R, Rosenfeld MG, Willson TM, Glass CK, Milburn MV (1998) Ligand binding and co-activator assembly of the peroxisome proliferator-activated receptor-γ. *Nature* 395: 137–143
41 Chen JD, Evans RM (1995) A transcriptional co-repressor that interacts with nuclear hormone receptors. *Nature* 377: 454–457
42 Horlein AJ, Naar AM, Heinzel T, Torchia J, Gloss B, Kurokawa R, Ryan A, Kamei Y, Soderstrom M, Glass CK et al (1995) Ligand-independent repression by the thyroid hormone receptor mediated by a nuclear receptor-co-repressor. *Nature* 377: 397–404
43 Smith CL, Nawaz Z, O'Malley BW (1997) Coactivator and corepressor regulation of the agonist/antagonist activity of the mixed antiestrogen, 4-hydroxytamoxifen. *Mol Endocrinol* 11: 657–666
44 Xu L, Lavinsky RM, Dasen JS, Flynn SE, McInerney EM, Mullen TM, Heinzel T, Szeto D, Korzus E, Kurokawa R et al (1998) Signal-specific co-activator domain requirements for Pit-1 activation. *Nature* 395: 301–306
45 Brent GA, Dunn MK, Harney JW, Gulick T, Larsen PR, Moore DD (1989) Thyroid hormone aporeceptor represses T3-inducible promoters and blocks activity of the retinoic acid receptor. *New Biologist* 1: 329–336
46 Baniahmad A, Kohne CA, Renkawitz R (1992) A transferable silencing domain is present in the thyroid hormone receptor, in the v-erbA oncogene product and in the retinoic acid receptor. *EMBO J* 11: 1015–1023
47 Kurokawa R, Soderstrom M, Horlein A, Halachmi S, Brown M, Rosenfeld MG, Glass CK (1995) Polarity-specific activities of retinoic acid receptors determined by a co-repressor. *Nature* 377: 451–454
48 Kurokawa R, DiRenzo J, Boehm M, Sugarman J, Gloss B, Rosenfeld MG, Heyman RA, Glass CK (1994) Regulation of retinoid signalling by receptor polarity and allosteric control of ligand binding. *Nature* 371: 528–531
49 Heinzel T, Lavinsky RM, Mullen T, Soderstrom M, Laherty CD, Torchia J, Yang W-M, Brard G, Ngo SD, Davie JR et al (1997) A complex containing N-CoR, mSin3 and histone deacetylase mediates transcriptional repression. *Nature* 387: 43–55
50 Nagy L, Kao H, Chakravarti D, Lin RJ, Hassig CA, Ayer DA, Schreiber SL, Evans RM (1997) Nuclear receptor repression mediated by a complex containing SMRT, mSin3A, and histone deacetylase. *Cell* 89: 373–380
51 Ayer DE, Lawrence QA, Eisenman RN (1995) Mad-Max transcriptional repression is mediated by ternary complex formation with mammalian homologs of yeast repressor Sin3. *Cell* 80: 767–776
52 Ogryzko VV, Schiltz RL, Russanova V, Howard BH, Nakatani Y (1996) The transcriptional coactivators p300 and CBP are histone acetyltransferases. *Cell* 87: 953–959
53 Wade PA, Wolffe AP (1997) Chromat*In*: Histone acetyltransferases in control. *Curr Biol* 7: R82–R84
54 Keirmaier A, Eilers M (1997) Transcriptional control: Calling in histone deacetylase. *Curr Biol* 7: R505–R507
55 Davie JR (1997) Nuclear matrix, dynamic histone acetylation and transcriptionally active chromatin. *Mol Biol Rep* 24: 197–207
56 Blackwood EM, Eisenman RN (1991) Max: a helix-loop-helix zipper protein that forms a sequence-specific DNA binding complex with Myc. *Science* 251: 1211–1217

57 Ayer DE, Kretzner L, Eisenman RN (1993) Mad: a heterodimeric partner for Max that antagonizes Myc transcriptional activity. *Cell* 72: 211–222
58 Schreiber-Agus N, Chin L, Chen K, Torres R, Rao G, Guida P, Skoultchi AI, DePinho RA (1995) An amino-terminal domain of Mxi1 mediates anti-Myc oncogenic activity and interacts with a homolog of the yeast transcriptional repressor SIN3. *Cell* 80: 777–786
59 Alland L, Muhle R, Hou HJr, Potes J, Chin L, Schreiber-Agus N, DePinho RA (1997) Role for N-CoR and histone deacetylase in Sin3-mediated transcriptional repression. *Nature* 387: 49–55
60 Ingraham HA, Chen RP, Mangalam HJ, Elsholtz HP, Flynn SE, Lin CR, Simmons DM, Swanson L, Rosenfeld MG (1988) A tissue-specific transcription factor containing a homeodomain specifies a pituitary phenotype. *Cell* 55: 519–529
61 Kapiloff MS, Farkash Y, Wegner M, Rosenfeld MG (1991) Variable effects of phosphorylation of Pit-1 dictated by the DNA response elements. *Science* 253: 786–789
62 Cassanova J, Helmer E, Selmi-Ruby S, Qi JS, Au-Fliegner M, Desai-Yajnik V, Koudinova N, Yarm F, Raaka BM, Samuels HH (1994) Functional evidence for ligand-dependent dissociation of thyroid hormone and retinoic acid receptors from an inhibitory cellular factor. *Mol Cell Biol* 14: 5756–5765
63 Bryant HU, Dere WH (1998) Selective estrogen receptor modulators: an alternative to hormone replacement therapy. *Proc Soc Exp Biol Med* 217: 45–52
64 Yang NN, Bryant HU, Hardikar S, Sato M, Galvin RJ, Glasebrook AL, Termine JD (1996) Estrogen and raloxifene stimulate transforming growth factor-β 3 gene expression in rat bone: a potential mechanism for estrogen- or raloxifene-mediated bone maintenance. *Endocrinology* 137: 2075–2084
65 Yang NN, Venugopalan M, Hardikar S, Glasebrook A (1996) Identification of an estrogen response element activated by metabolites of 17β-estradiol and raloxifene *Science* 273: 1222–1225
66 Forman BM, Tzameli I, Choi H-S, Chen J, Simha D, Seol W, Evans RM, Moore DD (1998) Androstane metabolites bind to and deactivate the nuclear receptor CAR-β. *Nature* 395: 612–615
67 Choi HS, Chung M, Tzameli I, Simha D, Lee YK, Seol W, Moore DD (1997) Differential transactivation by two isoforms of the orphan nuclearhormone receptor CAR. *J Biol Chem* 272: 23 565–23 571
68 Kovacs JAJ, Shear NH (1993) Adverse nonreproductive effects of retinoids in humans. *In*: G Koren (ed.): *Retinoids in Clinical Practice: The Risk-Benefit Ratio.* Dekker, New York, 241–260
69 Armstrong RB, Ashenfelter KO, Eckhoff C, Levin AA, Shapiro SS (1994) General and reproductive toxicology of retinoids. *In*: MB Sporn, AB Roberts, DS Goodman (eds): *The Retinoids: Biology, Chemistry and Medicine.* Raven Press, New York, 545–572
70 Standeven AM, Johnson AT, Escobar M, Chandraratna RAS (1996) Specific antagonist of retinoid toxicity in mice. *Toxicol Appl Pharmacol* 138: 169–175
71 White GM, Chen W, Yao J, Wolde-Tsadik G (1998) Recurrence rates after the first couse of isotretinoin. *Arch Dermatol* 134: 376–378
72 Cunliffe WJ, van de Kerkho PCM, Caputo R, Cavicchini S, Cooper A, Fyrand OL, Gollnick H, Layton AM, Leyden JJ, Mascaro J-M et al (1997) Roaccutane treatment guidelines: Results of an international survey. *Dermatology* 194: 351–357
73 Koren G, Pastuszak A, Ito S (1998) Drugs in pregnancy. *New Engl J Med* 338: 1128–1137
74 Ghazal P, LeBlanc JF, Angulo A (1997) Vitamin A regulation of viral growth. *Rev Med Virol* 7: 21–34
75 Sundar SK, Levine PH, Ablashi DV, Menzes J (1984) Retinoic acid and steroids inhibit Epstein-Barr virus-induced nuclear antigen, DNA synthesis and lymphocyte transformation. *Anticancer Res* 4: 415–418
76 Agarwal C, Hembree JR, Rorke EA, Eckert RL (1994) Interferon and retinoic acid suppress the growth of human papillomavirus type 16 immortalized cervical epithelial cells, but only interferon suppresses the level of the human papillomavirus transforming oncogenes. *Cancer Res* 54: 2108–2112
77 Poli G, Kinter AL, Justement JS, Bressler P, Kehrl JH, Fauci AS (1992) Retinoic acid mimics transforming growth factor β in the regulation of human immunodeficiency virus expression in monocytic cells. *Proc Natl Acad Sci USA* 89: 2689–2693
78 Semmel M, Macho A, Coulaud D, Alileche A, Plaisance S, Aguilar J, Jasmin C (1994) Effect of retinoic acid on HL-60 cells infected with human immunodeficiency virus type 1. *Blood* 84: 2480–2488
79 Orchard K, Lang G, Harris J, Collins M, Latchman D (1993) A palindromic element in the human immunodeficiency virus longterminal repeat binds retinoic acid receptors and can confer retinoic acidresponsiveness on a heterologous promoter. *J Acq Immun Defic Synd Hum R* 6: 440–445
80 Ladias JA (1994) Convergence of multiple nuclear receptor signalling pathways onto the long terminal repeat of human immunodeficiency virus-1. *J Biol Chem* 269: 5944–5951
81 Gonczol E, Andrews PW, Plotkin SA (1984) Cytomegalovirus replicates in differentiated but not inundifferentiated human embryonal carcinoma cells. *Science* 224: 159–161
82 Ghazal P, DeMattei C, Giulietti E, Kliewer SA, Umesono K, Evans RM (1992) Retinoic acid receptors initiate induction of the cytomegalovirus enhancer in embryonal cells. *Proc Natl Acad Sci USA* 89: 7630–7634
83 Angulo A, Suto C, Heyman RA, Ghazal P (1996) Characterization of the sequences of the human cytomegalovirus enhancer that mediate differential regulation by natural and synthetic retinoids. *Mol Endocrinol* 10: 781–793
84 Angulo A, Chandraratna RAS, LeBlanc JF, Ghazal P (1998) Ligand induction of retinoic acid receptors alters an acute infection by murine cytomegalovirus. *J Virol* 72: 4589–4600

85 Saunders NA, Jetten AM (1994) Control of growth regulatory and differentiation-specific genes in humanepi-dermal keratinocytes by interferon γ Antagonism by retinoic acid and transforming growth factor β 1. *J Biol Chem* 269: 2016–2022
86 Hohl D, de Viragh PA, Amiguet-Barras F, Gibbs S, Backendorf C, Huber M (1995) The small proline-rich proteins constitute a multigene family of differentially regulated cornified cell envelope precursor proteins. *J Invest Dermatol* 104: 902–909
87 Wilkinson MM, Busuttil A, Hayward C, Brock DJH, Dorin JR, van Heyningen V (1988) Expression pattern of two related cystic fibrosis-associated calcium-binding proteins in normal and abnormal tissues. *J Cell Sci* 91: 221–230
88 Kelly SE, Jones DB, Fleming S (1989) Calgranulin expression in inflammatory dermatoses. *J Pathol* 159: 17–21
89 Madsen P, Rasmussen HH, Leffers H, Honoré B, Celis JE (1992) Molecular cloning and expression of a novel keratinocyte protein (psoriasis-associated fatty acid-binding protein [PA-FABP]) that is highly up-regulated in psoriatic skin and that shares similarity to fatty acid-binding proteins. *J Invest Dermatol* 99: 299–305
90 Lackmann M, Rajasekariah P, Iismaa SE, Jones G, Cornish CJ, Hu S, Simpson RJ, Moritz RL, Geczy CL (1993) Identification of a chemotactic domain of the pro-inflammatory S100 protein CP-10. *J Immunol* 150: 2981–2991
91 Nagpal S, Thacher SM, Patel S, Friant S, Malhotra M, Shafer J, Krasinski G, Asano AT, Teng M, Duvic M et al (1996) Negative regulation of two hyperproliferative keratinocyte differentiation markers by a retinoic acid receptor-specific retinoid: insight into the mechanism of retinoid action in psoriasis. *Cell Growth Differ* 7: 1783–1791
92 DiSepio D, Malhotra M, Chandraratna RAS, Nagpal S (1997) Retinoic acid receptor-nuclear factor-interleukin 6 antagonism. *J Biol Chem* 272: 25 555–25 559
93 Thacher SM, Nagpal S, Klein ES, DiSepio D, Arefieg T, Krasinski G, Agawal C, Eckert RL, Chandraratna RAS (1998) Cell type and gene-specific activity of the retinoid inverse agonist AGN 193109: Divergent effects from agonist at retinoic acid receptor gamma in human keratinocytes. *Cell Growth Differ* 10: 235–262

Subject index